Innovative Technologies in Beverage Processing

Innovative Technologies in Beverage Processing

Edited by

Ingrid Aguiló-Aguayo
*IRTA, XaRTA-Postharvest, Edifici Fruitcentre,
Parc Científic i Tecnològic Agroalimentari de Lleida,
Lleida, Catalonia, Spain*

Lucía Plaza
Cincagroup, Huesca, Spain

This edition first published 2017 © 2017 John Wiley & Sons Ltd

Registered Offices
John Wiley & Sons Ltd, The Atrium, Southern Gate, Chichester, West Sussex, PO19 8SQ, UK

Editorial Office: 111 River Street, Hoboken, NJ 07030, USA
9600 Garsington Road, Oxford, OX4 2DQ, UK
The Atrium, Southern Gate, Chichester, West Sussex, PO19 8SQ, UK

For details of our global editorial offices, customer services, and more information about Wiley products visit us at www.wiley.com.

Wiley also publishes its books in a variety of electronic formats and by print-on-demand. Some content that appears in standard print versions of this book may not be available in other formats.

Library of Congress Cataloging-in-Publication Data

Names: Aguiló-Aguayo, Ingrid, 1981- editor. | Plaza, Lucía, 1972- editor.
Title: Innovative technologies in beverage processing / edited by Ingrid
    Aguiló-Aguayo, Lucía Plaza.
Description: Hoboken, NJ : John Wiley & Sons, 2017. | Includes
    bibliographical references and index.
Identifiers: LCCN 2017005486 (print) | LCCN 2017011095 (ebook) | ISBN
    9781118929377 (cloth) | ISBN 9781118929353 (Adobe PDF) | ISBN
    9781118929360 (ePub)
Subjects: LCSH: Beverage industry–Technological innovations. | Fruit juice
    industry.
Classification: LCC HD9348.A2 I565 2017 (print) | LCC HD9348.A2 (ebook) | DDC
    663/.63–dc23
LC record available at https://lccn.loc.gov/2017005486

Cover Design: Wiley
Cover Images: (Left to Right) © fcafotodigital/Gettyimages; © monticelllo/Gettyimages;
© Kostiantyn AblazovShutterstock

Set in 9.5/11.5pt, TimesTenLTstd-Roman by SPi Global, Chennai, India.
Printed and bound in Malaysia by Vivar Printing Sdn Bhd
10 9 8 7 6 5 4 3 2 1

Contents

About the IFST Advances in Food Science Book Series

The Institute of Food Science and Technology (IFST) is the leading qualifying body for food professionals in Europe and the only professional organisation in the UK concerned with all aspects of food science and technology. Its qualifications are internationally recognised as a sign of proficiency and integrity in the industry. Competence, integrity, and serving the public benefit lie at the heart of the IFST philosophy. IFST values the many elements that contribute to the efficient and responsible supply, manufacture and distribution of safe, wholesome, nutritious and affordable foods, with due regard for the environment, animal welfare and the rights of consumers.

IFST Advances in Food Science is a series of books dedicated to the most important and popular topics in food science and technology, highlighting major developments across all sectors of the global food industry. Each volume is a detailed and in-depth edited work, featuring contributions by recognized international experts, and which focuses on new developments in the field. Taken together, the series forms a comprehensive library of the latest food science research and practice, and provides valuable insights into the food processing techniques that are essential to the understanding and development of this rapidly evolving industry.

The IFST Advances series is edited by Dr Brijesh Tiwari, Senior Research Officer in the Department of Food Biosciences at the Teagasc Food Research Centre, Dublin, Ireland.

Books in the series

Ultrasound in Food Processing: Recent Advances
by Mar Villamiel (Editor), Jose V. Garcia-Perez (Editor), Antonia Montilla (Editor), Juan A. Carcel (Editor), Jose Benedito (Editor)

Emerging Technologies in Meat Processing: Production, Processing and Technology
by Enda J. Cummins (Editor), James G. Lyng (Editor)

Tropical Roots and Tubers: Production, Processing and Technology
by Harish K. Sharma (Editor), Nicolas Y. Njintang (Editor), Rekha S. Singhal (Editor), Pragati Kaushal (Editor)

Emerging Dairy Processing Technologies: Opportunities for the Dairy Industry
by Nivedita Datta, Peggy M. Tomasula

Nutraceutical and Functional Food Processing Technology
by Joyce I. Boye

Seafood Processing: Technology, Quality and Safety
by Ioannis S. Boziaris (Editor)

List of Contributors

Maribel Abadias, IRTA, XaRTA-Postharvest, Edifici Fruitcentre, Parc Científic i Tecnològic Agroalimentari de Lleida, Lleida, Catalonia, Spain

Ingrid Aguiló-Aguayo, IRTA, XaRTA-Postharvest, Edifici Fruitcentre, Parc Científic i Tecnològic Agroalimentari de Lleida, Lleida, Catalonia, Spain

Begoña de Ancos, Institute of Food Science, Technology and Nutrition (ICTAN), Spanish National Research Council (CSIC), Madrid, Spain

Gerard Aragonès, Department of Biochemistry and Biotechnology, Nutrigenomics Research Group, Universitat Rovira i Virgili, Tarragona, Spain

Anna Arola-Arnal, Department of Biochemistry and Biotechnology, Nutrigenomics Research Group, Universitat Rovira i Virgili, Tarragona, Spain

Lluís Arola, Technological Unit of Nutrition and Health, EURECAT-Technological Center of Catalonia, Reus, Spain; Department of Biochemistry and Biotechnology, Nutrigenomics Research Group, Universitat Rovira i Virgili, Tarragona, Spain

Francisco J. Barba, University of Copenhagen, Faculty of Science, Department of Food Science, Rolighedsvej, Frederiksberg, Copenhagen, Denmark; Department of Preventive Medicine and Public Health, Faculty of Pharmacy, Universitat de València, Burjassot, Spain

Jules Beekwilder, Wageningen UR, Plant Research International, Bioscience, Wageningen UR, Wageningen, The Netherlands

Cinta Bladé, Department of Biochemistry and Biotechnology, Nutrigenomics Research Group, Universitat Rovira i Virgili, Tarragona, Spain

Gloria Bobo, IRTA, XaRTA-Postharvest, Edifici Fruitcentre, Parc Científic i Tecnològic Agroalimentari de Lleida, Lleida, Catalonia, Spain

Dilek Boyacioglu, Department of Food Engineering Faculty of Chemical & Metallurgical Engineering, Istanbul Technical University, Maslak, Istanbul, Turkey

Esra Capanoglu, Department of Food Engineering, Faculty of Chemical & Metallurgical Engineering, Istanbul Technical University, Istanbul, Turkey

María Consuelo Pina-Pérez, Instituto de Agroquímica y Tecnología de Alimentos (IATA-CSIC), Paterna, Valencia, Spain; Centro Avanzado de Microbiología de Alimentos (CAMA), Universidad Politécnica de Valencia, Valencia, Spain

Navneet S. Deora, Associate Cereal Specialist, Nestle R&D India Centre Private Limited, Gurgaon, India

Aastha Deswal, Manager, Product Formulation, Bright LifeCare Private Limited, Gurgaon, India

Zamantha Escobedo-Avellaneda, Tecnológico de Monterrey, Escuela de Ingeniería y Ciencias, Centro de Biotecnología FEMSA, Tecnológico, Monterrey, Nuevo León, Mexico

Ciaran Fitzgerald, Department of Food BioSciences, Teagasc Food Research Centre Ashtown, Dublin, Ireland

Rebeca García-García, Tecnológico de Monterrey, Escuela de Ingeniería y Ciencias, Centro de Biotecnología FEMSA, Monterrey, Nuevo León, Mexico

Nabil Grimi, Sorbonne Université, Laboratoire Transformations Intégrées de la Matière Renouvelable (TIMR EA 4297), Centre de Recherche de Royallieu, Université de Technologie de Compiègne, Compiègne Cedex, France

Robert D. Hall, Wageningen UR, Plant Research International, Bioscience, Wageningen UR, Wageningen, The Netherlands

Mohammad Hossain, Department of Food BioSciences, Teagasc Food Research Centre Ashtown, Dublin, Ireland

Antonio Martínez, Instituto de Agroquímica y Tecnología de Alimentos (IATA-CSIC), Valencia, Paterna, Spain

Hari N. Mishra, Department of Agricultural and Food Engineering, Indian Institute of Technology, Kharagpur, India

Begoña Muguerza, Department of Biochemistry and Biotechnology, Nutrigenomics Research Group, Universitat Rovira i Virgili, Tarragona, Spain

Indrawati Oey, Department of Food Science, University of Otago, Dunedin, New Zealand

Lucía Plaza, IRTA, XaRTA-Postharvest, Edifici Fruitcentre, Parc Científic i Tecnològic Agroalimentari de Lleida, Lleida, Catalonia, Spain

Francesc Puiggròs, Technological Unit of Nutrition and Health, EURECAT-Technological Center of Catalonia, Reus, Spain

Francisco Quilez, Valencian School for Health Studies (EVES), Professional Training Unit, Juan de Garay, Valencia, Spain

Dilip K. Rai, Department of Food BioSciences, Teagasc Food Research Centre Ashtown, Dublin, Ireland

Alejandro Rivas, Instituto de Agroquímica y Tecnología de Alimentos (IATA-CSIC), Valencia, Paterna, Spain

Dolores Rodrigo, Instituto de Agroquímica y Tecnología de Alimentos (IATA-CSIC), Valencia, Paterna, Spain

Ma. Janeth Rodríguez-Roque, Universidad Autónoma de Chihuahua, Facultad de Ciencias Agrotecnológicas Cd. Universitaria s/n, Chihuahua, Mexico

Elena Roselló-Soto, Universitat de Valencia, Faculty of Pharmacy, Department of Preventive Medicine and Public Health, Toxicology and Forensic Medicine Department, Avda. Vicent Andrés Estellés, Burjassot, Valencia, Spain

Concepción Sánchez Moreno, Institute of Food Science, Technology and Nutrition (ICTAN), Spanish National Research Council (CSIC), Madrid, Spain

Rogelio Sánchez-Vega, Centro de Investigación en Alimentación y Desarrollo A.C., Avenida Río Conchos S/N, Parque Industrial, Chihuahua, Mexico

David Sepúlveda-Ahumada, Centro de Investigación en Alimentación y Desarrollo A.C., Avenida Río Conchos S/N, Parque Industrial, Chihuahua, Mexico

Manuel Suárez, Department of Biochemistry and Biotechnology, Nutrigenomics Research Group, Universitat Rovira i Virgili, Tarragona, Spain

Susana Suárez-Garcia, Department of Biochemistry and Biotechnology, Nutrigenomics Research Group, Universitat Rovira i Virgili, Tarragona, Spain

Gamze Toydemir, Department of Food Engineering, Faculty of Engineering, Alanya Alaaddin Keykubat University, Kestel-Alanya, Antalya, Turkey

Inmaculada Viñas, Food Technology Department, University of Lleida, XaRTA-Postharvest, Agrotecnio Center, Lleida, Catalonia, Spain

Jorge Welti-Chanes, Tecnológico de Monterrey, Escuela de Ingeniería y Ciencias, Centro de Biotecnología FEMSA, Monterrey, Nuevo León, Mexico

Sze Ying Leong, Department of Food Science, University of Otago, Dunedin, New Zealand; Riddet Institute, Palmerston North, New Zealand

Preface

In the twenty first century, the beverage industry faced changing times with new developments as a response to consumers' demand to attain new and more differentiated food products of high quality and with guarantee of safe standards. The non-alcoholic beverage market itself is composed of various segments, such as juices, vegetable blends, teas, dairy drinks and alternative beverages. The drinks of today require specific knowledge of exotic ingredients, novel processing techniques and a variety of functional ingredients. With such diverse group of commodities, a considerable research effort in the application of innovative technologies for manufacturing beverages has been made in the recent years. Several new processing technologies have emerged with regards to beverage processing. A number of novel thermal and non-thermal technologies have become available to ensure high-quality food retention levels, while extending the products' shelf life. On the other hand, the increasing demand for natural ingredients, improving health and appearance, is also attracting non-alcoholic beverages as the fastest growing segment on the functional food market. Consumer demands 'miracle beverages' that are not only safe and nutritious but also natural, economical to manufacture, convenient, great tasting, environmentally friendly and enhance health and well being. Advances in food science and technology are presenting exciting opportunities for the beverages sector.

The book is divided into three main parts. The first part is focused on juices processing, including conventional and novel processing techniques. The different chapters of this first section cover juices made from pome fruits, citrus fruits, prunus fruits, vegetables, exotic fruits, berries and juice blends. The second part highlights non-alcoholic beverages, including grain-based beverages, soups and functional beverages (covering novel ingredients and food regulations). Finally, the last part is devoted to waste and by-products generated in the juice and non-alcoholic beverage sector.

Innovative Technologies in Beverage Processing provides a comprehensive overview on the application of novel processing technologies in the non-alcoholic beverage product manufacture. The authors of the chapters are international leading experts in the field of emerging technologies for beverages processing. We would like to thank all of them for their valuable contribution to this book.

Ingrid Aguiló-Aguayo
Lucía Plaza
Lleida, 2017

List of Abbreviations

AA	amino acid
AOA	antioxidant activity
BBI	Bowman–Birk inhibitor
BP	blood pressure
CFU	colony-forming units
CH	conventional heating
CUT	come up time
CVD	cardiovascular disease
DPCD	dense phase carbon dioxide
ECG	epicatechin-3-gallate
EFSA	European Food Safety Authority
EGC	epigallocatechin
EGCG	epigallocatechin-3-gallate
ENT	enterobacteriaceae
EOs	electro-osmotic
EO	essential oils
EU	European Union
FAO	Food and Agricultural Organization of the United Nations
FCV-F9	feline calicivirus
FDA	food and Drug Administration
FHC	foods with health claims
FJ	fresh juice
FJ-SM	fruit juice–soy milk
FOSHU	foods for specified health uses
GAE	gallic acid equivalents
GLVs	green leaf-associated volatiles
GRAS	generally regarded as safe
GSE	grape seed extract
HHP	high hydrostatic pressure
HIPEF	high-intensity pulsed electric fields
HIPL	high-intensity pulsed light
HMF	hydroxymethylfurfural
HP	high pressure
HPH	high-pressure homogenization

HPHT	high-pressure–high-temperature
HPP	high-pressure processing
HTST	high-temperature–short time
ILSI	international of Life Sciences Institute
IR	irradiation
IU	international unit
KSTI	Kunitz soya bean trypsin inhibitor
LAB	lactic acid bacteria
L-NAME	NG-nitro-L-arginine methyl ester
LTHT	low-temperature–high-time
M	microwave
MAP	modified atmosphere packaging
MNV	murine norovirus
MW	microwave heating
NHCR	Nutrition and Health Claims Regulation
NO	nitric oxide
NSP	non-starch polysaccharides
OH	ohmic heating
OLC	orange–carrot–lemon
PASSCLAIM	process for the scientific support for claims on foods
PATP	pressure-assisted thermal processing
PATS	pressure-assisted thermal sterilization
PE	pectinesterase
PEF	pulsed electric fields
PG	polygalacturonase
PL	pulsed light
PME	pectinmethylesterase
POD	peroxidase
PPO	polyphenol oxidase
PSY	psychrotrophs
PUFAs	polyunsaturated fatty acids
R	radio frequency
RF	radiofrequency heating
SEM	scanning electron microscopy
SGS	soluble gas stabilisation
TBC	total bacterial count
TCD	total colour difference
TE	trolox equivalents
TIA	trypsin inhibitor activity
TP	total phenolic
TPC	total phenolic compounds
U	ultrasonic waves
UF	ultrafiltration
UHP	ultrahigh-pressure processing
UHPH	ultrahigh-pressure homogenisation
UHT	ultrahigh temperature
US	ultrasound
USA	United States of America
USD	dollars

UST	ultrasound treatment
UV	ultraviolet light
UV-C	ultraviolet light-C
UVT	ultraviolet treatment
VYLB	vegetable Yoghurt-like beverages
WHO	World Health Organization

Part I
Juice Processing

1

Pome Fruit Juices

Ingrid Aguiló-Aguayo[1]*, Lucía Plaza[1], Gloria Bobo[1], Maribel Abadias[1], and Inmaculada Viñas[2]

[1]*IRTA, XaRTA-Postharvest, Edifici Fruitcentre, Parc Científic I Tecnològic Agroalimentari de Lleida, Lleida, Catalonia, Spain*
[2]*Food Technology Department, University of Lleida, XaRTA-Postharvest, Agrotecnio Center, Lleida, Catalonia, Spain*

1.1 Introduction

Apple and pear are the two major commercial importance pome fruits that are grown in most temperate regions of the world. Apples (*Malus domestica*) have a strong antioxidant activity, which is mainly being attributed to the polyphenolic fraction. Pears (European pears: *Pyrus communis* L.; Asian pears: *Pyrus serotina* L.) are a good source of dietary fiber and vitamin C. Both apples and pears are often consumed fresh and also canned, dried, baked, freeze-dried, and as a cloudy or concentrate juice. The concentrated pome juices are usually obtained by extraction or pressing and later, clarification. The first step produces a juice of about 12°Brix, and, after concentration, a final product of about 70°Brix is obtained (Falguera *et al.*, 2013). However, its properties are also constantly changing when the juices are subjected to processing, storage, transport, marketing, and consumption (Rao, 1986). Through the years, the process of optimization for obtaining and preserve these products has been conducted in order to avoid undesirable quality changes. This chapter discusses the application of conventional and emerging techniques in the processing of pome fruit juices and their effect on the final quality of the product.

1.2 Conventional Processing Techniques

The juice production starts with handling, which, in the case of apples and pears, is with the use of water or conveyor belts. Rotten or moldy apples and pears should be removed. Later, washing is required to remove leaves and twigs and dirt or other

*Corresponding author: Ingrid Aguiló-Aguayo, Ingrid.Aguilo@irta.cat

Innovative Technologies in Beverage Processing, First Edition.
Edited by Ingrid Aguiló-Aguayo and Lucía Plaza.

water-soluble agricultural spray residues. Sanitizing process will then be carried out to avoid a high microbial load, undesirable flavors, or mycotoxin contamination (Barret *et al.*, 2005).

Prior to juice extraction, the fruit has to be pulped to release the trapped juices, and in the case of apples and pears, they need to be pressed at fairly high pressure to force the juice to flow through the cell structure (Downes, 1995). The fruits are milled to a pulp by a disintegration process that starts with a crushing step to break down the cell tissue. When fruits have been crushed, it undergoes a pressing step where the juice is extracted from the fruit by using conventional pack presses or horizontal rotatory presses. Juice extraction can also be performed by pectolytic enzymes, but apples and pears can normally be pressed fresh without any assistance. In general, juice-extraction process should be carried out as quickly as possible in order to minimize oxidation of the product. When juice has been extracted, clarification and filtration methods are generally conducted depending on the characteristics of the final product. For cloudy juices, clarification will not be necessary and only controlled centrifugation or a course filtration will be conducted to remove larger insoluble particles. To obtain a clarified juice, it will be necessary to remove the turbidity by a clarification step. Therefore, a complete depectinization by addition of pectinase enzymes, fine filtration, or high-speed centrifugation will be required (Barret *et al.*, 2005). To obtain concentrate juices, a multiple-stage evaporation process after clarification is carried out. The clarified juice has a soluble solid content of around 15 °Brix, and processing industries normally obtain juices with a content of 70 °Brix. Hence, in the concentration process, the juice changes its soluble solids content and temperature flows from one evaporator. This process, which often works at a temperature of 60 °C, not only preconcentrates the juice but also subjects the product to some reduction in quality and loss of nutrients (Falguera & Ibarz, 2014). Then, the concentrate is immediately cooled below 20 °C (NPCS, 2008). In order to destroy microorganisms and inactivate the enzymes that are present in all the obtained products, thermal pasteurization process is conducted in a flow-through heat exchanger by high-temperature–short-time (HTST) treatment, for example, 76.6–87.7 °C for holding time between 25 and 30 s (Moyer & Aitken, 1980). In-pack pasteurization directly processes into the preheated pack ensuring product integrity (e.g., held at 75 °C for 30 min) (NPCS, 2008). However, slight correction of acidity of pome fruit juices to reach equilibrium pH of 4.6 or below by addition of commercial citric acid is necessary before pasteurization process. Pome fruit juices can also be processed by an aseptic method, in which temperatures are risen well over 100 °C, but holding times are shortened to only few seconds. Aseptic technology, also known as ultrahigh-temperature (UHT) processing, involves the production of a sterile product by rapid heating at high temperatures, followed by a short holding time and ending with rapid cooling (Charles-Rodríguez, 2002).

1.2.1   Influence on Microbial Quality

Microflora present in fruit juices are normally associated to the surface of fruits during harvest and postharvest processing including transport, storage, and processing (Tournas *et al.*, 2006). The severity and extent of the thermal treatment will depend on factors such as type and heat resistance of target microorganisms, spore or enzyme present in food, pH, oxidation–reduction potential, or water activity of the food (Ramaswamy & Singh, 1997). The pH is also a critical factor, taking into account that foods may be broadly divided into high-acid (pH < 3.7) and low-acid (pH > 4.5) foods. Apple juices have the pH range between 3.35 and 4.00 serving as important barrier for microbial growth. Pears are not abundant in ascorbic acid (Kopera & Mitek, 2006) and some pear varieties have pH sometimes higher than 4.6 (Visser *et al.*, 1968). The acidic nature of apple or pear juices allows pasteurization, defining the use of temperatures close to 100 °C in order to inactivate spoilage microorganisms. However, when pH is greater than 4.6, spore heat resistance marks the process temperature that should be greater than 115 °C for extended shelf life. Therefore, the pH reduction by acid addition is a widely practice in the juice food industries.

Spoilage organisms of particular interest to fruit juice manufacturers are *Alicyclobacillus* and yeasts and molds. *Alicyclobacillus acidoterrestris* is acidophilic spore-forming, heat-resistant organism that may be found in fruit juices. This microorganism can survive commercial pasteurization processes commonly applied to apple and pear juices spoiling fruit juice by producing cloud loss, development of off-flavors, CO_2 production, or changes in color and appearance (Lawlor *et al.*, 2009). Saritas *et al.* (2011) evaluated the effect of ascorbic and cinnamic acid addition and different pasteurization parameters on the inactivation of *A. acidoterrestris*. They concluded that the increase in the acid concentrations led to a decline in the z-value of the microorganisms, which is greater in the clear apple juice than in the apple nectar.

Several studies have been conducting regarding the recommended pasteurization treatments in apple juice. Mazzotta (2001) recommended a general thermal process of 3 s at 71.1 °C for achieving a 5-log reduction of *Escherichia coli* O157:H7, *Salmonella*, and *Listeria monocytogenes* in apple juices adjusted to a pH of 3.9. They reported that pH ranges from 3.6 to 4.0 in apple juice had no significant impact on the heat resistance of *E. coli* O157:H7. However, the acid tolerance of some microorganisms including *E. coli* and *Salmonella* led to the use of chemical preservatives such as benzoic acid in pome fruit juices to help processors to increase, substantially, the safety of the pasteurized products (Albashan, 2009; Koodie & Dhople, 2001). According to the Food and Drug Administration (FDA) recommendations, the pasteurization process must ensure a 5-log reduction of the three stated vegetative bacterial pathogens (*E. coli* O157:H7, *Salmonella*, and *L. monocytogenes*) at pH values of 3.9. However, thermal destruction of the protozoan parasite *Cryptosporidium parvum* oocysts causing illness outbreaks associated with the consumption of apple juice must be taken into consideration. Published studies suggest that *C. parvum* might be more resistant to heat processing than the indicated three vegetative bacterial pathogens (Deng & Cliver, 2001). Therefore, FDA suggests a treatment at 71.7 °C for 15 s for apple juice at pH values of 4.0 or less to achieve the 5-log reduction for the three mentioned vegetative bacterial pathogens and for oocysts of *C. parvum* (FDA, 2004). For the case of pear juice with pH of 4.6 or less, FDA recommends to conduct the pasteurization process that guarantees the same 5-log reduction for the indicated microorganisms. Regarding patulin, a mycotoxin produced by certain species of *Penicillium*, *Aspergillus*, and *Byssochlamys* molds, the regulations limit their content in apple juice to no more than 50 µg/kg (Codex, 1999).

1.2.2 Influence on Nutritional Attributes

Apple and pear cultivars are well known for being good sources of antioxidant properties, especially for their content of polyphenolic compounds concentrated in their skin. Apples contain natural sugars, organic acids, dietary fiber, minerals, and vitamins, whereas pears are good sources of fiber and micronutrients such as chlorogenic acid, flavonols, and arbutin (Sinha, 2012). In general, pear juice is characterized by similar soluble solids content than apple juices, but in sugar-free extract, sorbitol content has higher contribution, which is estimated to be 10–25 g/l (Dietrich *et al.*, 2007).

Thermal processing inactivates spoiling microorganisms efficiently but may also degrade nutritional quality of foods (Qin *et al.*, 1995). According to Charles-Rodríguez (2002), conventional UHT treatment has apparent effects on pH, soluble solids, and acidity of the processed apple juice. Throughout the years, process optimization has been carried out in order to reduce processing times at high temperatures and avoid undesirable quality changes during the process (Awuah *et al.*, 2007). Nevertheless, thermal processing can still induce reactions that could affect color, odor, flavor, texture, and health-related compounds, which all could be linked to the change in phenolic profile (Niu *et al.*, 2010). Aguilar-Rosas *et al.* (2007) reported that HTST treatment caused a considerable loss of phenols (32.2%) in apple juices. These results agree with those reported by Spanos and Wrolstad (1992) who reported that total phenol concentration is reduced up to 50% in thermally treated apple juice at 80 °C for 15 min. Gardner *et al.* (2000) also observed a considerable loss of phenolics in pasteurized apple juice by thermal means. De Paepe *et al.* (2014) identified 42 compounds to be susceptible to thermal degradation, indicating that the most heat labile phenolic constituent in cloud apple juice was procyanidin subclass, according to the high degree of structural elucidation achieved. Jiang *et al.* (2016a) also observed an important decrease of 53.11% and 46.47% in the total phenolic and flavonoid content of pear juice concentrates from Hwasan and Niitaka cultivars. They indicated that the antioxidant levels were significantly affected during the pear juice production, especially during the pressing step in which press cake waste retains the seed and skin. A complete inactivation of enzymes (peroxidase, POD; pectin methylesterase, PME; and polyphenol oxidase, PPO) and microorganisms (total plate count, yeasts, and molds) was observed in conventional pasteurization at 95 °C, but this treatment also showed highest losses of ascorbic acid, total phenols, flavonoids, and antioxidant capacity (Saeeduddin *et al.*, 2015).

Moreover, during the clarification step of juice processing, a part of polyphenols is also lost(Dietrich, 2004; Hubert *et al.*, 2007). In contrast, the enzymatic and clarification stages are avoided in cloudy juices, which results in higher phenolic content in the final product. This means that cloudy juices have higher antioxidant activity and dietary fiber than clear ones (Oszmianski *et al.*, 2007). Willems and Low (2016) studied the effect of enzyme treatment and processing on the oligosaccharide profile of commercial pear juice samples. They observed that the majority of polysaccharide hydrolysis and oligosaccharide formation occurred during the enzymatic treatment at the pear-mashing stage, and the remaining processing steps had a minimal impact on the carbohydrate-base chromatographic profile of pear juice.

A reduction in nutritional quality could also be caused by browning reactions of fruits during processing and storage. Due to the general low acidity of pears, the addition of antibrowning agents to pear juice can lower the pH with a suitable sugar–acid ratio. According to Jiang *et al.* (2016b), addition of 0.20% ascorbic acid might solve the browning problem as well as improve the antioxidant capacity of pear juice.

1.2.3 Influence on Organoleptic Attributes

Thermal pasteurization of apple and pear juices is effective in preventing microbial spoilage and extends their shelf life. However, these treatments induce a negative impact on their natural flavor and color that is perceived by consumers (Mak *et al.*, 2001).

1.2.3.1 Changes in Aroma Profile Volatile composition and content in apple and pears is strongly influenced by the variety, maturity, and storage (Sinha, 2012). Different studies reported that ester compounds like ethyl butyrate, ethyl-2-methyl butyrate, butyl acetate, and hexyl acetate are key odorants giving the "apple" flavor; hexyl acetate the "sweet-fruit," and 1-butanol the "sweetish sensation" (Lavilla *et al.*, 1999; Mehinagic *et al.*, 2006; Sinha, 2012). Regarding pear aroma, compounds including hexanal, cinnamaldehyde, methyl and ethyl decadienoates, and farnesenes have been reported to be responsible for apple aroma (Riu-Aumatell *et al.*, 2005). Few studies have been focused on the volatile composition of pasteurized pome fruit juices. It is generally accepted that a conventional thermal juice pasteurization treatment induces a significant loss of the fresh flavor content (Braddock, 1999). In apple juice, a decrease in the aroma components including iso-butyl acetate, ethyl butyrate, ethyl 2-methyl butyrate, and hexanal was observed after pasteurization at 85 °C for 10 min (Su & Wiley, 1998). An HTST treatment at 90 °C for 30 s resulted in more than 50% loss in hexanal, ethyl acetate, ethyl butyrate, methyl butyrate, and acetic acid concentrations in apple juice (Aguilar-Rosas *et al.*, 2007). Kato *et al.* (2003) concluded that the odor attributes of pasteurized apple juices could not be predictable. However, they could determine that treatment temperature affected the apple juice volatile concentration more than the treatment time.

The effect of heat on apple and pear juice aromas could be stated during the juice concentration process. Volatiles are normally recovered and concentrated for later addition of the concentrated products, resulting in the loss of volatiles if incomplete restoration takes place (Renard & Maingonnat, 2012). In apple juices, conventional evaporation could induce a loss of more than 95% of *trans*-2-hexenal (Onsekizoglu *et al.*, 2010). Steinhaus *et al.* (2006) observed a distinct loss of the major juice aroma component β-damascenone during apple juice concentration, and high losses were also found for dimethyl sulfide and 1-octen-3-one.

1.2.3.2 Changes in Appearance (Color and Juice Clarity) Cloudiness and color stability are two of the main important quality characteristics for cloudy apple and pear juices, strongly influenced by the enzyme activities of pectinesterase (PE) and PPO. In general, HTST treatment can promote the inactivation of these enzymes but their action begins from initial processing steps such as the breakage of the fruit during the juice-extraction process (Krapfenbauer *et al.*, 2006). In addition to heat treatments, other methods including blanching, pectolytic enzyme treatments, or the use of antibrowning agents are some of the strategies to control browning and cloud destabilization (Fukutami *et al.*, 1986; Gierschener & Baumann, 1988; Castaldo & Loiudice, 1997; Quoc *et al.*, 2000; Riahi & Ramaswamy, 2003). Krapfenbauer *et al.* (2006) focused their study on the impact of HTST treatment combinations (60–90 °C for 20–100 s) on enzymatic browning and cloud stability of different apple juices considering eight apple varieties. They concluded that HTST at 80 °C inactivated PPO and reduced PE activity at 50%. In addition, they indicated that the best stability of cloud and color in relation to heat treatment was observed at 70 °C/100 s and 80 °C/20 s.

Sun (2012) reported that the influence of HTST treatment at 73, 80, or 83 °C for 27 s on apple juice induced values of total color difference (TCD) of 1.3. TCD reflects the overall color difference evaluation between the reference sample, normally that of initial fresh product, and the final processed product. The accumulation of brown color during thermal treatment of juices is attributed to nonenzymatic reactions involving caramelization and Maillard reactions (Ibarz et al., 1990). Nonenzymatic browning kinetics of concentrated pear juices from three pear varieties ("Alexandrine Douillard," "Flor de Invierno," and "Blanquilla") has been studied at three temperatures (90, 80, and 70 °C) and different soluble solid contents (52, 62, and 72 °C) (Ibarz et al., 1997). The "Alexandrine Doullard" pear juices showed less nonenzymatic browning in contrast to the "Flor de Invierno" and "Blanquilla" juices, but formation of 5-hydroxymethyl-2-furfuraldehyde followed first-order kinetics in all products. Burdurlu and Karadeniz (2003) evaluated the kinetics of nonenzymatic browning in *Golden Delicious* and "Amasya" apple juice concentrates for 4 months. They reported an increase in browning level of all apple juices according to a zero-order reaction kinetics. Hsu et al. (1990) observed higher degree of browning in juice and concentrate from "Barlett" than "Comice" and "Anjou" pears, which attributed to the high amino acid content.

1.3 Novel Processing Techniques

The application of promising novel processing techniques in the manufacture of pome fruit juices is being established in different steps of their production chain. On the other hand, the production process of pome fruit juices includes the steps of extraction and clarification. The yield of juice extracted by the application of pretreatments has been explored to ensure the highest possible quality of the product and increase the juice recovery. Clarification process allows to obtain a final product without turbidity and to preserve its organoleptic properties. Novel clarification processes have some advantages in relation to the conventional processes taking into account the final sensory and nutritional properties of the product. Moreover, novel food preservation approaches including physical methods based on nonthermal treatments, chemical methods such as natural food preservatives, and their combinations for extending their shelf life and also preserving natural organoleptic and nutritional value are result-oriented novel techniques currently applied in these products.

1.3.1 Improvement in Juice Extraction

Juice extraction is the process where the liquid phase is separated from the solid particles (Girard & Sinha, 2012). Juice extraction can be carried out using various mechanical processes, which may be achieved through diffusion extraction, decanter centrifuge, screw type juice extractor, fruit pulper, and by different types of presses (Sharma et al., 2016). Crushing operation prior to pressing can be divided into chopping and preparation for pressing. In addition, enzymatic treatment prior to mechanical extraction significantly improves juice recovery compared to any other extraction process (Sharma *et al.*, 2016). In this way, pectinase and polygalacturonase (PG) enzymes, which act on glycosidic linkage acting as fruit softening, can be used as a press aid adding to milled/crushed apples before the juice-extraction step (Sinha, 2012).

According to Bazhal and Vorobiev (2000), electrotechnologies such as electro-osmotic (EOs) treatments and pulsed electric fields (PEFs) could intensify the juice-extraction yield. EO treatments lead to addition or extraction of liquid from capillary-porous materials. The application of PEF as a pretreatment operation before pressing could allow significant increase in the juice yield and obtaining products with higher quality (Vorobiev *et al.*, 2004). Bazhal and Vorobiev (2000) carried out an experiment with *Golden Delicious* apple slices in a laboratory filter-press cell fitted with an appropriate electric treatment system. PEF treatments consisted of 1000 monopolar pulses of 1000 V with duration of 100 μs and a period of 10 ms, whereas EO treatments consisted of successively 50, 100, and 200 mA for 30 min each. They observed that both EO and PEF treatments caused significant increase in apple juice yield. However, the energy consumption was 50–100 times less for PEF treatments. Wu and Zhang (2015) evaluated the influence of PEF pretreatment on apple tissue structure, confirming an evident disorder in cells of fresh apple caused by the electroporation and the membrane damage of PEF. However, the pretreated samples did not collapse and maintained a desirable structure. Carbonell-Capella *et al.* (2016) studied the impact of apple pretreatment by PEF on juice extraction using the freezing-assisted pressing. They observed that freezing-assisted pressing at subzero temperatures was an effective tool in order to obtain an apple juice rich in bioactive compounds. In addition, the process was improved by the application of PEF pretreatment of apple tissue before freezing resulting in a reduction of both freezing and thawing time, and pressing was more effective.

1.3.2 Improvement in Juice Clarification

At industrial level, clarification of pome fruit juices is normally conducted mechanically or enzymatically. Mechanical clarification normally uses centrifuges and filtration devices by the first filtration phase in settling centrifuges, meanwhile decanters are used to remove fibers from cloudy juices (Welter *et al.*, 1991; Nagel, 1992). Conventional filtration methods are labor intensive and time consuming (Zhao *et al.*, 2014). New filtration methods such as ultrafiltration (UF) require lower labor and energy operation cost and have lower waste disposal in comparison to conventional methods (de Bruijn *et al.*, 2003; Zhao *et al.*, 2014). UF has shown excellent quality attributes for clarified apple juice (Álvarez *et al.*, 2000; De Bruijn *et al.*, 2002). Álvarez *et al.* (2000) proposed an integrated membrane process for producing apple juice and apple juice aroma concentrates. The process involved an integrated membrane reactor to clarify the raw juice, reverse osmosis to preconcentrate the juice up to 25 °Brix, prevaporation to recover and concentrate the aroma compounds, and a final evaporation step to concentrate apple juice up to 72 °Brix. The obtained juice was more clear and brilliant than apple juice produced by conventional methods, and the system had more economical advantages than the conventional process. The combination of UF (0.05 μm) with ceramic membrane and high hydrostatic pressure processing (HHP) at 500 MPa for 6 min to process fresh apple juice shows good results in terms of color, clarity, and total phenols retention, leading to microbiologically safe products for 60 days at 4 °C (Zhao *et al.*, 2014).

Pectic enzymes, that is, PME and PG, are often used in combination with amylases and cellulases in fruit and vegetable juice clarifications to obtain higher juice yields and clarity (Raju & Bawa, 2012). Fungal PG is normally used in the industrial processes for juice clarification. However, recently clarification of juice has been carried out through recombinant and nonrecombinant fungal strains. Singh and Gupta (2004) evaluated the effect of gelatin on the efficacy of a fungal pectinolytic enzyme preparation from *Aspergillus niger* van Tieghem for clarification of apple juice. They observed a complete clarification of the apple juice (200 ml) using 0.01% gelatin and 15 IU of enzyme preparation at 45 °C after a holding time of 6 h. In addition, the scaling up to 200 times results in about 143% more transmittance than the control juice, and the clarified juice stored at room temperature did not show any haze development after 2 months.

1.3.3   Preservation of Pome Fruit Juices by Innovative Technologies

1.3.3.1   Pulsed Electric Field Processing   Several researchers (Qin *et al.*, 1994; Ortega-Rivas *et al.*, 1998; Zárate-Rodríguez *et al.*, 2000; Aguilar-Rosas *et al.*, 2007) have successfully studied the application of PEF processing for extending the shelf life of apple juices. The application of high voltage induced structural changes in membranes of microbial cells based on electroporation, leading to microbial inactivation (Tsong, 1991; Barbosa-Cánovas *et al.*, 1999). Energy savings for PEF processing have been reported in comparison to conventional thermal processing in addition to be more environmentally friendly (Toepfl *et al.*, 2006). The PEF treatment has been reported to use less than 10% of the electric energy for heat treatment of apple juice (Qin *et al.*, 1994). Aguilar-Rosas *et al.* (2007) observed that PEF treatment at 35 kV/cm and a frequency of 1200 pulses per second in bipolar mode using pulses of 4 μs wide pasteurized an apple juice from *Golden delicious* fruits. They reported that PEF treatment proved to be efficient in microbial inactivation and induced better preservation in quality attributes than a conventional HTST pasteurization. Ortega-Rivas *et al.* (1998) compared UF and PEF in apple juice pasteurization reporting 6-log reductions in survivability of total aerobic microorganisms using the indigenous flora of the PEF-treated product with no differences in quality. Evrendilek *et al.* (2000) studied the microbial safety and shelf life of PEF-treated apple juice showing that PEF (35 kV/cm for 94 μs total treatment time) not only could extend the shelf life of the fresh product but also maintained a fresh flavor. They reported an inactivation of *E. coli* O157:H7 of 4.5-log by the PEF treatment and no changes in the natural food color and vitamin C of the product. Timmermans *et al.* (2014) conducted experiments using a continuous-flow PEF system at 20 kV/cm with variable frequencies to study the inactivation *of Salmonella panama, E. coli, L. monocytogenes*, and *Saccharomyces cerevisiae* in apple juice. They observed that under the same conditions, *S. cerevisiae* was the most sensitive microorganism, followed by *S. panama* and *E. coli*. A synergistic effect between the inlet temperatures above 35 °C and PEF treatment was demonstrated, suggesting that optimization of the PEF conditions to

reduce the energy input should aim for processing at higher inlet temperature to allow more effective inactivation per pulse. Applying PEF treatment at 30.76 kV/cm and using up to 21 pulses in apple juice achieved the 5-log reduction of *E. coli*, FDA-required standard for alternative pasteurization methods (Moody *et al.*, 2014). Little information is available regarding the effect of PEF on pear juices. Jin *et al.* (2008) evaluated the application of PEF at 30 kV/cm for 240 μs at 200 Hz and 10 °C on the microbial inactivity and quality of freshly squeezed pear juice. They reported an inactivation of *E. coli* and *S. cerevisiae* of 4.6 and 2.7-log after the PEF treatment, with no effect on the physicochemical properties and nutrition of the product.

Related to enzyme inactivation, Giner *et al.* (2001) reported a decrease in PPO activity up to 3.15% at 24.6 kV/cm for 6 ms of treatment time, whereas a total inactivation of 38% was obtained for pear PPO at 22.3 kV/cm for the same treatment time. Almost a complete inactivation of PPO and POD was achieved after treating apple juice at 35 kV/cm and 2 ms of pulse rise time (Bi *et al.*, 2013). The same PEF treatment conditions led to apple juices with significant higher lightness and yellowness than the control samples as well as preservation of initial total phenol content. No differences between PEF-treated apple juices samples (1, 3, 5 kV/cm, $n = 30$ pulses) and fresh samples were reported related to the overall apple juice composition described by pH, total soluble solids, total acidity, density, contents of sugar, malic acid, and pectin as well as polyphenol contents and antioxidant capacity (Schilling *et al.*, 2007). Charles-Rodríguez *et al.* (2007) also reported that PEF treatment seemed to retain more the color of natural apple juice than thermally pasteurized juices since less browning effect was observed. Harrison *et al.* (2001) reported that PEF-treated apple juice could be stored at 4 °C for 1 month with no change in the volatile flavor profile.

Combination of preheating at 40 °C and PEF treatment has been shown to induce an antimicrobial effect while avoiding detrimental effects on the quality of the product and inactivated PPO enzymes in apple juice (Heinz *et al.*, 2003; Ertugay *et al.*, 2013). Riener *et al.* (2008) reported that preheating at 50 °C and PEF treatment at 40 kV/cm for 100 μs achieved a PPO inactivation of 71% in PPO apple juice. Similar conditions combining preheating at 50 °C and PEF processing at 38.5 kV/cm and 300 pulses per second reduced a 70% of PPO activity in apple juice (Sanchez-Vega *et al.*, 2009). Synergistic lethal effects of PEF treatment and carvacrol have been reported in relation to increase the microbial inactivation in apple juices (Ait-Ouazzou *et al.*, 2013). PEF treatments consisting of 50 exponential decay pulses at 30 kV/cm induced less than 1-log cell cycle reduction, while the combination with 1.3 mM of carvacrol caused the inactivation of 5-log cell cycles with 20 pulses. Similar synergistic lethal effects on apple juice have been observed for the combination of essential oils including *Cyperus longus* L., *Eucalyptus globulus* L., *Juniperus phoenicea* L., *Mentha pulegium* L., *Rosmarinus officinalis* L., and *Thymus algeriensis* and mild PEF treatment (30 kV/cm, 25 pulses) (Ait-Ouazzou *et al.*, 2012). Mosqueda-Melgar *et al.* (2008) also reported that combinations of PEF treatment (35 kV/cm, 4 μs, <40 °C) with 0.1% cinnamon bark oil or 1.5% citric acid achieved more than 5-log reductions in microbial populations and extended the shelf life of apple and pear juices for 91 days at 5 °C.

The application of PEF and membrane UF as nonthermal preservation technique to process apple juice has been shown to be very efficient in preserving the quality attributes in terms of soluble solids, pH, acidity, and color ratio of the product (Zárate-Rodríguez *et al.*, 2000). Noci *et al.* (2008) studied the influence of ultraviolet irradiation (UV) and PEF on microbial inactivation, quality attributes, and enzymatic activity of fresh apple juice. They observed that application in combination UV + PEF or PEF + UV achieved similar reductions of 6.2- and 7.1-log cycles, respectively, of total bacterial counts. Both combinations also showed better retention of juice color and level of phenolic compounds as well as better reduction of PPO and POD activities in comparison to heat pasteurization. Walkling-Ribeiro *et al.* (2008) reported a higher reduction of *S. aureus* with a hurdle approach (UV; 46 °C (PEF inlet) and 58 °C (PEF outlet); PEF 40 kV/cm and 100 µs) in comparison to conventional pasteurization (9.5- vs 8.2-log, respectively) with little effect on the quality of apple juice.

1.3.3.2 High Hydrostatic Pressure Processing
HHP processing has been proved to meet the FDA requirement of a 5-log reduction of microorganisms in pome fruit juices without changes in the sensory and nutritional characteristics of the product due to the low processing temperature. Ramaswamy *et al.* (2003) achieved more than 5-log reduction of *E. coli* 29,055 after treatment of 400 MPa at 25 °C. Moody *et al.* (2014) also achieved 7-log reduction of *E. coli* after treatment of 400 MPa for 3 min. Evelyn *et al.* (2016) evaluated the efficacy of HHP at 600 MPa in combination with 75 °C to inactivate *Neosartorya fischeri*, a mold that spoils apple juice and can produce mycotoxins. They observed around 3.3-log reduction after the HHP-75 °C treatment for 10 min, suggesting that HHP could be an option for apple juice preservation. According to Juarez-Enriquez *et al.* (2015), HHP treatment consisting of 430 MPa for 7 min could extend shelf life of apple juice for 34 days with no changes in the physicochemical properties, nutritive value, or sensory attributes. In addition, PME and PPO PPO activities were controlled by the HHP treatment.

Combination of ultrafiltration (UF, 0.05 µm) and HHP treatments (500 MPa for 6 min) showed that total plate count and yeasts and molds decreased below 1-log cycle after the treatment of apple juice (Zhao *et al.*, 2014). In addition, the product showed lower browning degree, higher total phenols and clarity, and better retention of main volatile aroma compounds than HTST for 60 days of storage at 4 °C. The combination of HHP processing with nisin for the inactivation of *A. acidoterrestris* spores in apple has also been evaluated. HHP treatment of 200 MPa for 45 min with a nisin content of 250 IU/ml enabled total spore inactivation (Sokolowska *et al.*, 2012).

1.3.3.3 Other Innovative Technologies for Preservation Purposes
The application of other innovative physical methods has been explored to preserve pome fruit juices including ultrasound (US), dense phase carbon dioxide (DPCD) processing, ultraviolet light-C (UV-C), pulsed light (PL), high-pressure homogenization (HPH), or ohmic heating (OH).

The application of US has showed positive results for apple juice pasteurization meeting the criteria of the FDA, concerning a 5-log reduction of microbial cells in the fruit juices. Yuan *et al.* (2009) reported a 80% of reduction of *A. acidoterrestris* when treating apple juice at 24 kHz, 300 W for 60 min. Moody *et al.* (2014) also reported more than 6-log reduction after 5 min of US treatment at 400 W and 24 kHz, keeping the temperature at 60 °C. Bastianello *et al.* (2016) investigated the microbial shelf life and the volatile compounds of cloudy apple juices after US treatment (400 W and 24 kHz at 35 °C applying a process of 360 s/100 ml). They observed a sublethal injury of spoiler yeasts (*Candida parapsilosis* and *Rhodotorula glutinis*) 14 days post treatments, estimating the shelf life of the product to be around 21 days. The typical aroma compounds of untreated samples decreased in comparison to the US-treated juices. US treatment has shown to have some effect on PPO inactivation in apple juice at 23 °C and applying US power densities of 3300 W/l for 20 min (Silva *et al.*,

2015). Sun *et al.* (2015) also observed that US inhibited the browning of fresh apple (*Malus pumila Mill*, *cv. Red Fuji*) juice by applying an US intensity of 2 W/cm^2 at 15 °C for 10 min but decreased the total phenolic content and antioxidant capacity of the product. Abid *et al.* (2014) also observed that US treatment (20 kHz, 20 °C for 3 min) did not change the total soluble solids, pH, and acidity of apple juice, while significant increase in nonenzymatic browning and cloud values was observed. Ertugay and Başlar (2014) also observed that US treatment increased the cloudiness level up to 16.9 times and the cloud stability up to 9.8 times of apple juice. Saeeduddin *et al.* (2016) reported that US treatment at 25 kHz for 60 min at 25 °C exhibited optimum results in terms of physicochemical and microbial quality of pear juice. Saeeduddin *et al.* (2015) also observed that US-treated (20 kHz at 10 °C for 65 °C) pear juice quality was not affected by the US treatment, but an improvement of phenolic compounds and ascorbic acid was achieved in the US-treated product. Synergistic effects of US (30 min at 20 kHz) and continuous flow-through PL system (0.73 J/cm^2, 155 ml/min) for the treatment of apple juice were observed by Ferrario and Guerrero (2016). They reported that the combined treatment delayed yeast and mold recovery and prevented browning development during the storage of the product. The combination of the US and mild temperatures, called thermosonication, has been attempted as an alternative thermal treatment for the processing of apple juices. Abid *et al.* (2014) observed a quality enhancement of apple juice in terms of enzyme and microbial inactivation when using ultrasound with-probe sonicator (20 kHz for 10 min at 60 °C).

Other nonthermal methods include the DPCD, based on a cold pasteurization in which the product is in contact with either (pressurized below 50 MPa) sub- or supercritical CO$_2$ for certain time (Porto *et al.*, 2010). Only one study has been carried out in apple juice resulting in a pasteurization effect, minimizing the quality loss in terms of nutritional value and volatile compounds applying 15 MPa at 35 °C for 15 min or 25 MPa at 35 °C for 15 min (Porto *et al.*, 2010).

The UV-C (200–800 nm) has been used as nonthermal method to successfully reduce the microbial load of apple juice. The good UV-C penetration in clear apple juice treated with UV dose of 1377 J/l showed a 7.42-log reduction of inoculated *E. coli* with no color changes (Keyser *et al.*, 2008). Char *et al.* (2010) observed that UV-C treatments in a UV-C device, consisting of a 90-cm long UV-C lamp (100 W) placed inside a glass tube leaving an annular flow space (0.2 l/min, 40 °C), effectively inactivated microbial populations (*E. coli* and *S. cerevisiae*) in apple juice.

PL technology involves applying intense and short pulses of white light that have antimicrobial action due to the presence of a UV component within the broad spectrum of light flash (Maftei *et al.*, 2014). Ferrario *et al.* (2015) reported that PL treatments of 3 pulses per second and fluencies ranging 2.4–71.6 J/cm^2 induced 3.8-log reductions in clear apple juice. According to Caminiti *et al.* (2012), treatments above 2.66 J/cm^2 achieved reduction levels below 1-log CFU/ml for both *E. coli* and *L. innocua* in apple juice. Treating with doses up to 10.62 J/cm^2 were in the same range as fresh samples in relation to the sensory evaluation of the products. Microbial reductions of 4-log-cycles for *E. coli* and 2.98-log-cycles for *L. innocua* were observed after applying PL fluences of 4 J/cm^2 in apple juices (Pataro *et al.*, 2011). Reductions up to 3.76-log CFU/ml of *Penicillium expansum* were reported after applying 0.4 J/cm^2 per pulse, 40 flashes, and a depth of the juice of 6 mm with some darkening color changes (Maftei *et al.*, 2014). Funes *et al.* (2013) suggested that PL treatments would be a potential alternative method to reduce patulin contamination in apple products including juices, since a significant decrease in patulin levels (up to 22%)

Table 1.1 Application of antimicrobial chemical treatments in pome fruit juices

Product	Treatment	Storage conditions	Quality changes	Source
Apple juice	Bacteriocin (brevicin from *Lactobacillus brevis* NS01)		Increase shelf life and clarification comparing to chemical agents (potassium sorbate or sodium benzoate)	Duraisamy *et al.* (2015)
Apple juice (12 °Brix, pH 4.0)	4.0 mM Sorbic acid, 1.5 mM cinnamic acid, 20 mM vanillin, 10 mM ferulic acid or 12 mM *p*-coumaric acid, 20 min at temperatures from 46 to 55 °C	4 °C	Complete inhibition of all initial yeast inoculum (10^3 CFU/ml). Addition of antimicrobial prevented spoilage during storage	Wang *et al.* (2016)
Apple juice	Rosemary commercial extracts V20 and V40 at 7.8 mg/ml and 3.9 mg/ml, respectively		Reduce growth of alicyclobacilli vegetative cells but did not show sporicidal effects on alicyclobacilli. Do not change sensory properties	Piskernik *et al.* (2016)
Apple juice	Propolis 0.1–0.2 mg/ml and heat treatment at 51 °C		Strong synergistic and lethal effects against *E. coli* O157:H7 Sakai and sensorially acceptable product	Luis-Villaroya *et al.* (2015)
Apple juice	Propolis extract at 1.2% and 5% concentrations		Antimicrobial activity against *E. coli* and *E. coli* O157:H7	Sagdic *et al.* (2007)
Apple juice (pH 3.82, 11.3 °Brix)	Grape seed extract (GSE) (0–1.9% v/v)	37 °C	Inhibition of growth of *A. acidoterrestris* cells and spore germination/outgrowth	Molva and Baysal (2015a)
Apple juice (pH 6.0)	Yerba mate extract 40 mg/ml		4.5-log reduction of *E. coli*	Burris *et al.* (2012)
Apple juice	Chitosan 5 g/l	25 °C	Yeast-free conditions for 14 days	Roller and Covill (1999)

Product	Treatment	Temperature	Effect	Reference
Red apple juice	Six essential oils extracted from Shieh, kafoor and Neem (0.5%)	4°C	Reduction of PPO. Increased the shelf life up to 4 weeks	Eissa et al. (2012)
Apple juice (pH 3.82, 11.3°Brix)	Commercial pomegranate extract (POMELLA®, PE)	37°C	Inhibition of growth of A. acidoterrestris cells and spore germination/outgrowth	Molva and Baysal (2015b)
Apple juice	Caprylic acid (0.6 mmol/l or 0.8 mmol/l) at 50°C for 5 min		Complete inactivation of inoculated E. coli (7.25–7.34-log CFU/ml)	Kim and Rhee (2015)
Apple juice	Carvacrol and p-cymene (0.25–1.25 mM)	4°C	Undetectable levels of E. coli O157:H7 for 19 days	Kiskó and Roller (2005)
Apple juice	0.3% Cinnamon with 0.1% sodium benzoate	8 and 25°C	Inactivation of 5.2-log CFU/ml of E. coli O157:H7 in 11 days	Ceylan et al. (2004)
Apple juice (pH 3.3 and 3.8)	Vanillin (3000 ppm)		Reduction ranging from 4-log cycles after a 4–8 h exposure at 30°C	Corte et al. (2004)
Apple juice	2.5% (v/v) Malic acid	5°C	Reduction of more than 5-log cycles of L. monocytogenes, S. enteritidis and E. coli O157:H7 after 24 h inactivation	Raybaudi-Massilia et al. (2009)
Pear juice	2.5% (v/v) Malic acid	5°C	Reduction of more than 5-log cycles of L. monocytogenes, S. enteritidis and E. coli O157:H7 after 24 h inactivation	Raybaudi-Massilia et al. (2009)
Apple juice (pH 6.0)	Yerba mate extract 40 mg/ml		4.5-log Reduction of E. coli	Burris et al. (2012)

was observed after applying PL doses of $35.8\,J/cm^2$. In terms of nutritious quality, Orlowska *et al.* (2013) observed that PL treatments consisting of 31 J per pulse only reduced 1.30% vitamin C content.

On the other hand, HPH technology is considered to be a promising nonthermal technology, which involves combination of spatial pressure and velocity gradients, turbulence, impingement, cavitation, and viscous shear (Vasantha & Yu, 2012). Homogenization pressures from 100 to 200 MPa induced significant inactivation of *E. coli* K-12 in apple juice (Kumar *et al.*, 2009). However, synergistic effects of the HPH treatment and chitosan at 0.1% concentration resulted in enhancing microbial inactivation. Pathanibul *et al.* (2009) reported more than 5-log reductions of *E. coli* and *L. innocua* when apple juice was exposed to homogenization pressures higher than 250 MPa. They observed interaction effects of 10 IU nisin in combination with the HPH treatment on *L. innocua*. Bevilacqua *et al.* (2012) observed a reduction by 2–4-log CFU/ml of *Saccharomyces bayanus* by applying homogenization pressures of 20 mPa in combination with 900 ppm of limonene and 2 ppm of citrus extract, respectively.

The potential of OH on the treatment of juices has been explored as an alternative method for sterilization or pasteurization. Jakób *et al.* (2010) observed that kinetic parameters related to inactivation of PME and POD in apple juice remained the same after OH compared to conventional indirect heating. Park and Kang (2013) studied the effect of electric field-induced (60 V/cm) OH for inactivation of *E. coli* O157:H7, *Salmonella enterica* serovar Typhimurium, and *L. monocytogenes* in apple juice. They observed reduction between 3.40 and 3.59-log for all the three foodborne pathogens when processing the juice at 60 °C for 30 s. However, combinations at 58 and 60 °C with electric field increased the inactivation effect 2–3 times higher than those obtained after conventional thermal treatment. On the other hand, high lethal rate (6.39-log reduction) of *E. coli* K12 was obtained in apple juice when combining ultraviolet radiation and OH at 65 °C, leading to cell disruption and inhibition of cell replication (Lee *et al.*, 2013).

The consumer demand for more natural products has led to finding natural methods for extending the shelf life of pome fruit juices. In this way, the application of natural antimicrobials including bacteriocins, lactoperoxidase, herbs, and spices containing essential oils or organic acids, also called GRAS (generally recognized as safe) substances, has shown feasibility for use in apple juice (Table 1.1).

Other alternative chemical methods such as the application of gaseous ozone (3 l/min flow rate and $2-3\,g/m^3$) treatments have been shown to be effective in the reduction of *E. coli* O157:H7, *S. typhimurium*, and *L. monocytogenes* (Song *et al.*, 2015).

1.4   Conclusion and Future Trends

The application of several novel processing technologies has been explored in particular in apple juices. However, the research in other pome fruit juices such as pear juices is still scarce. The application of nonthermal technologies is of particular interest in pome fruit juices since preservation of color and flavor characteristics in these products has been achieved. However, the interest for improving the quality characteristics and health-related compounds such as polyphenols in pome fruit juices is still a challenge. One approach would be to introduce new pome fruit cultivars and consider them as new varieties for juice making.

Acknowledgments

Dr Aguiló-Aguayo thanks the Spanish Government for the FPDI-2013-15583. L. Plaza thanks the National Institute for Agronomic Research (INIA) for a DOC-INIA research contract. Authors acknowledge financial support from CERCA Programme/Generalitat de Catalunya.

References

Abid, M., Jabbar, S., Wu, T., Hashim, M., Hu, B., Saeeduddin, M. & Zeng, X. (2014) Qualitative assessment of sonicated apple juice during storage. *Journal of Food Processing and Preservation* **39**, 6, 1299–1308.

Aguilar-Rosas, S., Ballinas-Casarrubias, M., Nevarez-Moorillon, G., Martín-Belloso, O. & Ortega-Rivas, E. (2007) Thermal and pulsed electric fields pasteurization of apple juice: effects on physicochemical properties and flavour compounds. *Journal of Food Engineering* **83**, 41–46.

Ait-Ouazzou, A., Espina, L., Cherrat, L., Hassani, M., Laglaoui, A., Conchello, P. & Pagán, R. (2012) Synergistic combination of essential oils from Morocco and physical treatments for microbial inactivation. *Innovative Food Science and Emerging Technologies* **16**, 283–290.

Ait-Ouazzou, A., Espina, L., García-Gonzalo, D. & Pagán, R. (2013) Synergistic combination of physical treatments and carvacrol for *Escherichia coli* O157:H7 inactivation in apple, mango, orange, and tomato juices. *Food Control* **32**, 1, 159–167.

Albashan, M. (2009) Acid tolerance of *Escherichia coli* O157:H7 serotype and *Salmonella typhi* (A Group D Serotype) and their survival in apple and orange juices. *World Journal of Medical Sciences* **4**, 1, 33–40.

Álvarez, S., Riera, F., Álvarez, R., Coca, J., Cuperus, F., Bouwer, S., van Gemert, R., Veldsink, J., Giorno, L., Donato, L., Todisco, S., Drioli, E., Olsson, J., Trägardh, G., Gaeta, S. & Panyor, L. (2000) A new integrated membrane process for producing clarified apple juice and apple juice aroma concentrate. *Journal of Food Engineering* **46**, 2, 109–125.

Awuah, G., Ramaswamy, H. & Economides, A. (2007) Thermal processing and quality: principles and overview. *Chemical Engineering and Processing: Process Intensification* **46**, 584–602.

Barbosa-Cánovas, G., Góngora-Nieto, M., Pothakamury, U. & Swanson, B. (1999) *Preservation of Foods with Pulsed Electric Fields*. Academic Press, San Diego.

Barret, D., Somogyi, L. & Ramaswamy, H. (eds) (2005) *Processing Fruits: Science and Technology*. CRC Press, Boca Raton.

Bastianello, E., Montemurro, F., Fasolato, L., Balzan, S., Marchesini, G., Contiero, B., Cardazzo, B. & Novelli, E. (2016) Volatile compounds and microbial development in sonicated cloudy apple juices: preliminary results. *CyTA – Journal of Food* **14**, 1, 65–73.

Bazhal, M. & Vorobiev, E. (2000) Electrical treatment of apple cossettes for intensifying juice pressing. *Journal of the Science of Food and Agriculture* **80**, 1668–1674.

Bevilacqua, A., Corbo, M. & Sinigaglia, M. (2012) Use of natural antimicrobials and high-pressure homogenization to control the growth of *Saccharomyces bayanus* in apple juice. *Food Control* **24**, 12, 109–115.

Bi, X., Liu, F., Rao, L., Li, J., Liu, B., Liao, X. & Wu, J. (2013) Effects of electric field strength and pulse rise time on physicochemical and sensory properties of apple juice by pulsed electric field. *Innovative Food Science & Emerging Technologies* **17**, 85–92.

Braddock, R. (1999). Single-strength juices and concentrate. In *Handbook of Citrus By-Products and Processing Technology* (ed R. J. Braddock), pp. 53–58. John Wiley, New York.

Burdurlu, H. & Karadeniz, F. (2003) Effect of storage on nonenzymatic browning of apple juice concentrates. *Food Chemistry* **80**, 91–97.

Burris, K., Davidson, P., Stewart, Jr, C., Zivanovic, S. & Harte, F. (2012) Aqueous extracts of yerba mate (*Ilex paraguariensis*) as a natural antimicrobial against *Escherichia coli* O157:H7 in a microbiological medium and pH 6.0 apple juice. *Journal of Food Protection* **75**, 4, 753–757.

Caminiti, I., Palgan, I., Muñoz, A., Noci, F., Whyte, P., Morgan, D., Cronin, D. & Lyng, J. (2012) The effect of ultraviolet light on microbial inactivation and quality attributes of apple juice. *Food and Bioprocess Technology* **5**, 2, 680–686.

Carbonell-Capella, J., Parniakov, O., Barba, F.J., Grimi, N., Bals, O., Lebovka, N. & Vorobiev, E. (2016) "Ice" juice from apples obtained by pressing at subzero temperatures of apples pretreated by pulsed electric fields. *Innovative Food Science & Emerging Technologies* **33**, 187–194.

Castaldo, D. & Loiudice, R. (1997) Presence of residual pectin methylesterase activity in thermally stabilized industrial fruit preparations. *LWT – Food Science and Technology* **30**, 479–484.

Ceylan, E., Fung, D. & Sabah, J. (2004) Antimicrobial activity and synergistic effect of cinnamon with sodium benzoate or potassium sorbate in controlling *Escherichia coli* O157:H7 in apple juice. *Journal of Food Science* **69**, 4, 102–106.

Char, C., Mitilinaki, E., Guerrero, S. & Alzamora, S. (2010) Use of high-intensity ultrasound and UV-C light to inactivate some microorganisms in fruit juices. *Food and Bioprocess Technology* **3**, 6, 797–803.

Charles-Rodríguez, A. (2002) *Comparación de proceso térmico y uso de pulsos eléctricos en la pasteurización de jugo de manzana.* MSc Thesis. Chihuahua, Mexico, Autonomous University of Chihuahua.

Charles-Rodríguez, A., Nevárez-Moorillon, G., Zhang, Q. & Ortega-Rivas, E. (2007) Comparison of thermal processing and pulsed electric fields treatment in pasteurization of apple juice. *Food and Bioproducts Processing* **85**, 93–97.

Codex Alimentarius Commission. (1999) Codex Committee on food additives and contaminants. Position paper on patulin, 22–29 March 1999.

Corte, F., Fabrizio, S., Salvatori, D. & Alzamora, S. (2004) Survival of *Listeria innocua* in apple juice as affected by vanillin or potassium sorbate. *Journal of Food Safety* **24**, 1, 1–15.

De Bruijn, J., Venegas, A. & Bórquez, R. (2002) Influence of crossflow ultrafiltration on membrane fouling and apple juice quality. *Desalination* **148**, 131–136.

De Bruijn, J., Venegas, A., Martínez, J. & Bórquez, R. (2003) Ultrafiltration performance of Carbosep membranes for the clarification of apple juice. *LWT – Food Science and Technology* **36**, 4, 397–406.

De Paepe, D., Valkenborg, D., Noten, B., Servaes, K., Diels, L., De Loose, M., Van Droogenbroeck, B. & Voorspoels, S. (2014) Variability of the phenolic profiles in the fruits from old, recent and new apple cultivars cultivated in Belgium. *Metabolomics* **11**, 3, 1–14.

Deng, M. & Cliver, D. (2001) Inactivation of *Cryptosporidium parvum* oocysts in cider by flash pasteurization. *Journal of Food Protection* **64**, 4, 523–527.

Dietrich, H. (2004) Bioactive compounds in fruit and juice. *Fruit Processing* **1**, 50–55.

Dietrich, H., Früger-Steden, E., Dieter Patz, C., Rheinberger, A. and Hopf, I. (2007) Increase of sorbitol in pear and apple juice by water stress, a consequence of climatic change? *Science & Research* 348–355.

Downes, J. (1995) Equipment for extraction and processing of soft and pome fruit juices. In *Production and Packaging of Non-Carbonated Fruit Juices and Fruit Beverages*. (ed P.R. Ashurst), pp 197–220. Springer, US.

Duraisamy, S., Kasi, M., Balakrishnan, S., Al-Sohaibani, S. & Ramasamy, G. (2015) Optimization of *Lactobacillus brevis* NS01 Brevicin production and its application in apple juice biopreservation using food-grade clarifying agent silica as a carrier. *Food and Bioprocess Technology* **8**, 1750–1761.

Eissa, H., El-Seideek, L., Ibrahim, N. & Emam, W. (2012) Utilisation of some natural medical plant (NMP) extracts as antibacterial, antifungal and antibrowning in red apple juice preservation. *Journal of Applied Sciences Research* **8**, 5, 2821–2831.

Ertugay, M. & Başlar, M. (2014) The effect of ultrasonic treatments on cloudy quality-related quality parameters in apple juice. *Innovative Food Science & Emerging Technologies* **26**, 226–231.

Ertugay, M.F., BaŞlar, M. & Ortakci, F. (2013) Effect of pulsed electric field treatment on polyphenol oxidase, total phenolic compounds, and microbial growth of apple juice. *Turkish Journal of Agricultural and Forestry* **37**, 772–780.

Evelyn, Kim, H. & Silva, F. (2016) Modeling the inactivation of *Neosartorya fischeri* ascospores in apple juice by high pressure, power ultrasound and thermal processing. *Food Control* **59**, 530–537.

Evrendilek, G., Jin, Z., Ruhlman, K., Qiu, X., Zhang, Q. & Richter, E. (2000) Microbial safety and shelf-life of apple juice and cider processed by bench and pilot scale PEF systems. *Innovative Food Science & Emerging Technologies* **1**, 77–86.

Falguera, V. & Ibarz, A. (eds) (2014) *Juice Processing: Quality, Safety and Value-Added Opportunities*. CRC Press, Boca Raton.

Falguera, V., Vicente, M., Garvín, A. & Ibarz, A. (2013) Flow behavior of clarified pear and apple juices at subzero temperatures. *Journal of Food Processing and Preservation* **37**, 133–138.

FDA U.S. Food Drug and Administration (2004) *Guidance for Industry: Juice HACCP Hazards and Controls Guidance First Edition; Final Guidance*, [Online], Available: http://www.fda.gov/Food/GuidanceRegulation/GuidanceDocumentsRegulatoryInformation/Juice/ucm072557.htm#ftn7 [11 Aug 2016].

Ferrario, M., Alzamora, S., Guerrero, S. (2015) Study of the inactivation of spoilage microorganisms in apple juice by pulsed light and ultrasound. *Food Microbiology* **46**, 635–642.

Ferrario, M. & Guerrero, S. (2016) Effect of a continuous flow-through pulsed light system combined with ultrasound on microbial survivability, color and sensory shelf life of apple juice. *Innovative Food Science & Emerging Technologies*, **34**, 214–224.

Fukutami, K., Sano, K., Yamayuchi, T. & Ogana, H. (1986) Production of turbid apple juice and apple puree. *Basic patent*, JP 62259568 A2 871111.

Funes, G., Gómez, P., Resnik, S. & Alzamora, S. (2013) Application of pulsed light to patulin reduction in McIlvaine buffer and apple products. *Food Control* **30**, 2, 405–410.

Gardner, P., White, T., McPhail, D. & Duthie G. (2000) The relative contributions of vitamin C, carotenoids and phenolics to the antioxidant potential of fruit juices. *Food Chemistry* **68**, 4, 471–474.

Gierschener, K. & Baumann, G. (1988) New method of producing stable cloudy fruit juices by the action of pectolytic enzymes. *Industrie Obst-Gemueseverarbeitung* **54**, 217–218.

Giner, J., Gimeno, V., Barbosa-Canovas, G. & Martin, O. (2001) Effects of pulsed electric field processing on apple and pear polyphenoloxidases. *Food Science and Technology International* **7**, 339–345.

Girard, K. & Sinha, N. (2012) Cranberry, blueberry, currant, and gooseberry. In *Handbook of Fruits and Fruit Processing* (eds N. Sinha, J. Sidhu, J. Barta, J. Wu & M.P. Cano), 2nd edition, pp. 399–418. Wiley-Blackwell, Iowa, USA.

Harrison, S., Chang, F., Boylston, T., Barbosa-Cánovas, G. & Swanson, B. (2001) Shelf stability, sensory analysis, and volatile flavor profile of raw apple juice after pulsed electric field, hydrostatic pressure, or heat exchanger processing. In *Pulsed Electric Fields in Food Processing: Fundamental Aspects and Applications* (eds G.V. Barbosa-Cánovas & Q.H. Zhang), pp. 241–257. Technomic Publishing Co. Inc., Lancaster, USA.

Heinz, V., Toepfl, S. & Knorr, D. (2003) Impact of temperature on lethality and energy efficiency of apple juice pasteurization by pulsed electric fields treatment. *Innovative Food Science & Emerging Technologies* **4**, 167–175.

Hsu, J., Heatherbell, D. & Yorgey, B. (1990) Effects of variety, maturity and processing on pear juice quality and protein stability. *Journal of Food Science* **55**, 6, 1610–1613.

Hubert, B., Baron, A., Le Quere, J. & Renard, C. (2007) Influence of prefermentary clarification on the composition of apple musts. *Journal of Agricultural and Food Chemistry* **55**, 13, 5118–5122.

Ibarz, I., Gonzales, C., Esplugas, S., & Miguelsanz, R. (1990) Non-enzymatic browning kinetics of clarified peach juice at different temperatures. *Confructa* **34**, 152–159.

Ibarz, A., Martín, O. & Barbosa-Cánovas, G. (1997) Non-enzymatic browning kinetics of concentrated pear juice (Note). *Food Science and Technology International* **3**, 213–218.

Jakób, A., Bryjak, J., Wójtowicz, H., Illeová, V., Annus, J. & Polakovič, M. (2010) Inactivation kinetics of food enzymes during ohmic heating. *Food Chemistry* **123**, 2, 369–376.

Jiang, G., Kim, Y., Nam, S., Yim, S. & Eun, J. (2016b) Enzymatic browning inhibition and antioxidant activity of pear juice from a new cultivar of Asian pear (*Pyrus pyrifolia Nakai* cv. *Sinhwa*) with different concentrations of ascorbic acid. *Food Science and Biotechnology* **25**, 1, 153–158.

Jiang, G., Kim, Y., Nam, S., Yim, S., Gwak, H. & Eun, J. (2016a) Changes in total phenolic and flavonoid content and antioxidative activities during production of juice concentrate from Asian pears (*Pyrus pyrifolia Nakai*). *Food Science and Biotechnology* **25**, 47–51.

Jin, Z., Ruijin, Y., Wei, Z., Songshan, M. & Yanfang, W. (2008) Effect of pulsed electric fields on inactivation of microbe and quality of freshly-squeezed pear juice. *Transactions of the Chinese Society of Agricultural Engineering* **24**, 6, 239–244.

Juarez-Enriquez, E., Salmeron-Ochoa, I., Gutierrez-Mendez, N., Ramaswamy, H. & Ortega-Rivas, E. (2015) Shelf life studies on apple juice pasteurised by ultrahigh hydrostatic pressure. *LWT – Food Science and Technology* **62**, 1, 915–919.

Kato, T., Shimoda, M., Suzuki, J., Kawaraya, A., Igura, N. & Hayakawa, I. (2003) Changes in the odors of squeezed apple juice during thermal processing. *Food Research International* **36**, 8, 777–785.

Keyser, M., Muller, I., Cilliers, F., Nel, W. & Gouws, P. (2008). Ultraviolet radiation as a non-thermal treatment for the inactivation of microorganisms in fruit juice. *Innovative Food Science & Emerging Technologies* **9**, 3, 348–354.

Kim, S. & Rhee, M. (2015) Use of caprylic acid to control pathogens (*Escherichia coli* O157:H7 and *Salmonella enterica* serovar *Typhimurium*) in apple juice at mild heat temperature. *Journal of Applied Microbiology* **119**, 5, 1317–1323.

Kiskó, G. & Roller, S. (2005) Carvacrol and *p*-cymene inactivate *Escherichia coli* O157:H7 in apple juice. *BMC Microbiology* **5**, 36.

Koodie, L. & Dhople, A. (2001) Acid tolerance of *Escherichia coli* O157: H7 and its survival in apple juice. *Microbios* **104**, 409, 167–175.

Kopera, M. & Mitek, M. (2006) Effect of addition L-ascorbic acid in pulp on polyphenols content in pears juices. *Zywnosc-Nauka Technologia Jakosc* **2**, 47, 116–123.

Krapfenbauer, G., Kinner, M., Gössinger, M., Schönlechner, R. & Berghofer, E. (2006) Effect of thermal treatment on the quality of cloudy apple juice. *Journal of Agricultural and Food Chemistry* **54**, 5453–5460.

Kumar, S., Thippareddi, H., Subbiah, J., Zivanovic, S., Davinson, P. & Harte, F. (2009) Inactivation of *Escherichia coli* K-12 in apple juice using combination of high pressure homogenization and chitosan. *Journal of Food Science* **74**, 8–14.

Lavilla, T., Puy, J., Lopez, M., Recasens, I. & Vendrell M. (1999) Relationships between volatile production, fruit quality, and sensory evaluation in Granny Smith apples stored in different controlled-atmosphere treatments by means of multivariate analysis. *Journal of Agriculture and Food Chemistry* **47**, 3791–3803.

Lawlor, K., Schuman, J., Simpson, P. & Taormina P. (2009) Microbiological spoilage of beverages. In *Compendium of the Microbiological Spoilage of Foods and Beverages* (eds W. H. Sperber & M. P. Doyle) Food Microbiology and Food Safety series, pp. 245–284. Springer, New York, USA.

Lee, S., Yamada, K. & Jun, S. (2013) Ultraviolet radiation assisted with ohmic current for microbial inactivation in apple juice. *Transactions of the ASABE* **56**, 3, 1085–1091.

Luis-Villaroya, A., Espina, L., García-Gonzalo, D., Bayarri, S. Pérez, C. & Pagán, R. (2015) Bioactive properties of a propolis-based dietary supplement and its use in combination with mild heat for apple juice preservation. *International Journal of Food Microbiology* **205**, 90–97.

Maftei, N., Ramos-Villarroel, A., Nicolau, A., Martín-Belloso, O. & Soliva-Fortuny, R. (2014) Influence of processing parameters on the pulsed-light inactivation of *Penicillium expansum* in apple juice. *Food Control* **41**, 27–31.

Mak, P., Ingham, B. & Ingham, S. (2001) Validation of apple cider pasteurization treatments against *Escherichia coli* O157:H7, *Salmonella*, and *Listeria monocytogenes*. *Journal of Food Protection* **64**, 1679–1689.

Mazzotta, A. (2001) Thermal inactivation of stationary-phase and acid-adapted *Escherichia coli* O157:H7, *Salmonella*, and *Listeria monocytogenes* in fruit juices. *Journal of Food Protection* **64**, 3, 315–320.

Mehinagic, E., Royer, G., Symoneaux, R., Jourjon, F. & Prost, C. (2006) Characterization of odor-active volatiles in apples: influence of cultivars and maturity stage. *Journal of Agricultural and Food Chemistry* **54**, 7, 2678–2687.

Molva, C. & Baysal, A. (2015a) Antimicrobial activity of grape seed extract on *Alicyclobacillus acidoterrestris* DSM 3922 vegetative cells and spores in apple juice. *LWT – Food Science and Technology* **60**, 1, 238–245.

Molva, C. & Baysal, A. (2015b) Evaluation of bioactivity of pomegranate fruit extract against *Alicyclobacillus acidoterrestris* DSM 3922 vegetative cells and spores in apple juice. *LWT – Food Science and Technology* **62**, 2, 989–995.

Moody, A., Marx, G., Swanson, B. & Bermúdez-Aguirre, D. (2014) A comprehensive study on the inactivation of *Escherichia coli* under nonthermal technologies: high hydrostatic pressure, pulsed electric fields and ultrasound. *Food Control* **37**, 305–314.

Mosqueda-Melgar, J., Raybaudi-Massilia, R. & Martín-Belloso, O. (2008) Non-thermal pasteurization of fruit juices by combining high-intensity pulsed electric fields with natural antimicrobials. *Innovative Food Science & Emerging Technologies* **9**, 3, 328–340.

Moyer, J. & Aitken, H. (1980) Apple juice. In *Fruit and Vegetable Juice Processing Technology* (eds P. E. Nelson & D. K. Tressler), pp. 212–267. AVI Publishing Co., Westport, CT, USA.

Nagel B. (1992) Continuous production of high quality cloudy apple juices. *Fruit Processing* **2**, 1, 3–5.

Niu, S., Xu, Z., Fang, Y., Zhang, L., Yang, Y., Liao, X., & Hu, X. (2010) Comparative study on cloudy apple juice qualities from apple slices treated by high pressure carbon dioxide and mild heat. *Innovative Food Science & Emerging Technologies* **11**, 91–97.

Noci, F., Riener, J., Walkling-Ribeiro, M., Cronin, D., Morgan, D. & Lyng, J. (2008) Ultraviolet irradiation and pulsed electric fields (PEF) in a hurdle strategy for the preservation of fresh apple juice. *Journal of Food Engineering* **85**, 1, 141–146.

NPCS Board of Consultants & Engineers (2008) *The Complete Technology Book on Alcoholic and Non-Alcoholic Beverages (Fruit Juices, Whisky, Beer, Rum and Wine)*. Asia Pacific Business Press Inc., India.

Onsekizoglu, P., Bahceci, K. & Acar, M. (2010) Clarification and the concentration of apple juice using membrane processes: a comparative quality assessment. *Journal of Membrane Science* **352**, 160–165

Orlowska, M., Koutchma, T., Grapperhaus, M., Gallagher, J., Schaefer, R. & Defelice, C. (2013) Continuous and pulsed ultraviolet light for nonthermal treatment of liquid foods. Part 1: effects on quality of fructose solution, apple juice, and milk. *Food and Bioprocess Technology* **6**, 6, 1580–1592.

Ortega-Rivas, E., Zárate-Rodríguez, E. & Barbosa-Cánovas, G. (1998) Apple juice pasteurization using ultrafiltration and pulsed electric fields. *Food and Bioproducts Processing* **76**, 4, 193–198.

Oszmianski, J., Wolniak, M., Wojdylo, A. & Wawer, I. (2007) Comparative study of polyphenolic content and antiradical activity of cloudy and clear apple juices. *Journal of the Science of Food and Agriculture* **87**, 4, 573–579.

Park, I. & Kang D. (2013) Effect of electropermeabilization by ohmic heating for inactivation of *Escherichia coli* O157:H7, *Salmonella enterica* serovar *Typhimurium*, and *Listeria monocyto*genes in buffered peptone water and apple juice. *Applied and Environmental Microbiology* **79**, 23, 7122–7129.

Pataro, G., Muñoz, A., Palgan, I., Noci, F., Ferrari, G. & Lyng, J. (2011) Bacterial inactivation in fruit juices using a continuous flow Pulsed Light (PL) system. *Food Research International* **44**, 6, 1642–1648.

Pathanibul, P., Taylor, T., Davidson, P. & Harte, F. (2009) Inactivation of *Escherichia coli* and *Listeria innocua* in apple and carrot juices using high pressure homogenization and nisin. *International Journal of Food Microbiology* **129**, 3, 316–320.

Piskernik, S., Klanĉnik, A., Demšar, L., Mozina, S. & Jeršek, B. (2016) Control of *Alicyclobacillus* spp. vegetative cells and spores in apple juices with rosemary extracts. *Food Control* **60**, 205–214.

Porto, C., Decorti, D. & Tubaro, F. (2010) Original article: effects of continuous dense-phase CO_2 system on antioxidant capacity and volatile compounds of apple juice. *International Journal of Food Science and Technology* **45**, 9, 1821–1827.

Qin, B., Pothakamury, H., Vega-Mercado, H., Martin, O., Barbosa-Cánovas, G. & Swanson, B. (1995) Food pasteurization using high intensity pulsed electric fields. *Journal of Food Technology* **49**, 12, 55–60.

Qin, B., Zhang, Q., Barbosa-Cánovas, G., Swanson, B. G. & Pedrow, P. (1994) Inactivation of microorganisms by pulsed electric fields of different voltage waveforms. *IEEE Transactions on Dielectric Electric Insulation* **1**, 1047–1050.

Quoc, A., Lamarche, F. & Makhlouf, J. (2000) Acceleration of pH variation in cloudy apple juice using electrodialysis with bipolar membranes. *Journal of Agricultural and Food Chemistry* **48**, 2160–2166.

Raju, P. & Bawa, A. (2012) Food Additives in fruit processing. In *Handbook of Fruits and Fruit Processing* (eds N.K. Sinha, J.S. Sidhu, J. Barta, J.S. Wu & M.P. Cano). 2nd edition, pp. 189–214. Blackwell Publishing, Iowa, USA.

Ramaswamy, H., Riahi, E. & Idziak, E. (2003) High-pressure destruction kinetics of *E. Coli* (29055) in apple juice. *Journal of Food Science* **68**, 5, 1750–1756.

Ramaswamy, H. & Singh, R. (1997) Sterilization process engineering. In *Handbook of Food Engineering Practice* (eds K. J. Valentas, E. Rotstein & R. P. Singh), pp 37–69. CRC Press, New York.

Rao, M. (1986) Rheological properties of fluid foods. In *Engineering Properties of Foods* (eds. M. A. Rao & S. S. H. Rizvi), pp. 1–47. Marcel Dekker Inc., New York.

Raybaudi-Massilia, R., Mosqueda-Melgar J. & Martín-Belloso, O. (2009) Antimicrobial activity of malic acid against *Listeria monocytogenes*, *Salmonella enteritidis* and *Escherichia coli* O157:H7 in apple, pear and melon juices. *Food Control* **20**, 2, 105–112.

Renard, C. & Maingonnat, J. (2012) Thermal processing of fruits and fruit juices. In *Thermal Food Processing: New Technologies and Quality Issues*, (ed D.W. Sun), 2nd edition, pp. 413–440. CRC Press, Boca Raton.

Riahi, E. & Ramaswamy, H. (2003) High-pressure processing of apple juice: kinetics of pectinesterase inactivation. *Biotechnology Progress* **19**, 908–914.

Riener, J., Noci, F., Cronin, D., Morgan, D. & Lyng J. (2008) Combined effect of temperature and pulsed electric fields on apple juice peroxidase and polyphenoloxidase inactivation. *Food Chemistry* **109**, 402–407.

Riu-Aumatell, M., Lopez-Tamames, E. & Buxaderas, S. (2005) Assessment of the volatile composition of juices of apricot, peach, and pear according to two pectolytic treatments. *Journal of Agricultural and Food Chemistry* **53**, 7837–7843.

Roller, S. & Covill, N. (1999) The antifungal properties of chitosan in laboratory media and apple juice. *International Journal of Food Microbiology* **47**, 67–77.

Saeeduddin, M., Abid, M., Jabbar, S., Hu, B., Hashim, M., Khan, M., Xie, M., Wu, T. & Zeng, X. (2016) Physicochemical parameters, bioactive compounds and microbial quality of sonicated pear juice. *International Journal of Food Science and Technology* **51**, 7, 1552–1559.

Saeeduddin, M., Abid, M., Jabbar, S., Wu, T., Hashim, M., Awad, F., Hu, B., Lei, S. & Zeng, X. (2015) Quality assessment of pear juice under ultrasound and commercial pasteurization processing conditions. *LWT – Food Science and Technology* **64**, 1, 452–458.

Sagdic, O., Silici, S. & Yetim, H. (2007) Fate of *Escherichia coli* and *E. coli* O157:H7 in apple juice treated with propolis extract. *Annals of Microbiology* **57**, 3, 345–348.

Sanchez-Vega, R., Mujica-Paz, H., Marquez-Melendez, R., Ngadi, M., Ortega-Rivas, E. (2009) Enzyme inactivation on apple juice treated by ultrapasteurization and pulsed electric fields technology. *Journal of Food Processing and Preservation* **33**, 486–499.

Saritas, A., Acar, J. & Bahçeci, K. (2011) The effects of ascorbic and cinnamic acid addition and different pasteurization parameters on the inactivation of *alicyclobacillus acidoterrestris* in apple juice. *In:* 6th International CIGR Technical Symposium – Towards a Sustainable Food Chain: Food Process, Bioprocessing and Food Quality Management, Nantes, France.

Schilling, S., Alber, T., Toepfl, S., Nedihart, S., Knorr, D., Schieber, A. & Carle, R. (2007) Effects of pulsed electric field treatment of apple mash on juice yield and quality attributes of apple juices. *Innovative Food Science & Emerging Technologies* **8**, 1, 127–134.

Sharma, H., Patel, H. & Sugandha (2016) Enzymatic extraction and clarification of juice from various fruits: a review. *Critical Reviews in Food Science and Nutrition*, 5 January, http://dx.doi.org/10.1080/10408398.2014.977434.

Silva, L., Almeida, P., Rodrigues, S. & Fernandes, F. (2015) Inactivation of polyphenoloxidase and peroxidase in apple cubes and in apple juice subjected to high intensity power ultrasound processing. *Journal of Food Processing and Preservation* **39**, 6, 2081–2087.

Singh S. & Gupta, R. (2004) Apple juice clarification using fungal pectinolytic enzyme and gelatin. *Indian Journal of Biotechnology* **3**, 573–576.

Sinha, N. (2012) Apples and pears: production, physicochemical and nutritional quality, and major products. In *Handbook of Fruits and Fruit Processing.* (Eds N. K. Sinha, J. S. Sidhu, J. Barta, S. B. Wu & M. P. Cano), 2nd edition, pp. 367–384. Wiley-Blackwell, Iowa, USA.

Sokolowska, B., Skapska, S., Fonberg-Broczek, M., Niezgoda, J., Chotkiewicz, M., Dekowska, A. & Rzoska, S. (2012) The combined effect of high pressure and nisin or lysozyme on the inactivation of *Alicyclobacillus acidoterrestris* spores in apple juice. *High Pressure Research* **32**, 1, 119–127.

Song, W., Shing, J., Ryu, S. & Kang, D. (2015) Inactivation of *Escherichia coli* O157:H7, *Salmonella Typhimurium* and *Listeria monocytogenes* in apple juice at different pH levels by gaseous ozone treatment. *Journal of Applied Microbiology* **119**, 2, 465–474.

Spanos, A. & Wrolstad E. (1992) Phenolics of apple, pear, and white grape juices and their changes with processing and storage. A review. *Journal of Agricultural and Food Chemistry* **40**, 1478–1487.

Steinhaus, M., Bogen, J. & Schieberle, P. (2006) Key aroma compounds in apple juice – changes during juice concentration. In: *Flavour Science: Recent Advances and Trends* (eds W.L.P. Bredie & M.A. Petersen), pp 189–192. Elsevier BV, Amsterdam, The Netherlands.

Su, S. & Wiley, R. (1998) Changes in apple juice flavor compounds during processing. *Journal of Food Science* **63**, 4, 681–691.

Sun, D. (ed) (2012) *Thermal Food Processing: New Technologies and Quality Issues*, CRC Press, Boca Raton.

Sun, Y., Zhong, L., Cao, L., Lin, W. & Ye, X. (2015) Sonication inhibited browning but decreased polyphenols contents and antioxidant activity of fresh apple (*malus pumila mill, cv. Red Fuji*) juice. *Journal of Food Science and Technology* **52**, 12, 8336–8342.

Timmermans, R., Nierop Groot, M., Nederhoff, A., van Boekel, M., Matser, A. & Mastwijk, H. (2014) Pulsed electric field processing of different fruit juices: impact of pH and temperature on inactivation of spoilage and pathogenic micro-organisms. *International Journal of Food Microbiology* **173**, 105–111.

Toepfl, S., Mathys, A., Heinz, V. & Knorr, D. (2006) Review: potential of high hydrostatic pressure and pulsed electric fields for energy efficient and environmentally friendly food processing. *Food Reviews International* **22**, 405–423.

Tournas, V., Heeres, J. & Burgess L. (2006) Moulds and yeasts in fruit salads and fruit juices. *Food Microbiology* **23**, 7, 684–688.

Tsong, T. (1991) Electroporation of cell membranes. *Biophysical Journal* **60**, 297–306.

Vasantha, R.H. & Yu, L. (2012) Emerging preservation methods for fruit juices and beverages. In *Food Additive* (ed Y. El-Samragy), pp. 65–86. InTechOpen, Croatia.

Visser, T., Schaap, A. & De Vries, P. (1968) Acidity and sweetness in apple and pear. *Euphytica* **17**, 153–167.

Vorobiev, E., Jemai, A. & Bouzrara, H. (2004) Pulsed electric field-assisted extraction of juice from food plants. In *Novel Food Processing Technologies* (eds G.V. Barbosa-Canovas, M.P. Cano, M.S. Tapia), pp. 105–130. CRC Press, Boca Raton.

Walkling-Ribeiro, M., Noci, F., Cronin, D., Riener, J., Lyng, J. & Morgan, D. (2008) Reduction of *Staphylococcus aureus* and quality changes in apple juice processed by ultraviolet irradiation, pre-heating and pulsed electric fields. *Journal of Food Engineering* **89**, 3, 267–273.

Wang, H., Hu, Z., Long, F., Guo, C., Niu, C., Yuan, Y. & Yue, T. (2016) The effects of stress factors on the growth of spoilage yeasts isolated from apple-related environments in apple juice. *Journal of Food Safety* **36**, 2, 162–171.

Welter, C., Hartmann, E. & Frei, M. (1991) Production of very light colored cloudy juices. *Flüssiges Obst* **58**, 5, 230–233.

Willems, J. & Low, N. (2016) Oligosaccharide formation during commercial pear juice processing. *Food Chemistry* **204**, 84–93.

Wu, Y. & Zhang, D. (2015). Effect of pulsed electric field pretreatment on apples microstructures. *International Agricultural Engineering Journal* **24**, 4, 80–87.

Yuan, Y., Hu, Y., Yue, T., Chen, T. & Lo, Y. (2009) Effect of ultrasonic treatments on thermoacidophilic *Alicyclobacillus acidoterrestris* in apple juice. *Journal of Food Processing and Preservation* **33**, 3, 370–383.

Zárate-Rodríguez, E., Ortega-Rivas, E. & Barbosa Cánovas, G. (2000) Quality changes in apple juice as related to nonthermal processing. *Journal of Food Quality* **23**, 337–349.

Zhao, L., Wang, Y., Qiu, D. & Liao, X. (2014) Effect of ultrafiltration combined with high-pressure processing on safety and quality features of fresh apple juice. *Food and Bioprocess Technology* **7**, 11, 3246–3258.

2

Citrus Fruit Juices

Maria Consuelo Pina-Pérez[1,2], Alejandro Rivas[1], Antonio Martínez[1], and Dolores Rodrigo[1]*

[1]*Instituto de Agroquímica y Tecnología de Alimentos (IATA-CSIC), Paterna, Valencia, Spain*
[2]*Centro Avanzado de Microbiología de Alimentos (CAMA), Universidad Politécnica de Valencia, Valencia, Spain*

2.1 Introduction

Fruit juices are foods rich in bioactive compounds. Citrus juices in particular, especially orange, mandarin, grapefruit, and lemon, are rich in vitamin C, vitamin A (in the form of carotenoids, β-carotene being the most abundant), flavonoids (vitamin factors that have a synergic action with vitamin C), and folic acid (the concentration of which is very low compared with other vitamins, but sufficient to supply 25% of the recommended daily intake) (Primo Yúfera, 1997). Recent epidemiological studies and other investigations have demonstrated that these bioactive compounds (limonoids, flavonoids, and carotenoids) have a broad range of physiological effects and may contribute to the associations between citrus juice consumption and prevention of chronic diseases such as cardiovascular disease and cancer, among others (Valko *et al.*, 2006; Liu *et al.*, 2012).

The soluble solids in citrus juices consist mainly of sugars (saccharose, glucose, and fructose) and acids (mainly citric and malic). Notable features are the high concentration of potassium (149 mg/100 ml and 182 mg/100 ml, in orange and lemon, respectively) and low concentration of sodium (0.75 and 1.78 mg/100 ml, respectively), making them suitable foods for diets poor in Na and high in K (Primo Yúfera, 1997).

Among the spoilage microflora encountered in citrus juices, the more common ones include lactic acid bacteria (*Lactobacillus* and *Leuconostoc* species), fermentative yeasts (*Saccharomyces cerevisiae*), and spore-forming molds, owing to their capability of growing at low pH values (<4.0). Lactic acid bacteria produce characteristic off-flavors due to the production of diacetyl as a metabolic end product. Yeasts frequently cause spoilage due to ethanolic fermentation. Spoilage of fruit juices can

*Corresponding author: Dolores Rodrigo, lolesra@iata.csic.es

Innovative Technologies in Beverage Processing, First Edition.
Edited by Ingrid Aguiló-Aguayo and Lucía Plaza.
© 2017 John Wiley & Sons Ltd. Published 2017 by John Wiley & Sons Ltd.

also result in an increase in viscosity and production of hydrogen sulfide and other off-odors (Basak *et al.*, 2002).

It is recognized that fruit juices and fruit-based beverages are microbiologically safe because of their low pH. However, some strains of *Escherichia coli* O157, *Salmonella*, and *Shigella* species are acid-resistant and can survive for long periods in acidic environments at low temperatures (Miller & Kasper, 1994). There are various possible causes for the presence of these microorganisms in the final product. Postprocessing recontamination, high raw material contamination, or high processing resistance are some of the ways of introducing contamination, leading to the survival of a small proportion of cells, and causing food safety concern due to their low infective dose.

The characteristic turbidity of citrus juices is mainly provided by pectin, present in the liquid in dissolved form (soluble pectin), giving it viscosity and body, or bound to cellulose (insoluble pectin) that comes mainly from the pulp. These pectins suffer the effect of the enzyme pectin methylesterase (PME), producing loss of turbidity and therefore loss of commercial value. In the industry, the conventional way of inactivating this kind of enzyme is by pasteurization. However, to achieve inactivation that does not affect the quality of the juice, it is necessary to apply treatments of such intensity (temperatures over 90 °C) that they affect organoleptic characteristics (giving a "cooked" taste) and nutritional characteristics (affecting vitamin content).

Because of this loss of quality, various alternatives to conventional pasteurization have been developed in order to obtain a microbiologically safe product while maintaining the maximum organoleptic and nutritional quality. These strategies can be divided into (i) development of alternative thermal technologies, such as ohmic heating and microwaves; (ii) development of nonthermal technologies, particularly pulsed electric fields (PEF); and high-pressure processing (HPP).

2.2 Conventional Preservation Processing Techniques

2.2.1 Effect on Microbial Quality

As commented earlier, the low pH of this kind of juice reduces the type of microorganisms capable of proliferating, mainly to filamentous fungi and lactic bacteria (LAB). Furthermore, some bacteria of the genus *Alicyclobacillus* have been confirmed as causing spoilage of citrus juice in the industry because their spores are resistant to high temperatures in acidic media (pH 3.7) (Yamazaki *et al.*, 1996). *Alicyclobacillus acidoterrestris* is a spore-forming, nonpathogenic, and thermoacidophilic microorganism that presents a problem in the fruit industry worldwide because it has been associated with various spoilage incidents in shelf-stable orange juices, with the consequent economic losses involved (Jensen, 1999). Spoilage is characterized by a bad taste and/or flavor detected at an approximate microorganism concentration of 10^5–10^6 CFU/ml (Pettipher *et al.*, 1997). The high heat resistance of *Alicyclobacillus* spores has been studied in various cases. Maldonado *et al.* (2008) studied heat resistance in lemon juice concentrate and found that the D value depended on pH, degrees Brix, and the kind of juice (clarified or unclarified). The results show the capability of spores of the microorganism to survive in pasteurization treatments that are used in the industry (82 °C, 2 min), and even at high temperatures (95 °C), the D value ranges between 6 and 8.5 min. The high heat resistance of these spores has been verified in other kinds of citrus juice, such as orange juice (maximum D_{95} values of 8.70 min) (Eiroa *et al.*, 1999; Silva *et al.*, 1999; Komitopoulou *et al.*, 1999) and grapefruit juice ($D_{95} = 1.85$ min) (Komitopoulou *et al.*, 1999). The type of strain of the microorganism also affects its heat resistance, with the D_{95} value varying between 2.50 and 8.70 min (Eiroa *et al.*, 1999). Therefore, it is difficult to eliminate these spores using standard heat treatment (e.g., 95 °C, 10 min) without causing damage to the commercial product.

Another group of microorganisms that can survive in acidic media consists of lactic bacteria (*Leuconostoc* and *Lactobacillus* genera). These microorganisms are sensitive to temperature, so the heat treatments usually applied in the industry would be sufficient to eliminate microorganisms of this kind, but they can be a problem in recontamination of products that have already been pasteurized. Inactivation of up to 2.5 decimal reductions of *Lactobacillus plantarum* has been achieved in orange juice by applying mild heat treatments (60 °C, 40 s) (Alwazeer *et al.*, 2002).

Yet another type of flora responsible for spoilage of citrus juice is fermentative yeast, particularly *S. cerevisiae* and *Zygosaccharomyces bailii*, both of which come from contamination of the surface of the fruit. These yeasts are capable of growing in acidic media even in the presence of preservatives (sorbic or benzoic acid) and, in the case of *Z. bailii*, in beverages with low water activity (syrups). The vegetative forms of these yeasts have low resistance to pasteurization (more than 6 and 2 decimal reductions for 60 °C, 40 s and 55 °C, 40 s) (Alwazeer *et al.*, 2002). However, ascospores of *Z. bailii* have higher resistance to heat treatment, with D_{50} values of 10.4 min, than vegetative cells, with D_{50} values of 1.97 min (Raso *et al.*, 1998b).

Traditionally, it was thought that the acidic pH (1.8–4.1) of citrus juices restricted the survival of pathogenic microorganisms. However, outbreaks associated with the consumption of fresh juice have been observed, mainly owing to the presence of *Salmonella*, which indicates the existence of strains of pathogens that are resistant to acidic pHs (Hlady *et al.*, 1998; Khan *et al.*, 2007). Good manufacturing practices and handling together with pasteurization treatments, given that these microorganisms are sensitive to heat treatment ($D_{58} = 0.1$ min) (Alvarez-Ordoñez *et al.*, 2009), are strategies that are sufficient to obtain juice that is safe from a microbiological viewpoint.

2.2.2 Effect on Quality-Related Enzymes

The juice cloud, which is composed of finely divided particles of pectin, cellulose, hemicellulose, proteins, and lipids in suspension, is considered a desirable characteristic of orange juice; however, it shows a loss of cloudiness and concentrate gelation a short time after squeezing, which has been associated with PME activity. Given that the heat resistance of this enzyme (for inactivation >90%) is higher than that of most spoilage microorganisms present in citrus juices, PME is used as a reference for pasteurization (Tribess & Tadini, 2006). PME has two forms, one thermostable and the other thermolabile, and severe heat treatment is required to inactivate the thermostable form (90 °C, 1 min or 95 °C, 30 s) (Cameron *et al.*, 1998).

2.2.3 Effect on Nutritional Quality

The use of the intense heat treatments that are required to inactivate the thermostable forms of PME and the ascospores (*Z. bailii*) and bacterial spores (*A. acidoterrestris*) that are present causes loss of nutritional and organoleptic quality. The main micronutrient present in citrus juices is vitamin C because of its nutritional value and its antioxidant capacity. This vitamin is very sensitive to the presence of oxygen and light and to heat treatment, causing oxidation and degradation of it during processing and storage of citrus juice (Nagy & Smoot, 1977; Robertson & Samaniego, 1986; Fiore *et al.*, 2005; Burdulu *et al.*, 2006).

Carotenoids are the main compounds that are responsible for providing the color of fruit juices, and some of them also have provitamin A activity (Meléndez-Martínez *et al.*, 2007). These compounds are sensitive to light, oxygen, and heat. Processing affects the profile of carotenoids in orange juice. It has been observed that the pasteurization process (90 °C, 30 s) produces a decrease in the levels of violaxanthin (46.4%), antheraxanthin (24.8%), and carotenoids with provitamin A activity (Lessin *et al.*, 1997; Lee & Coates, 2003). However, in red grapefruit juice, conventional heat treatments (80–95 °C, 15–30 s) do not affect concentrations of the main carotenoids in the juice (β-carotene and lycopene) (Lee & Coates, 1999).

2.2.4   Effect on Organoleptic Quality

In addition to degradation of nutritional compounds in the juice (mainly vitamin C and carotenoids), heat treatment also causes loss of organoleptic quality, that is, distancing from the characteristics of fresh juice. Heat treatment (90 °C, 30 s) caused a change of color in sweet orange juice, which led to the juice color becoming lighter and more saturated (Lee & Coates, 2003). In the case of red grapefruit juice, pasteurization (91 °C) brought about a change in the values of *b* and chroma, causing a slight color shift toward lighter and brighter (Lee & Coates, 1999).

The flavor of citrus juices is easily changed by heat during processing or storage (Moshonas & Shaw, 2000; Shaw *et al.*, 2000). Irreversible damage to citrus juice flavor results from chemical reactions initiated or occurring during the heating process (Braddock, 1999; Yeom *et al.*, 2000).

2.3   Novel Processing Techniques

2.3.1   Changes in Conventional Methods

Heat treatment is the preservation technology most commonly used by the food industry. However, since it was first applied in the nineteenth century it has gradually evolved, especially during the second half of the past century. The traditional techniques used to apply heat are based on conduction and convention mechanisms, in which energy is transmitted by means of hot media such as steam, vapor, hot water, and hot air. Conventional thermal techniques have some disadvantages: (i) loss of energy, (ii) reduction of heat transfer, and (iii) overheating of foods. The type of heating applied, from the surface to the interior, which is necessary in order to achieve the food safety objectives at the coldest point of the food, leads to overheating of the food, causing corresponding losses of nutritional and organoleptic quality.

These drawbacks have led to the development of emerging heating techniques in the area of food preservation, microwaves and ohmic heating being two of these new techniques. While in conventional heat treatment, energy is transmitted from the surface to the interior by a heat gradient, these new technologies are characterized by producing heating inside the food, with a consequent benefit in terms of energy efficiency, and therefore, ultimately, better preservation of the quality of the food treated (Tewari & Juneja, 2007).

2.3.2   Ohmic Heating

Ohmic heating (OH) is a method that is based on the passage of electrical current through a food product that serves as an electrical resistance. Electrical energy applied is instantly converted inside the food to heat, the amount of which is directly related to the current induced by the voltage gradient in the field and the electrical conductivity of the food.

This technology is characterized by its high heating speed, depending on the resistance of the food to the flow of the current and the voltage gradient. In this kind of technology, the uniformity of the heating is greater than in conventional heat treatments. The heat transfer mechanism is not conductive or convective, and this reduces the possibility of a cold point, thermal damage, and nutritional losses, and increases overall lethality in the food.

Ohmic treatment can be applied to foods with an electrical conductivity between 0.01 and 10 S/m (Evrendilek *et al.*, 2012), a range that includes citrus juices (Icier & Ilicali, 2005). In order to introduce this technology on an industrial level, it is necessary to have a profound knowledge of the evolution of the conductivity of the food during ohmic treatment because it is on this that the temperature attained depends.

2.3.2.1   Effect on Microbial Quality   Few studies have been conducted on inactivation of microorganisms in citrus juice by OH (Table 2.1). Sagong *et al.* (2011) obtained inactivation of *E. coli* O157:H7, *S. typhimurium*, and *Listeria monocytogenes* exceeding $5D$ in orange juice (10–20 V/cm, 0–540 s; maximum temperature not greater than 80 °C).

Other scientific studies have concentrated on determining whether the application of a low-level electric field affects the survival of microorganisms present in juices. For *A. acidoterrestris* spores, Baysal and Icier (2010) obtained (30 V/cm) lower D values (58.48, 12.24, and 5.97 min) than those obtained using the equivalent conventional heat treatments (70, 80, and 90 °C) (83.33, 15.11, and 7.84 min). However, Leizerson and Shimoni (2005a) did not find differences in inactivation of natural flora in orange juice when it was treated by OH or conventionally (up to 90, 120, and 150 °C; F-values: 1.89×10^{-5}, 1.89×10^{-2}, 18.9).

2.3.2.2   Effect on Quality-Related Enzymes   Just as with the inactivation of microorganisms, scientific studies do not show a clear synergy between heat and electric current in the inactivation of PME. Demirdöven and Baysal (2014) observed greater inactivation of PME in orange juice when it was treated by OH (42 V/cm, 69 °C), obtaining inactivation of 96%, than when conventional treatments were applied (95 °C, 60 s; 83.3%), whereas Leizerson and Shimoni (2005a) did not find differences between the two technologies.

2.3.2.3   Effect on Organoleptic and Nutritional Quality   The concentration of vitamin C in orange juice is one of its most important attributes for the consumer. Leizerson and Shimoni (2005b) did not observe differences in the concentration of ascorbic acid in orange juice when it was treated by OH or traditional technologies. However, Vikram *et al.* (2005) carried out a comparative study of the kinetics of vitamin C degradation after it had been sterilized using different methods of heating, and they observed that OH (50, 60, 75, and 90 °C) was the technology that preserved the vitamin C content better, obtaining D-values (95.96, 58.55, 23.72, and 14.76 min) higher than

Table 2.1 Studies related to the application of ohmic heating to citrus juice

Product	Aspect	Conditions	Results	References
Orange juice	PME	OH: 42 V/cm, 69 °C CH: 95 °C, 60 s	OH[a]: 96% inactivation D_{min}: 0.751–0.742 min CH[b]: 83.3% inactivation	Demirdöven and Baysal (2014)
	Ascorbic acid		D_{min}: 1.07 min OH: 43.08–45.20 mg/ml CH: 42.09 mg/ml	
Orange juice	E. coli 0157:H7, Salmonella typhimurium, L. monocytogenes	10–20 V/cm 0–540 s	Over 5 log reductions	Sagong et al. (2011)
Orange juice	Alicyclobacillus acidoterrestris spores	OH: 30 V/cm + 70, 80, 90 °C CH: 70, 80, 90 °C	D-values: 58.48, 12.24, and 5.97 min D-values: 83.33, 15.11, and 7.84 min	Baysal and Icier (2010)
Orange juice	Natural flora	Up to 90, 120, 150 °C (F-values: 1.89×10^{-5}, 1.89×10^{-2}, 18.9)	2–3 log reductions No difference compared with CH	Leizerson and Shimoni (2005a,b)
	PME		90–98% inactivation No difference compared with CH	
	Vitamin C		15% loss No difference compared with CH	
	Browning		0.25–0.43 browning level No difference compared with CH	
	Flavor components: limonene, pinene, myrcene, octanal, and decanal Sensory evaluation		Ohmic treatment preserves better than conventional No difference between fresh and ohmic treated	
Orange juice	Vitamin C	OH: 50, 60, 75, 90 °C CH: 50, 60, 75, 90 °C	D-values: 95.96, 58.55, 23.72, and 14.76 min D-values: 65.67, 49.81, 27.02, and 12.91 min	Vikram et al. (2005)

[a]OH, ohmic heating

those obtained using conventional heating (65.67, 49.81, 27.02, and 12.91 min). Better preservation of vitamin C by OH (42 V/cm, 69 °C) than by conventional treatment (95 °C, 60 s) was also found by Demirdöven and Baysal (2014).

Leizerson and Shimoni (2005a) observed that ohmically heated orange juice maintained higher amounts of the five representative flavor compounds (decanal, octanal, limonene, pinene, and myrcene) than heat-pasteurized juice did. In addition, sensory evaluation tests showed no difference between fresh and ohmically heated orange juice. The authors also observed that the sensory shelf life of ohmically treated juice was almost two times longer than that of conventionally pasteurized juice.

These results, and those obtained for other foods, support the view that OH achieves microorganism inactivation that is as great as or greater than that achieved by equivalent conventional heat treatments and that it preserves the sensory and nutritional properties of fresh juices to a greater extent, and therefore, it could be considered as an alternative to conventional heat treatment.

2.3.3 Microwave Heating

Microwaves are nonionizing, time-varying electromagnetic waves of radiant energy with frequencies ranging from 300 MHz to 300 GHz. Owing to possible interference with telecommunications, microwave ovens operate at 915 and 2450 MHz.

When microwaves impinge on a dielectric material, such as a food, part of the energy is transmitted, part is reflected, and part is absorbed by the material, where it is dissipated as heat. The heating of the food is due mainly to ionic polarization and to rotation of the dipoles.

The mechanisms of microwave heating (MW) produce rapid heating of food, and the time needed to reach the desired temperature is less than that required by conventional heating systems. In products with high water content such as juices, the energy is absorbed very fast, leading to the product being heated quickly (Venkatesh & Raghavan, 2004). Other advantages of microwaves are the saving in energy, the high precision in control of the process, and the improved nutritional and sensory quality of the food with respect to conventional technologies.

2.3.3.1 Effect on Microbial Quality

Studies of the effectiveness of MW against pathogens and spoilage are limited (Table 2.2). The products treated are mainly vegetable-based beverages (Benlloch-Tinoco et al., 2014a,b) and milk-related products (Pina-Pérez et al., 2013a). Regarding fruit-based beverages, acidic juices have also scarcely been studied (Tajchakavit & Ramaswamy, 1995; Tajchakavit et al., 1998; Gentry & Roberts, 2005; Puligundla et al., 2013). In general, the z values obtained by MW processing of juices are slightly lower than values obtained by conventional heating, for example, in the inactivation kinetics of spoilage microorganisms such as *L. plantarum* (4.5 °C vs 15.9 °C) and *S. cerevisiae* (7 °C vs 13.4 °C). In the specific case of orange juices, Nikdel et al. (1993) used MW continuous flow processing to pasteurize orange juice, taking PME and *L. plantarum* inactivation levels as indicators. The results showed the greater effectiveness of MW treatment versus conventional heating in equivalent thermal conditions. This superiority has been reported by other authors and might indicate the possibility of some enhanced effects associated with microwaves (Banik et al., 2003). The research group of Dr Zadyraka

Table 2.2 Studies related to the application of microwaves to citrus juice

Type of juice	Aspect	Conditions	Results	References
Orange juice	PME Color	MW[a]: 60, 70, 75, 85 °C	$Z = 22.1$ °C For 60 °C and 5 min, no differences in color parameters	Cinquanta et al. (2010)
	Total carotenoid content Ascorbic acid Vitamin C		Differences for 70, 75, and 85 °C 60 °C, 2.5 min (3% loss); 70 °C, 1 min (13% loss) 3–3.9% loss for 70–85 °C, 1 min 18% loss 70 °C, 5 min 12% loss 60 °C, 5 min	
Orange juice	PME Ascorbic acid Browning	89.7 and 96.4 °C (CH[b] and MW)	No differences between technologies (>97% inactivation) No loss in both technologies No difference between the two technologies	Villamiel et al. (1998)
Grapefruit juice	Color Particle size distribution Flow behavior Density Turbidity	MW: 900 W, 30 s; 80 °C CH: 80 °C, 11 s	MW and CH: lower levels of lightness, but higher b^* and C^* values. But differences between MW/CH and FJ[c]. MW: no difference compared with FJ CH: less than MW and FJ Both reduce juice consistency Both reduce juice density Both increase juice turbidity, more with MW	Igual et al. (2014)
Grapefruit juice	Ascorbic acid and vitamin C Organic acids Total phenols Antioxidant capacity	MW: 900 W, 30 s; 80 °C CH: 80 °C, 11 s	Better preservation of ascorbic acid, total phenols, and antioxidant activity than CH	Igual et al. (2010)

[a] MW, microwave heating.
[b] CH, conventional heating.
[c] FJ, fresh juice.

at Queen Margaret University in Edinburgh evaluated the feasibility of microwave volumetric heating (MVH) on the pasteurization of orange juice, obtaining promising results (Zadyraka, 2012). According to the results of that research group, the shelf life of the product was increased (up to 28 days) without detrimental effects on the flavor, nutritional content, and antioxidant capacity of valuable orange juice compounds.

2.3.3.2 Effect on Quality-Related Enzymes

MW is a thermal technology, and therefore, in principle, it should be valid for treating citrus juices. The studies that have been conducted so far have concentrated on aspects of inactivation of PME and on aspects of organoleptic and nutritional quality (Tajchakavit & Ramaswamy, 2007). Cinquanta *et al.* (2010) treated orange juice in a pilot-scale microwave unit, controlling the temperature of the process in real time (up to 85 °C). The enzyme inactivation levels achieved ($z = 22.1$ °C for 70 °C) were suitable for the marketing of orange juice (5.5×10^{-5} PME units, 70 °C, 1 min) (Kimball, 1991).

In a comparison of conventional heat treatments and MW treatments (similar final temperatures), Villamiel *et al.* (1998) achieved inactivation values greater than 97%, with no differences between the two technologies.

2.3.3.3 Effect on Nutritional Quality

Studies conducted by Igual *et al.* (2010) (900 W, 30 s, 80 °C) revealed that MW-pasteurized grapefruit juice has better preservation of ascorbic acid, total phenols, and antioxidant activity than traditional pasteurization. However, Villamiel *et al.* (1998) did not observe a loss of ascorbic acid content in grapefruit juice when it was treated by conventional heating or by MW.

2.3.3.4 Effect on Organoleptic Quality

One of the organoleptic characteristics that suffer most modification when a citrus juice is pasteurized is color. A parameter to evaluate whether there are differences in the color of a food is ΔE, which indicates the magnitude of the color difference between juices at initial time and after thermal treatment. When 70 °C and 1 min was applied to orange juice (Cinquanta *et al.*, 2010), no variations in color were observed ($\Delta E = 0.77$), taking the value of $\Delta E = 2$ as the limit for the visual detection of color changes (Francis & Clydesdale, 1975). In grapefruit juice, independently of the technology, Igual *et al.* (2014) observed lower levels of lightness but higher b^* and C^* values. However, despite these differences in color parameters, there was no difference in color between the treated juices and the fresh juice ($\Delta E < 2$). The same authors studied other physical properties (particle size distribution, flow behavior, density, and turbidity), verifying that there were no differences between the two technologies with the exception of turbidity, which increased more with MW, an aspect that, in principle, is not prejudicial in a cloudy juice such as orange juice. Igual *et al.* (2014) concluded that no strong changes in the physical properties of the juice occurred as a result of microwave heating.

These studies, together with studies on the effect of MW on other kinds of juice, show the benefit of using this type of technology as a method for preserving citrus juice, always provided that the temperature of the product is controlled in order to avoid overheating that would affect its quality.

2.4 Processing Citrus by Innovative Methods

2.4.1 High-Pressure Processing

HPP is a nonthermal food processing technology in which food, liquid or solid, and usually packed, is subjected to hydrostatic pressures (at commercial level, from 400 to 650 MPa) by a noncompressible fluid (usually water), generally at moderate temperatures. The food is placed in a pressure vessel that is filled with the fluid. The fluid is usually compressed by a pump or pressure intensifier, and the hydrostatic pressure is distributed uniformly throughout the pressure vessel and equally in all directions of the food surfaces, avoiding disuniformities due to processing. Typical treatment times applied once constant pressure is achieved last for some minutes (between 3 and 7 min at commercial level).

In general, HPP tends not to destroy covalent bonds between atoms of the constituent molecules, as the energy used during the treatment is relatively low, and the process affects hydrogen bonds and ionic and hydrophobic interactions in macromolecules. Accordingly, HPP is less aggressive than heat treatment, so the food retains more of the flavor, texture, nutrients, and quality attributes of the product before processing.

2.4.1.1 Effect on Microbial Quality As commented earlier, yeasts and fungi are notable among the spoilage flora in citrus juices because of their resistance to acidity. Both *S. cerevisiae* and *Z. bailii* have the ability to form spores on the surface of the fruit, where there are conditions of low concentrations of sugars or ethanol. During extraction of the fruit, these ascospores may contaminate it, eventually producing physicochemical changes that would cause spoilage of the beverage. Vegetative cells of these organisms are easily inactivated by pressure treatment. Raso *et al.* (1998a) and Parish (1998a) reported inactivation close to 5 log cycles in *Z. cerevisiae* and *S. cerevisiae*, with pressures ranging between 300 and 350 MPa for 5–10 min at room temperature. However, the same authors needed times of 30 min at 300 MPa, or pressures of 500 MPa for 5 min, to reduce *Z. cerevisiae* and *S. cerevisiae* by between 1 and 3 log cycles (Raso *et al.*, 1998a, Parish, 1998a).

Silva *et al.* (2012) studied the use of high pressure (200–600 MPa) in combination with mild temperature (45–65 °C) for 1–15 min to inactivate *A. acidoterrestris* spores in orange juice. Spores were inactivated at 45 °C and 600 MPa (1.2 log cycles), 65 °C and 200 MPa (1.9 log cycles), and 65 °C and 600 MPa, at which the greatest inactivation (2.5 log cycles) was achieved. Similar *A. acidoterrestris* spore inactivation was described by Hartyáni *et al.* (2013) at 600 MPa, 10 min, and 50 °C. The same authors studied whether an electronic nose was able to show differences in changes of aroma that were connected with decreases in the number of microorganisms during 4 weeks of storage at 4 °C. Changes in the volatile compounds were observed when the treatment temperature increased. The differences were most obvious in the case of a 60 °C treatment temperature. Although the electronic nose was able to show the differences in those experimental conditions (600 MPa, 10 min, 60 °C, and 28 days of storage at 4 °C), it would be interesting to check and compare the results with human olfactory experiments.

Because of their resistance to acidity, lactic acid bacteria are microorganisms that also tend to spoil citrus juices. *Leuconostoc mesenteroides* is one of them. Basak *et al.* (2002) studied inactivation of this bacterium by HPP in fresh orange juice and orange concentrate (42 °Brix). A microbial reduction of 5 log cycles was achieved after a treatment of 350 MPa for 15 min in the fresh juice and of 2.5 log cycles under the same conditions in the concentrated juice. *L. plantarum* is another of the microorganisms that can reduce the shelf life of citrus juices because of its ability to grow in acidic media. The inactivation kinetics of *L. plantarum* in a mandarin juice treated by HPP was fitted to a Weibull distribution function (Carreno *et al.*, 2011). A synergistic effect was observed between temperature (45 °C) and pressure (400 MPa) at 1 min treatment time, reaching a maximum inactivation of 6.12 log cycles. The Weibull model accurately described microorganism inactivation after HPP processing.

With regard to pathogenic flora, *E. coli* is one of the microorganisms that may be found in citrus juices because of its resistance to low pH values. It belongs to the family of *Enterobacteriaceae*, Gram-negative microorganisms that can survive in conditions of anaerobiosis and that mostly develop in the intestinal tract of numerous animals (Madigan *et al.*, 2014). Specifically, the pathogenic strain O157:H7 has been isolated from citrus juices on several occasions and has been related to various outbreaks owing to its low infective dose. The resistance to pressure of this strain of *E. coli* has been studied in depth with regard to processing conditions (treatment time and pressure), temperature employed, and physicochemical properties of the medium (basically, the pH of the juice) (Alpas & Bozoglu, 2000; Bayindirli *et al.*, 2006; Whitney *et al.*, 2007; Pina-Pérez *et al.*, 2010). Summing up the studies that have been reported, it is possible to achieve inactivation of 5 log cycles with a pressure of 600 MPa and 4–6 min of treatment at room temperature, which is the pressure currently applied in HPP-processed foods that are being marketed. Studies have also been conducted recently on combined application of various technologies (hurdle technology) and their possible synergic effect on inactivation of *E. coli* O157:H7. Espina *et al.* (2013a) studied the joint application of HPP (300 MPa, 20 min) and essential oils, (+)-limonene, achieving inactivation of over 5 log cycles.

Some strains of *Salmonella* are also capable of surviving in conditions of acidity, and therefore, they are present in citrus juices if there are bad hygiene conditions, contaminated raw matter, or recontamination after processing. As in the previous case, several studies have been conducted on the resistance to pressure of various strains of *Salmonella* (Alpas & Bozoglu, 2000; Teo *et al.*, 2001; Bull *et al.*, 2005; Bayindirli *et al.*, 2006). To sum up, a treatment of 550–615 MPa for 2–5 min is sufficient to inactivate at least 5 log cycles of the strains of *Salmonella* that have been studied.

Illness-causing organisms that are ubiquitous in nature, such as *L. monocytogenes*, have also been identified as possible contaminants in juice. The use of contaminated products to make the juice, and the ability of some of these pathogens to survive in acidic and refrigerated foods such as juices, together with the use of inadequate controls for these pathogens during juice processing, are believed to be among the causative factors for these outbreaks. Various authors (Alpas & Bozoglu, 2000; Jordan *et al.*, 2001) studied *Listeria* inactivation by HPP. A high-pressure treatment of 600 MPa for 5 min at room temperature is typically sufficient to obtain a 5 log reduction of this microorganism.

2.4.1.2 Effects on Quality-Related Enzymes

Enzymes are biocatalysts that are essential in the physiology and metabolism of plants. However, most of them remain active after harvesting, which could be desirable if ripening takes place during postharvest, but at the same time, it may also produce detrimental changes in quality properties such as color, flavor, texture, or nutritional characteristics.

The most important enzyme related to citrus juice quality is PME (EC 3.1.1.11) (PME), which catalyzes the de-esterification of pectin to acidic pectin with a lower degree of esterification and methanol. It is a texture-related enzyme, whose main action in citrus fruits depends on pulp content. It destabilizes the suspension formed by pectin mycelia (cloud), leading to a clarified product with low commercial value. The effect of HPP on PME has been extensively studied. Treatments lower than 500 MPa at room temperature seem not to inactivate PME in orange or grapefruit juice (Goodner *et al.*, 1998; Nienaber & Shellhammer, 2001a, b; Bisconsin-Junior *et al.*, 2014) or in a smoothie (50% orange juice, 20% milk, 30% water) (Sampedro *et al.*, 2008). However, Polydera *et al.* (2004) described inactivation of a PME labile fraction obtained from Greek orange juice at 100 MPa and 30 °C. Processing conditions of 600–700 MPa for 3–5 min combined with mild temperatures (50–60 °C) seem to be effective in inactivating native PME (80–90% maximum inactivation achieved) (Nienaber & Shellhammer, 2001a; Sampedro *et al.*, 2008; Bisconsin-Junior *et al.*, 2014). In general, although in real, practical conditions HPP does not result in complete inactivation of PME, good cloud stability has been reported in many cases (Terefe *et al.*, 2014).

Peroxidase (EC 1.11.1.7) (POD) is found in almost all living organisms. It catalyzes the oxidation of phenolic compounds in the presence of hydrogen peroxide, leading to the formation of brown degradation products, so it is involved in color and flavor degradation of horticultural products. In citric juices, it is related to loss of flavor quality. There are not many articles that have studied the resistance to pressure of POD in citrus juices, probably because PME is more important than POD in the quality of the juice. Cano *et al.* (1997) achieved only 50% POD inactivation in orange juice after 400 MPa for 15 min at 32 °C. Other authors found 16.1% POD residual activity after 500 MPa for 5 min in a fruit extract matrix (15% orange, 20% mango, and 65% papaya). When a leaf infusion of *Stevia rebaudiana* (2.5% w/v) was added to the beverage, Stevia enhanced the inactivation percentage obtained by HPP, achieving complete enzyme inactivation (Barba *et al.*, 2014).

2.4.1.3 Effect on Nutritional Quality

Citrus juices are a rich source of phytochemicals with biological activity. Their activity has been related to the scavenging of free radicals and nonradical reactive oxygen species, which have been identified as the major cause of cellular toxic processes, including oxidative damage to proteins and DNA, membrane lipid oxidation, enzyme pathway inhibition, and gene mutation (Oms-Oliu *et al.*, 2012). HPP treatment is expected to be less detrimental than thermal treatment to low-molecular-weight food compounds such as pigments and vitamins.

Vitamin C is the most important water-soluble nutrient, and it is related to the antioxidant capacity of citrus juices (Cortés *et al.*, 2008). Many authors have reported that vitamin C is retained better after HPP treatment than after the equivalent thermal treatment. Several authors (Sánchez-Moreno *et al.*, 2005; Plaza *et al.*, 2006) studied the effect of HPP treatment (100–400 MPa, 1–5 min at 30–60 °C) and thermal treatments (70 °C, 30 s, and 90 °C, 1 min) on orange juice stored at 4 °C for 40 days. They reported 91% retention after treatment (400 MPa, 1 min, 40 °C or 100 MPa, 5 min, 60 °C). Losses during refrigerated storage (24% and 32% after HPP and thermal processing, respectively) were attributed to the different levels of POD and ascorbate oxidase inactivation, which could degrade L-ascorbic acid by an oxidative process. The antioxidant activity was not affected after HPP and low thermal pasteurization processing, whereas 90 °C, 1 min thermal treatment reduced antioxidant activity by 6.5%.

No changes in vitamin B1, B2, B3, and B6 contents were observed after pressurizing orange juices (Donsi *et al.*, 1996).

As far as flavanones are concerned, HPP treatment (400 MPa, 1 min, 40 °C) increased the extractability of each individual flavanone with regard to the untreated orange juice and therefore the total flavanone content (15.46%) (Plaza *et al.*, 2011). The treatment increased the naringenin and hesperidin contents by 20% and 40%, respectively, in comparison with untreated juice. However, the same study reported a 50% decrease in flavanone content during 20 days of refrigerated storage, mainly related to the residual POD and PPO enzyme activity.

Carotenoids are widespread pigments, with bioactive functionality. Studies conducted by de Ancos *et al.* (2002) and Sánchez-Moreno *et al.* (2003) showed increases of 23% and 43% in total carotenoid content after pressure treatment at 100 and 350 MPa, respectively, probably owing to release of carotenoids from the orange cloud after denaturation of protein–carotenoid complexes induced by pressure. With regard to individual carotenoids, β-carotene increased by 50%, α-carotene by 60%, β-cryptoxanthin by 42%, and α-cryptoxanthin by 63% after 350 MPa for 5 min at 30 °C. As for stability during refrigerated storage, Plaza *et al.* (2011) reported that HPP-treated orange juice showed a higher carotenoid content than heat-pasteurized juice. Consequently, vitamin A values showed an increase of over 30% compared with the value of the untreated sample. Inactivation of enzymes that caused carotenoid losses during storage and improved extraction as a consequence of the treatment are the reasons presented by the authors to explain the results (de Ancos *et al.*, 2002).

Folates are vitamins with special importance for women during pregnancy. However, very few data have been published in relation to the effects of HPP treatment and folate stability. Butz *et al.* (2004) studied the stability of three main types of folates found in orange juice (tetrahydrofolate, 5-methyltetrahydrofolate, and 5-formyltetrahydrofolate) after HPP treatment (600 MPa at 25 and 80 °C). At 25 °C, the losses ranged from 10% to 40%, and at 80 °C from 25% to 95%, after treatment for 24 min. The pressure sensitivity was as follows: 5-methyltetrahydrofolate > 5-formyltetrahydrofolate > tetrahydrofolate.

2.4.1.4 Effect on Organoleptic Quality

The main driving force for research in nonthermal technologies has been the increasing consumer demand for fresh-like, healthy products with a long shelf life. Flavor is one of the most characteristic properties of fresh juices. The changes due to HPP in the analytical profile of volatile compounds related to fruit juice aroma have been deeply studied. In general, HPP treatment retains the volatile profile of juices better than an equivalent thermal treatment. Twenty volatile compounds were analyzed in the storage of orange juice at 4 and 10 °C. Considerable reductions were found in thermally and HPP-treated samples as compared with fresh juice. The compounds showing the greatest changes were octanal, citral, ethyl butanoate, and limonene, and the final concentration of the compounds was 6–38% lower than the initial level (Baxter *et al.*, 2005).

Because of the difficulty of human testing of these novel products, an electronic nose and tongue were applied to detect possible differences between fresh and HPP-treated (600 MPa, 10 min) citrus juices (orange, grapefruit, and tangerine) (Hartyáni *et al.*, 2011). The results showed that the electronic devices were able to differentiate between treated and untreated samples, making them promising tools for investigating the effect of treatment on organoleptic properties.

2.4.2   Pulsed Electric Fields

PEFs technology is specifically applied to pumpable liquid and semiliquid foods with low conductivity, such as juices, milk-based beverages, liquid whole egg, purees, smoothies, yogurt, liquors, and other beverages or infusions. In PEF technology, pulses of high intensity, of the order of 10–40 kV/cm, and short duration (from microseconds to milliseconds) are applied to inactivate enzymes and microorganisms that can affect the chemical stability and microbiological safety of the final product (Min *et al.*, 2007; Toepfl *et al.*, 2007). PEF processing has been studied by a number of researchers across a wide range of liquid foods (Raso *et al.*, 1998a, b; Jin & Zhang, 1999; Heinz *et al.*, 2002; Rodrigo *et al.*, 2003; Plaza *et al.*, 2011; Pina-Pérez *et al.*, 2013b), as a pasteurization process where the temperature remains below 50 °C during the treatment. After treatment, the food is packaged aseptically and stored under refrigeration. As a pasteurization process, additional control measures are consequently required to achieve food safety objectives, reducing 5*D* of the pathogens of greatest concern, such as *E. coli* O157:H7, *S. typhimurium*, and *L. monocytogenes* (Riener *et al.*, 2009; Pina-Pérez *et al.*, 2012). Innovative functional beverages have been processed by PEF technology without significant depletion of their valuable bioactive composition. Specifically, this technology has been successfully applied mainly to acidic products such as fruit juices and also to fruit–milk mixed beverages (García *et al.*, 2005; Rivas *et al.*, 2006, 2007; Gurtler *et al.*, 2010, 2011), citrus juices being among the substrates most treated by PEF technology (Cortés *et al.*, 2006; Sampedro *et al.*, 2009; Monfort *et al.*, 2010). Moreover, combined barriers have been studied to improve the effectiveness of this technology, the application of moderate temperature, and the use of natural antimicrobials being the most important ones, exerting synergistic effects on the microbial and enzymatic inactivation achieved (Pina-Pérez *et al.*, 2012, 2013b).

This section gives an overview of the main aspects regarding current achievements in PEF processing of citrus juices, alone or in combination with the concept of hurdle technology.

2.4.2.1   Effect on Microbial Quality
The acceptance of PEF technology as an alternative to conventional thermal pasteurization of fruit juices requires the accomplishment of successful destruction of more than 5-$\log_{10}$ of pathogenic microorganisms. *Salmonella* spp. are generally accepted as the pertinent pathogens in citrus juice (Parish, 1998b; Diana *et al.*, 2012), and it is mandatory to reduce them by 99.999% or 5-$\log_{10}$ CFU/ml (USFDA, 2013). Several authors have demonstrated the effectiveness of PEF against *S. typhimurium* in different substrates and under a wide range of conditions (Sampedro *et al.*, 2011; Pina-Pérez *et al.*, 2012). Specifically, in orange juice substrate, a maximum inactivation of 5.9$\log_{10}$ was achieved with a PEF treatment of 90 kV/cm for 50 µs at 55 °C (Liang *et al.*, 2002). Gurtler *et al.* (2010) studied the inactivation of two strains of *S. typhimurium* (UK-1 and 14028) in orange juice, achieving reduction levels between 2.05 and 3.54 $\log_{10}$ cycles after application of PEF treatment at 22 kV/cm for 59 µs (45 °C outlet temperature). The impact of temperature on PEF lethality has been systematically studied by several research groups in citrus juices (McDonald *et al.*, 2000; Wouters *et al.*, 2001; Barbosa-Cánovas & Sepúlveda, 2005). The combination of PEF technology with mild heat is generally related to a high level of inactivation, attributed to the temperature-related phase transition of the membrane phospholipids from gel to liquid crystalline, which causes

membranes to lose their elastic properties and become more easily disrupted by PEF (Stanley, 1991).

According to Clark (2006), *Listeria* spp. are some of the most PEF resistant microorganisms. Owing to the psychrotrophic nature of *L. monocytogenes* and its ability to survive in acidic conditions, the bacterium can grow in the environment and become endemic within processing facilities, which can ultimately introduce the pathogen into domestic/food service environments. Specifically, in industrial plants with liquid flow processes, such as beverage production, the biofilm formation capability of *L. monocytogenes* has been recognized as the main cause of persistent contamination of pumpable foods (Szlavik *et al.*, 2012). In spite of this, reduction levels close to 3.9 $\log_{10}$ cycles have been obtained in *L. monocytogenes* surrogate by PEF processing of orange juice (pH 3.5) at 40 kV/cm, 100 μs, 56 °C (McNamee *et al.*, 2010). The low pH sensitivity of Gram-positive bacteria treated by PEF was pointed out by Gómez *et al.* (2005). PEF treatment at 28 kV/cm, 400 μs in acidic juices achieved *L. monocytogenes* reductions of up to 6 $\log_{10}$ cycles when the pH value remained below 3.5 and inactivation values close to 3 $\log_{10}$ cycles when the pH was $\geq$5.

The pH value is an important factor that affects PEF effectiveness, specifically in citrus juices. According to Buckow *et al.* (2013), the combined effect of pH (ranging from 2.9 to 4.2) and PEF treatment conditions in orange substrates can be summarized as producing pathogen inactivation levels ranging from 1.59 $\log_{10}$ cycles (*E. coli* O157:H7; PEF = 22 kV/cm, 59 μs, 45 °C; pH 3.4) to 6.0 $\log_{10}$ cycles for *S. aureus* (PEF = 40 kV/cm, 150 μs, 56 °C; pH 3.7). In general, application of combined conditions of PEF (25–35 kV/cm) and low pH (3.5) could produce an additional reduction of 1–3 $\log_{10}$ cycles of pathogenic bacteria within 24 h in refrigerated storage (Somolinos *et al.*, 2008; Saldaña *et al.*, 2010).

Regarding the effectiveness of PEF against the pathogen *E. coli* O157:H7, inactivation levels of 1.0, 2.4, and 3.79 $\log_{10}$ were achieved with treatments at 13.1, 19.7, and 23.7 kV/cm (75 μs), respectively, when mild temperature was applied in combination (55 °C) (Gurtler *et al.*, 2010).

Recently, the addition of natural antimicrobials has improved the effectiveness of PEF by avoiding the use of mild temperatures that might affect the flavor or aroma of processed citrus juices. In some cases, owing to the resistance of some pathogens, it is necessary to apply very intense PEF treatments (high electric field strength and energy inputs), application of which on an industrial scale has shown several technical limitations (Barbosa-Cánovas & Sepúlveda, 2005). Although many techniques are applied in combination with PEF to enhance its effectiveness, the use of natural or chemical preservatives in combination with PEF treatments in order to achieve additive or synergistic effects is one of the hurdles most frequently used in PEF processing. Nisin is one of the antimicrobials most commonly added to fruit juices. It has been added to orange-based beverages to test its possible antimicrobial potential against natural juice microflora and foodborne pathogens (Rupasinghe & Yu, 2012). The addition of nisin (0.1 mg/ml) to pasteurized, freshly squeezed orange juice led to a reduction of 2.95 $\log_{10}$ cycles in the initial load of *S. typhimurium* when the juice was processed by nonthermal PEF technology (30 pulses of 90 kV/cm) (Liang *et al.*, 2002). According to the results of Hodgins *et al.* (2002), combined treatments of PEF (80 kV/cm), mild heat (44 °C), and addition of nisin (100 U/ml) to orange juice achieved natural microflora inactivation levels of up to 6 $\log_{10}$ cycles. More intensive treatments, 90 kV/cm, 100 μs, mild heat of 55 °C, and addition of nisin (100 U/ml), were required to inactivate a minimum of 7 $\log_{10}$ cycles of *S. typhimurium* in orange juice matrices. In orange juice, specifically, a mixture of nisin (27.5 U/ml) and lysozyme (690 U/ml) added to an orange juice processed by applying 90 kV/cm, 50 pulses (55 °C) increased the cell viability lost in the *Salmonella* spp. population by an additional 0.04–2.75 $\log_{10}$ cycles (Liang *et al.*, 2002). Synergistic effects were observed by McNamee *et al.* (2010) following the incorporation of nisin (2.5 ppm) to orange juice treated by PEF (40 kV/cm, 100 µs), achieving inactivation levels of up to 5 $\log_{10}$ cycles in *E. coli* and *L. innocua* populations.

Combinations of heat (54 °C, 10 min) or PEF treatment (30 kV/cm, 25 pulses) with the addition of limonene (200 ml/l) to orange juice were studied by Espina *et al.* (2013b), who confirmed that synergistic effects of four extra $\log_{10}$ cycles are achievable by the application of this hurdle technology. Similarly, Arroyo *et al.* (2010) confirmed the synergistic effect against *Cronobacter sakazakii* of applying PEF and adding citral to an orange juice beverage, achieving a reduction of up to 2 extra $\log_{10}$ cycles in bacterial counts.

With regard to the microbiological safety of citrus-based smoothies, the application of PEF could be important as part of a trend toward health-promoting nutrition based on these tasty beverages in which, by using PEF treatment, organoleptic and bioactive properties are preserved like fresh characteristics. In recent years, smoothies have rapidly increased in popularity, HPP and PEF being the technologies recently suggested providing safe innovative products. Walking-Ribeiro *et al.* (2010) studied the effectiveness of PEF after processing a smoothie-type beverage prepared from juiced fresh fruits (pineapples, bananas, apples, and oranges) and coconut milk. A combination of mild heat (55 °C) and PEF treatment at 34 kV/cm, 60 µs resulted in reductions of 3.4 and 4 $\log_{10}$ cycles in total aerobic bacteria and total counts of yeasts and molds, respectively, increasing shelf life of the product by seven additional days compared with traditional heat pasteurization treatment (72 °C, 15 s). Other combined hurdles have been used in the treatment of citrus-based beverages, as reported by Palgan *et al.* (2012). They treated a smoothie based on milk and fruit juices (35% orange) by a combination of PEF (34 kV/cm, 32 µs) and manothermosonication (20 kHz, maximum amplitude of 31 µm, and intensity of 40 W/cm^2), achieving a mean reduction of 5.6$\log_{10}$ cycles in the *L. innocua* population, thereby exceeding the 5$\log_{10}$ cycles minimum requirement specified by the United States Food and Drug Administration.

In addition to ensuring food safety, PEF treatments have been proposed to prevent the development of spoiling microorganisms, thereby extending product shelf life. Several studies have focused on the inactivation of natural microflora in PEF-treated citrus juices and their subsequent shelf life. Many authors have reported the achievement of a ≥5 log reduction in the naturally occurring microflora of naturally squeezed orange juice (Dunn & Pearlman, 1987; Zhang *et al.*, 1995; Zhang, 1997; Yeom *et al.*, 2000), by means of the application of PEF treatments of 20–35 kV/cm and 35–60 µs, extending the shelf life of processed products to 1 week at a storage temperature of 37 °C and 4 weeks at 22 °C. Other orange juice products, such as blended orange and carrot juice, have also been treated by PEF. Rodrigo *et al.* (2003) studied inactivation of the natural microbial flora of a filtered, fresh orange–carrot juice mixture (80% orange juice–20% carrot juice) maintained at −40 °C prior to PEF treatment (OSU-4D, 6 cofield chambers in series). The initial count of microbial flora was 10^4 CFU/ml, and after PEF treatment ($E = 40$ kV/cm, $t = 130$ µs), the final count was 30 CFU/ml. The initial count of molds and yeasts was 10^4 CFU/ml, which reduced to 5 CFU/ml after PEF treatment ($E = 25$ kV/cm, $t = 340$ µs).

Lactic acid bacteria are normal flora in orange juice and present no health hazards. These microorganisms are Gram-positive, non-spore-forming, microaerophilic bacteria that mainly produce lactic acid from fermentable carbohydrates. Their metabolic activity of converting glucose to lactic acid through homofermentation, the bifidus pathway, and heterofermentation causes spoilage of juice and leads to the end of its shelf life. Among the most studied spoilage microorganisms inactivated by PEF, the following are noteworthy: *L. plantarum*, *L. mesenteroides*, and *Lactobacillus brevis*. McDonald *et al.* (2000) demonstrated the effectiveness of PEF for inactivation of spoilage microorganisms in orange juice, applying electric field strengths of 30 and 50 kV/cm for 12 μs at an outlet temperature of 55 °C. PEF treatment at 30 or 50 kV/cm inactivated 4.75 $\log_{10}$ and 6.2 $\log_{10}$ cycles, respectively, of *L. mesenteroides* in orange juice. In the same substrate, according to Soliva-Fortuny *et al.* (2005), *L. brevis* inoculated at an initial load of 10^8 CFU/ml was reduced by up to a maximum of 5.8 $\log_{10}$ cycles as a result of application of PEF (35 kV/cm for 1000 μs using 4-μs pulse width in bipolar mode). Treatment of *L. casei* in orange juice at 20 kV/cm for 70 μs and 55 °C outlet temperature achieved an inactivation level of 0.6 $\log_{10}$ CFU/ml, compared with inactivation levels for *L. plantarum*, *L. fermentum*, and *L. lactis* of 3.07, 3.22, and 4.75 $\log_{10}$, respectively. According to Élez-Martínez *et al.* (2005), *L. brevis* was inactivated to 5.8 $\log_{10}$ CFU/ml in orange juice during PEF processing at 35 kV/cm for 1000 μs at less than 32 °C (4-μs pulse width in bipolar mode). Inactivation levels of 2.5–7.8$\log_{10}$ cycles were obtained in orange juices treated by PEF at 35–50 kV/cm, 100 μs, 40–50 °C, in spoilage *Pichia fermentans*, *L. mesenteroides*, and *S. cerevisiae* by McNamee *et al.* (2010) and McDonald *et al.* (2000).

PEF inactivation of spoilage microorganisms in orange juice was also demonstrated by Sampedro *et al.* (2009) and Timmermans *et al.* (2011). According to studies conducted by Sampedro *et al.* (2009), PEF treatment (30 kV/cm for 50 μs) was effective in prolonging the shelf life of an orange juice–skim milk mixed beverage to 2.5 weeks at a slightly abusive storage temperature (8–10 °C), the reduction in bacterial counts and in yeasts and molds being close to 4.5–5 $\log_{10}$ cycles. A similar PEF treatment (23 kV/cm, 36 μs, inlet temperature 38 °C, outlet temperature 58 °C) was applied by Timmermans *et al.* (2011), resulting in microbial counts of levels less than the detection limit for up to 2 months of refrigerated (4 °C) storage. According to several authors, the shelf life of PEF-processed orange juices under refrigeration (4 °C) increases from 6 weeks (Sharma *et al.*, 1998; Yeom *et al.*, 2000) to 5 months (Jia *et al.*, 1999), depending on the PEF treatment conditions.

2.4.2.2 Effect on Quality-Related Enzymes

In general, enzymes require more intensive PEF treatment conditions than microorganisms to be reduced significantly (Ho *et al.*, 1997; Terefe *et al.*, 2015). Orange PME is partially inactivated by PEF processing at high temperatures. PME inactivation levels were found to range from 10% (Van Loey *et al.*, 2002, Walking-Ribeiro *et al.*, 2009) at 20 kV/cm, 4000 μs, approximately 25 °C, up to levels ≥90% using PEF treatments of 25–30 kV/cm, 300–400 μs, with maximum treatment temperatures of 60–72 °C (Yeom *et al.*, 2002; Espachs-Barroso *et al.*, 2006; Sentandreu *et al.*, 2006). According to Yeom *et al.* (2000), a PEF treatment of 35 kV/cm for 59 μs inactivated 88% of PME activity. This level of inactivation was irreversible after storage at 4 and 22 °C for 112 days.

Not only in orange juice but also in mixed beverages, levels of PME inactivation between 80% and 90% were achieved by PEF (Rodrigo *et al.*, 2003; Rivas *et al.*, 2006; Sampedro *et al.*, 2009). A PEF-treated orange–carrot juice mixed beverage was processed by PEF at 25 kV/cm, 330 μs, achieving 81% PME inactivation (Rivas *et al.*, 2006). Sampedro *et al.* (2009) processed an orange juice–skim milk mixed beverage under PEF conditions of 35 kV/cm for 59 μs, achieving a PME inactivation level (90%) equivalent to a thermal treatment of 85 °C, 66 s, with the added value of better preserved organoleptic values. A blended orange–carrot juice was also processed by Rodrigo *et al.* (2003), achieving levels of inactivation close to 80% with PEF conditions of 25 kV/cm, 340 μs, 63 °C. Nevertheless, the treatment times proposed are also much longer than the practical commercial limits (<200 μs) (Buckow *et al.*, 2013). In spite of the resistant PME fraction remaining in orange juice treated by PEF, citrus juices featured good cloud stability during storage, comparable to thermally pasteurized juices (Hodgins *et al.*, 2002). However, the type of enzyme, the enzyme source, and enzyme concentration have a significant effect on the level of enzyme inactivation by PEF processing. In this connection, orange peroxidase (POD) inactivation levels between 93% and 100% have been obtained in orange juice by Élez-Martínez *et al.* (2005) using PEF treatments of 35 kV/cm for 1000–1500 μs in combination with temperatures ≥40 °C.

2.4.2.3   Effect on Nutritional Quality   Changes in the concentrations of phytochemicals resulting from processing and storage can greatly compromise the quality, and ultimately the acceptance, of the product. Specifically, citrus fruits contain a range of key nutrients such as vitamin C, vitamin A, carotenes of various kinds (e.g., β-carotene, lutein, and zeaxanthin), folate, and fiber, as well as very many non-nutrient phytochemicals, including classes such as flavonoids, glucarates, coumarins, monoterpenes, triterpenes, and phenolic acids, and individual components such as hesperidin, naringin, tangeritin, limonene, nomilin, perillyl alcohol, myricetin, quercetin, sinensetin, tangeretin, and nobiletin (Baghurst, 2003). In particular, orange has over 170 different phytochemicals and more than 60 flavonoids. However, phytochemicals are highly sensitive to processing conditions. Furthermore, various examples are presented to illustrate the potential of PEF technology when aiming at preserving the health-promoting features of plant-based foods, such as citrus juices in this case.

Polyphenols (Sánchez-Moreno *et al.*, 2005, 2009) and vitamins (Rivas *et al.*, 2007; Zulueta *et al.*, 2010a,b) are the two most studied groups of phytochemical components with an impact on human health that have been processed by PEF.

In the vitamins group, vitamin C has been studied most in PEF-processed beverages (apple juice, strawberry juice, orange–carrot juice, tomato juice, orange juice, and fruit juice–soy milk blends) as a reference, owing to its heat lability. It is generally recognized that there is higher retention of ascorbic acid in PEF-processed juices than in thermally processed juices (Min *et al.*, 2003), just after treatment and during refrigerated storage (4 °C). According to Oms-Oliu *et al.* (2012), more acidic conditions are known to stabilize vitamin C, the retention of vitamin C being above 80% in PEF-processed (35 kV/cm for 2050 μs at 250 Hz applying mono- or bipolar 7-μs pulses) orange or orange–carrot. Specifically, in orange juice, according to Yeom *et al.* (2000), greater amounts of vitamin C and the five representative flavor compounds in the product were highly retained by PEF treatment (35 kV/cm for 59 μs) during storage at 4 °C compared with heat-pasteurized juice (94.6 °C, 30 s) ($p < 0.05$). Moreover, the PEF-treated orange juice had a lower browning index and higher whiteness (L) values than the heat-pasteurized orange juice during storage at 4 °C (Yeom *et al.*, 2000). According to Torregrosa *et al.* (2006), the vitamin C content in a PEF-treated orange juice was 55 mg/100 ml after processing, which equals the content present in the fresh product. Zulueta *et al.* (2010a) studied the kinetic degradation of vitamin C after PEF processing of an orange juice–skim milk mixed beverage. The degradation of vitamin C followed an exponential decay: the higher the kinetic constants, the higher the electric field applied in the range 15–40 kV/cm. During storage at 4 °C after processing, the degradation of vitamin C followed first-order kinetics, being equivalent to the degradation of vitamin C that occurs in a thermally processed orange juice (90 °C, 20 s).

Total carotenoid content is also better preserved in PEF-treated juices than under equivalent thermal pasteurization conditions. Cortés *et al.* (2006) observed that the total carotenoid concentration in orange juice increased slightly after intense PEF treatments of 35 and 40 kV/cm for 30–240 µs. Similarly, Torregrosa *et al.* (2005) observed that the carotenoid concentration of an orange–carrot juice mixture rose as treatment time increased with PEF treatments of 25 and 30 kV/cm. A slightly increased concentration of carotenoid compounds was observed in an orange juice–skim milk beverage after processing at 15 kV/cm (40–700 µs) with respect to the same substrate treated under thermal conditions (90 °C, 20 s) (Zulueta *et al.*, 2010a).

According to the studies of Morales-De la Peña *et al.* (2011), in PEF-treated fruit juice–soymilk beverages, the shorter the treatment time that was applied, the better the preservation of vitamin C that was achieved. They reported that vitamin C was better maintained in an 800-µs PEF-treated fruit juice–soymilk beverage (46.4%) than in the same substrate treated at 1400 µs (22.6%); the thermally treated beverage showed the lowest retention of vitamin C (6.7%) (Morales-De la Peña *et al.*, 2011). However, other authors did not find any difference in vitamin A content between PEF and thermal treatments. According to Sánchez-Moreno *et al.* (2005), PEF processing (35 kV/cm, 750 µs) and thermal treatments (70 °C, 30 s, and 90 °C, 30 s) did not have any effect on the vitamin A content of an orange juice. According to Plaza *et al.* (2011), a comparison among PEF, HPP, and mild pasteurization (70 °C, 30 s) with regard to carotenoid content revealed that after processing HPP was the technique that best preserved the vitamin C content, while there was no difference in orange juice vitamin C content between the thermal and PEF pasteurization processes. Furthermore, according to Cortés *et al.* (2009), xanthophyll content was reduced by PEF. The antheraxanthin concentration decreased during storage (10 °C) of a PEF-processed orange juice (30 kV/cm, 100 µs) to undetectable levels from the 6th week onward. However, the opposite was observed by Zulueta *et al.* (2010b). PEF treatment was a better process for preserving the bioactive compounds lutein and zeaxanthin compared with thermal treatment (90 °C, 20 s), which significantly reduced the concentrations of both xanthophylls, lutein (22.8%) and zeaxanthin (22.5%).

Regarding phenolic compounds, according to other authors (Odriozola-Serrano *et al.*, 2008), it seems that no significant differences were observed between PEF-treated and thermally treated juices, but high retention of these antioxidant compounds was observed during storage. Similarly, in an orange juice treated at 35 kV/cm, 450 µs, with 800-Hz bipolar pulses, Sánchez-Moreno *et al.* (2009) observed that the total flavonone content was invariable. The antioxidant capacity (mmol TE/L) and total phenolic content (mg GE/L) of PEF-processed fruit juices were significantly higher than the values obtained for the same matrices thermally pasteurized (Schilling *et al.*, 2008). In a blended fruit juice–soymilk beverage, the content of the phenolic compound hesperidin showed a huge rise after PEF treatment (35 kV/cm with 4-µs bipolar pulses at 200 Hz for 800 or 1400 µs), resulting in a significant increase in total phenol concentration. In addition, the total phenol concentration seemed to be highly stable during refrigerated storage (Morales-De la Peña *et al.*, 2011). The effects observed in phenolic compounds after PEF processing may be due to (i) biochemical reactions that could have occurred during PEF processing, leading to the formation of new phenolic compounds; (ii) PEF might have caused significant effects on cell membranes or in phenolic complexes with other compounds, releasing some free phenolic acids or flavonoids; (iii) PEF may inactivate PPO, preventing further loss of phenolic compounds; and (iv) PEF treatment might have induced favorable conditions for an increase in PAL activity, resulting in an enhancement of phenolic concentration in the beverage. Flavonoids are the most common and most widely distributed group of plant phenolics. Among them, flavones, flavonols, flavanols, flavanones, anthocyanins, and isoflavones are particularly common in fruits. Similarly, Sánchez-Moreno *et al.* (2005) evaluated the effect of a PEF treatment at 35 kV/cm for 750 μs (with 4-μs bipolar pulses at 800 Hz) on the flavanone content of orange juice. No changes were observed in total flavanones, or in the individual flavanone glycosides and their aglycones, hesperetin and naringenin.

It was observed that PEF processing of orange juice not only achieved better preservation of nutritional factors but also improved their bioavailability. According to Sánchez-Moreno *et al.* (2004), an *in vivo* study regarding the consumption of PEF-treated orange juice in a dose equivalent to an intake of about 185 mg/day of ascorbic acid, revealed that PEF treatment retained vitamin C bioavailability and antioxidant properties and had an impact on human health indicators such as plasma vitamin C and plasma 8-epiPGF2α concentrations (indicator of lipid peroxidation).

The main energy-yielding nutrient in citrus is carbohydrate; citrus contains the simple carbohydrates (sugars) fructose, glucose, and sucrose, as well as citric acid, which can also provide a small amount of energy. Citrus fruits also contain nonstarch polysaccharides (NSP), commonly known as dietary fiber, which is a complex carbohydrate with important health benefits. PEF treatment (35 kV/cm with 4-μs bipolar pulses at 200 Hz for 800 or 1400 μs) of an orange juice–soy milk mixed beverage was evaluated in terms of nutritional compound retention by Morales-De la Peña *et al.* (2010a). The treated beverage showed a higher concentration of glucosides than aglycones, genistin being the most abundant glucoside, immediately after PEF treatment. Regarding the lipid profile after PEF, Zulueta *et al.* (2007) reported that PEF treatment did not significantly change the amounts of total fat, saturated fatty acids, monounsaturated fatty acids, and polyunsaturated fatty acids contained in an orange juice–milk-based beverage fortified with $n-3$ fatty acids and oleic acid. However, according to Morales-De la Peña *et al.* (2011), higher concentrations of linoleic, oleic, linolenic, palmitic, and stearic acids were found in an orange juice–soy milk beverage after PEF treatment (35 kV/cm with 4-μs bipolar pulses at 200 Hz for 800 and 1400 μs) compared with the same substrate processed thermally (90 °C for 60 s).

2.4.2.4 Effect on Organoleptic Quality

Pink grapefruit, lemon, and tangerine juices were treated by Cserhalmi *et al.* (2006), who analyzed physical and chemical characteristics. According to their research, the acid contents (malic, citric, ascorbic, and fumaric acids) in PEF-treated orange juice (2-μs bipolar pulse duration, 28 kV/cm, −50 pulses) remained invariable between the untreated and treated samples. From a physicochemical point of view, according to Cserhalmi *et al.* (2006), there was no significant difference in Brix-, pH, conductivity, and viscosity between PEF-treated (2-μs bipolar pulse duration, 28 kV/cm, −50 pulses) and untreated citrus juices. Organoleptic characteristics also remained similar to the fresh product. In most of the samples, there was no or only a slightly visible difference in color. Moreover, closely related to the acceptability of the product, the most important volatile aroma compounds in citrus juices did not change significantly after PEF. This is in agreement with the studies of Yeom *et al.* (2000) regarding PEF-treated orange juice (35 kV/cm, 59 μs), in which they observed retention of volatile components (α-pinene, myrcene, octanal, D-limonene, and decanal). In general, according to several authors (Vega-Mercado *et al.*, 1997; Cserhalmi *et al.*, 2006), changes in flavor components of PEF-treated and heat-treated citrus juices showed 16% and 61% losses, respectively, compared with fresh citrus juices.

Another factor affecting the organoleptic quality of PEF-processed citrus juices is the packaging material. According to Ayhan *et al.* (2001), the retention of orange juice aroma compounds, color, and vitamin C was significantly increased up to a shelf life of >16 weeks in glass and PET packagings at 4 °C after PEF treatment (35 kV/cm for 59 μs). Regarding the significant influence of packaging on flavor and volatile compounds of citrus juices, recent studies have demonstrated the effectiveness of PEF technology in preserving valencene (28 kV/cm, 100 μs, ~34 °C), naringenin (35 kV/cm, 750 μs, ~50 °C), hesperidin (35 kV/cm, 750 μs, ~50 °C), and other characteristic citrus compounds by using PE cups at 4 °C or opaque tubes with nitrogen headspace (Hartyáni *et al.*, 2011; Plaza *et al.*, 2011).

2.5 Conclusions and Future Trends

The consumption of fruit juice, especially of citrus-based juice, is very important in the diet of the population, particularly in developing countries, not only because of their much-appreciated flavor but also because they are important sources of vitamins, phytochemicals, and other bioactive compounds. However, the traditional thermal preservation processes have a series of limitations that, in the case of citrus juices, lead to a deterioration in the nutritional and organoleptic quality of the processed product. In view of consumer preference for processed foods with fresh-like characteristics, the future of food preservation is moving toward "à la carte processing," consisting in designing a specific treatment for each food. In fact, there are foods for which traditional technologies are still, and will continue to be, the best processing alternative. In spite of this, niche markets have appeared for emerging technologies due to their potentiality to be applied to certain products, such as citrus juices, giving rise to foods that are microbiologically safe and healthy, and that retain the properties of the fresh product to a greater extent, that is, that better satisfy consumer demands.

References

Alpas, H. & Bozoglu, F. (2000) The combined effect of high hydrostatic pressure, heat and bacteriocins on inactivation of foodborne pathogens in milk and orange juice. *World Journal of Microbiology and Biotechnology* **16**, 387–392.

Alvarez-Ordoñez, A., Fernández, A., Bernardo, A. & López, M. (2009) A comparative study of thermal and acid inactivation kinetics in fruit juices of *Salmonella enterica* serovar *Typhimurium* and *Salmonella enterica* serovar *Senftenberg* grown at acidic conditions. *Foodborne Pathogens and Disease* **6**, 1147–1155.

Alwazeer, D., Cachon, R. & Divies, C. (2002) Behavior of *Lactobacillus plantarum* and *Saccharomyces cerevisiae* in fresh and thermally processed orange juice. *Journal of Food Protection* **65**, 1586–1589.

Arroyo, C., Somolinos, M., Cebrián, G., Condón, S. & Pagán, R. (2010) Pulsed electric fields cause sublethal injuries in the outer membrane of Enterobacter sakazakii facilitating the antimicrobial activity of citral. *Letters in Applied Microbiology* **51**, 525–531.

Ayhan, Z., Yeom, H.W., Zhang, Q.H., & Min, D.B. (2001) Flavor, color, and vitamin C retention of pulsed electric field processed orange juice in different packaging materials. *Journal of Agriculture and Food Chemistry* **49**, 669–674.

Baghurst, K. (2003) The health benefits of citrus fruits. *Horticulture Australia* 1–248.

Banik, S., Bandyopadhyay, S. & Ganguly, S. (2003) Bioeffects of microwave – a brief review. *Bioresource Technology* **87**, 155–159.

Barba, F. J., Criado, M., Belda-Galbis, C., Esteve, M. & Rodrigo, D. (2014) *Stevia rebaudiana* Bertoni as a natural antioxidant/antimicrobial for high pressure processed fruit extract: processing parameter optimization. *Food Chemistry* **148**, 261–267.

Barbosa-Cánovas, G.V. & Sepúlveda, D. (2005) Present status and the future of PEF technology. In: *Novel Food Processing Technologies.* (eds. G.V. Barbosa-Cánovas, M.S. Tapia & M.P. Cano) pp. 1–44 CRC Press, Boca Raton, Florida, Marcel Dekker.

Basak, S., Ramaswamy, H. & Piette, J. (2002) High pressure destruction kinetics of *Leuconostoc mesenteroides* and *Saccharomyces cerevisiae* in a single strength and concentrated orange juice. *Innovative Food Science and Emerging Technologies* **3**, 223–231.

Baxter, I. A., Easton, K., Schneebeli, K. & Whitreld, F. B. (2005). High pressure processing of Australian navel orange juices: sensory analysis and volatile flavor profiling. *Innovative Food Science & Emerging Technologies* **6**, 372–387.

Bayindirli, A., Alpas, H., Bozoglu, F. & Hizal, M. (2006) Efficiency of high pressure treatment on inactivation of pathogenic microorganisms and enzymes in apple, orange, apricot and sour cherry juices. *Food Control* **17**, 52–58.

Baysal, A. H. & Icier, F. (2010) Inactivation kinetics of *Alicyclobacillus acidoterrestris* spores in orange juice by ohmic heating: effects of voltage gradient and temperature on inactivation. *Journal of Food Protection* **73**, 299–304.

Benlloch-Tinoco, M., Martínez Navarrete, N. & Rodrigo, D. (2014b) Impact of temperature on lethality of kiwifruit puree pasteurization by thermal and microwave processing. *Food Control* **37**, 336–342.

Benlloch-Tinoco, M., Pina-Pérez, M.C., Martínez-Navarrete, N. & Rodrigo, D. (2014a) *Listeria monocytogenes* inactivation kinetics under microwave and conventional thermal processing in a kiwifruit puree. *Innovative Food Science & Emerging Technologies* **22**, 131–136.

Bisconsin-Junior, A., Rosenthal, A. & Moteiro, M. (2014) Optimisation of high hydrostatic pressure processing of pra rio orange juice. *Food and Bioprocess Technology* **7**, 1670–1677.

Braddock, R. (1999) *Handbook of Citrus By-products and Processing Technology.* John Wiley & Sons.

Buckow, R., Ng, S. & Toepfl, S. (2013) Pulsed electric field processing of orange juice: a review on microbial, enzymatic, nutritional, and sensory quality and stability. *Comprehensive Reviews in Food Science and Food Safety* **12**, 455–468.

Bull, M., Szabo, E., Cole, M. B. & Stewart, C. M. (2005) Toward validation of process criteria for high-pressure processing of orange juice with predictive models. *Journal of Food Protection* **68**, 949–954.

Burdulu, H. S., Koca, N. & Karadeniz, F. (2006). Degradation of vitamin C in citrus juice concentrates during storage. *Journal of Food Engineering* **74**, 211–216.

Butz, P., Serfert, Y., García, A. F., Dietrich, S., Lindauer, R., Bognar, A. & Tauscher, B. (2004) Influence of high-pressure treatment at 25 degrees C and 80 degrees C on folates in orange juice and model media. *Journal of Food Science* **69**, S117–S121.

Cameron, R. G., Baker, R. A. & Grohmann, K. (1998) Multiple forms of pectin methylesterase from citrus peel and their effects on juice cloud stability. *Journal of Food Science* **63**, 253–256.

Cano, M., Hernandez, A. & De Ancos, B. (1997) High pressure and temperature effects on enzyme inactivation in strawberry and orange products. *Journal of Food Science* **62**, 85–88.

Carreno, J. M., Gurrea, M. C., Sampedro, F. & Carbonell, J. V. (2011). Effect of high hydrostatic pressure and high-pressure homogenisation on *Lactobacillus plantarum* inactivation kinetics and quality parameters of mandarin juice. *European Food Research and Technology* **232**, 265–274.

Cinquanta, L., Albanese, D., Cuccurullo, G. & Di Matteo, M. (2010) Effect on orange juice of batch pasteurization in an improved pilot-scale microwave oven. *Journal of Food Science* **75**, E46–E50.

Clark, P. (2006) Pulsed electric fields processing. *Food Technology Processing* **60**, 66–67.

Cortés, C., Esteve, M.J. & Frígola, A. (2008) Effect of refrigerated storage on ascorbic acid content of orange juice treated by pulsed electric fields and thermal pasteurization. *European Food Research and Technology* **227**, 629–635.

Cortés, C., Esteve, M.J. & Frígola, A. (2009) Anteroxanthin concentration during refrigerated storage in orange juice treated by PEF. *Czech Journal of Food Sciences* **27**, 1–3.

Cortés, C., Torregrosa, F., Esteve, M.J. & Frígola, A. (2006) Carotenoid profile modification during refrigerated storage in untreated and pasteurized orange juice and orange juice treated with high-intensity pulsed electric fields. *Journal of Agriculture and Food Chemistry* **54**, 6247–6254.

Cserhalmi, Z., Sass-Kiss, A., Tóth-Markus, M. & Lechner, N. (2006) Study of pulsed electric field treated citrus juices. *Innovative Food Science and Emerging Technologies* **7**, 49–54.

De Ancos, B., Sgroppo, S., Plaza, L. & Cano, M. P. (2002) Possible nutritional and health-related value promotion in orange juice preserved by high-pressure treatment. *Journal of the Science of Food and Agriculture* **82**, 790–796.

Demirdöven, A. & Baysal, T. (2014) Optimization of ohmic heating applications for pectin methylesterase inactivation in orange juice. *Journal of Food Science and Technology* **51**, 1817–1826.

Diana, J.E., Pui, C.F. & Son, R. (2012) Enumeration of *Salmonella* spp. *Salmonella typhi* and *Salmonella typhimurium* in fruit juices. *International Food Research Journal* **19**, 51–56.

Donsi, G., Ferrari, G. & Dimatteo, M. (1996) High pressure stabilization of orange juice: evaluation of the effects of process conditions. *Italian Journal of Food Science* **8**, 99–106.

Dunn, J. E. & Pearlman, J. S. (1987) Methods and apparatus for extending the shelf life of fluid products. US Patent, 4,695,472.

Eiroa, M. N. U., Junqueira, V. C. A. & Schmidt, F. L. (1999) Alicyclobacillus in orange juice: occurrence and heat resistance of spores. *Journal of Food Protection* **62**, 883–886.

Élez-Martínez, P., Escolà-Hernández, J., Soliva-Fortuny, R.C. & Martín-Belloso, O. (2005) Inactivation of *Lactobacillus brevis* in orange juice by high-intensity pulsed electric fields. *Food Microbiology*, **22**, 311–319.

Espachs-Barroso, A., Van Loey, A., Hendrickx, M. & Martín-Belloso, O. (2006) Inactivation of plant pectin methylesterase by thermal or high intensity pulsed electric field treatments. *Innovative Food Sciences & Emerging Technologies* **7**, 40–48.

Espina, L., García-Gonzalo, D., Laglaoui, A., Mackey, B. M. & Pagán, R. (2013a) Synergistic combinations of high hydrostatic pressure and essential oils or their constituents and their use in preservation of fruit juices. *International Journal of Food Microbiology* **161**, 23–30.

Espina, L., Gelaw, T.K., De Lamo-Castellví, S., Pagán, R. & García-Gonzalo, D. (2013b) Mechanism of bacterial inactivation by (+)-limonene and its potential use in food preservation combined processes. *PLoS ONE* **8**(2): e56769. doi: 10.1371/journal.pone.0056769.

Evrendilek, G. A., Baysal, T., Icier, F., Yildiz, H., Demirdoven, A. & Bozkurt, H. (2012) Processing of fruits and fruit juices by novel electrotechnologies. *Food Engineering Reviews* **4**, 68–87.

Fiore, A., La Fauci, L., Cervellati, R., Guerra, M. C., Speroni, E., Costa, S., Galvano, G., De Lorenzo, A., Bacchelli, V., Fogliano, V. & Galvano, F. (2005) Antioxidant activity of pasteurized and sterilized commercial red orange juices. *Molecular Nutrition & Food Research* **49**, 1129–1135.

Francis, F. J. & Clydesdale, F. M. (1975) *Food Colorimetry: Theory and Applications.* AVI Publ. Co., Westport, CT.

García, D., Hassani, M., Mañas, P., Condón, S. & Pagán, R. (2005) Inactivation of *Escherichia coli* O157:H7 during storage under refrigeration of apple juice treated by pulsed electric fields. *Journal of Food Safety* **25**, 30–42.

Gentry, T.S. & Roberts, J.S. (2005) Design and evaluation of a continuous flow microwave pasteurization system for apple cider. *LWT – Food Science and Technology* **38**, 227–238.

Gómez, N., García, D., Álvarez, I., Condón, S. & Raso, J. (2005) Modelling inactivation of Listeria monocytogenes by pulsed electric fields in media of different pH. *International Journal of Food Microbiology* **103**, 199–206.

Goodner, J., Braddock, R.J. & Parish, M. (1998) Inactivation of pectinesterase in orange and grapefruit juices by high pressure. *Journal of Agricultural and Food Chemistry* **46**, 1997–2000.

Gurtler, J.B., Bailey, R.B., Geveke, D.J. & Zhang, H.Q. (2011) Pulsed electric field inactivation of *E. coli* O157:H7 and non-pathogenic surrogate *E. coli* in strawberry juice as influenced by sodium benzoate, potassium sorbate, and citric acid. *Food Control* **22**, 1689–1694.

Gurtler, J.B., Rivera, R.B., Zhang, H.Q. & Geveke, D.J. (2010) Selection of surrogate bacteria in place of *E. coli* O157:H7 and *Salmonella typhimurium* for pulsed electric field treatment of orange juice. *International Journal of Food Microbiology* **139**, 1–8.

Hartyáni, P., Dalmadi, I., Cserhalmi, Z., Kántor, D.B., Tóth-Markus, M., & Sass-Kiss, A. (2011). Physical–chemical and sensory properties of pulsed electric field and high hydrostatic pressure treated citrus juices. *Innovative Food Science & Emerging Technologies*, **12**, 255–260.

Hartyáni, P., Dalmadi, I. & Knorr, D. (2013) Electronic nose investigation of *Alicyclobacillus acidoterrestris* inoculated apple and orange juice treated by high hydrostatic pressure. *Food Control* **32**, 262–269.

Heinz, V., Álvarez, I., Angersbach, A. & Knorr, D. (2002) Preservation of liquid foods by high intensity pulsed electric field-basic concept for process design. *Trends in Food Science and Technology* **12**, 103–111.

Hlady, G., Cook, K., Dobbs, T., Hlady, W. G., Wells, J., Barrett, T., Puhr, N., Lancette, G., Bodager, D., Toth, B., Genese, C., Highsmith, A., Pilot, K., Finelli, L. & Swerd-low, D. (1998) Outbreak of *Salmonella Serotype* Hartford infections associated with unpasteurized orange juice. *JAMA: the Journal of the American Medical Association* **280**, 1504–1509.

Ho, S.Y., Mittal, G.S. & Griffiths, M.W. (1997) Effects of high field electric pulses on the activity of selected enzymes. *Journal of Food Engineering* **31**, 69–84.

Hodgins, A.M., Mittal, G.S. & Griffiths, M.W. (2002) Pasteurization of fresh orange juice using low-energy pulsed electrical field. *Journal of Food Science* **67**, 2294–2299.

Icier, F. & Ilicali, C. (2005) The effects of concentration on electrical conductivity of orange juice concentrates during ohmic heating. *European Food Research and Technology* **220**, 406–414.

Igual, M., Contreras, C., Camacho, M. M. & Martínez-Navarrete, N. (2014) Effect of thermal treatment and storage conditions on the physical and sensory properties of grapefruit juice. *Food and Bioprocess Technology* **7**, 191–203.

Igual, M., García-Martínez, E., Camacho, M. M. & Martínez-Navarrete, N. (2010) Effect of thermal treatment and storage on the stability of organic acids and the functional value of grapefruit juice. *Food Chemistry*, **118**, 291–299.

Jensen, N. (1999) *Alicyclobacillus* – a new challenge for the food industry. *Food Australia* **51**, 33–36.

Jia, M., Zhang, Q.H., & Min D.B. (1999). Pulsed electric field processing effects on flavor compounds and microorganisms of orange juice. *Food Chemistry*, **65**, 445–451.

Jin, Z.T. & Zhang, Q.H. (1999) Pulsed electric field inactivation of microorganisms and preservation of quality of cranberry juice. *Journal of Food Process Preservation* **23**, 481–497.

Jordan, S., Pascual, C., Bracey, E. & Mackey, B. M. (2001) Inactivation and injury of pressure-resistant strains of *Escherichia coli* O157:H7 and *Listeria monocytogenes* in fruit juices. *Journal of Applied Microbiology* **91**, 463–469.

Khan, A., Melvin, C. & Dagdag, E. (2007) Identification and molecular characterization of *Salmonella* spp. from unpasteurized orange juices and identification of new serotype *Salmonella* strain *S. enterica* serovar Tempe. *Food microbiology* **24**, 539–543.

Kimball, D.A. (1991). *Citrus Processing-Quality Control and Technology*, pp. 117–243. Van Nostrand Reinhold, New York.

Komitopoulou, E., Boziaris, I. S., Davies, E. A., Delves-Broughton, J. & Adams, M. R. (1999) Alicyclobacillus acidoterrestris in fruit juices and its control by nisin. *International Journal of Food Science and Technology* **34**, 81–85.

Lee, H. S. & Coates, G. A. (1999) Thermal pasteurization effects on color of red grapefruit juices. *Journal of Food Science* **64**, 663–666.

Lee, H. S. & Coates, G. A. (2003) Effect of thermal pasteurization on Valencia orange juice color and pigments. *Lebensmittel-Wissenschaft Und-Technologie-Food Science and Technology* **36**, 153–156.

Leizerson, S. & Shimoni, E. (2005a) Effect of ultrahigh-temperature continuous Ohmic heating treatment on fresh orange juice. *Journal of Agricultural and Food Chemistry* **53**, 3519–3524.

Leizerson, S. & Shimoni, E. (2005b). Stability and sensory shelf life of orange juice pasteurized by continuous ohmic heating. *Journal of Agricultural and Food Chemistry* **53**, 4012–4018.

Lessin, W. J., Catigani, G. L. & Schwartz, S. J. (1997) Quantification of cis-trans isomers of provitamin a carotenoids in fresh and processed fruits and vegetables. *Journal of Agricultural and Food Chemistry* **45**, 3728–3732.

Liang, Z., Mittal, G.S. & Griffiths, M.W. (2002) Inactivation of *Salmonella typhimurium* in orange juice containing antimicrobial agents by pulsed electric field. *Journal of Food Protection* **65**, 1081–1087.

Liu, Y., Heying, E. & Tanumihardjo, S. A. (2012) History, global distribution, and nutritional importance of citrus fruits. *Comprehensive Reviews in Food Science and Food Safety* **11**, 530–545.

Madigan, M. T., Martinko, J. M., Bender, K. S., Buckley, D. H., Stahl, D. A. & Brock, T. (2014) *Biology of Microorganisms*, Benjamin Cummings, New York.

Maldonado, M. C., Belfiore, C. & Navarro, A. R. (2008) Temperature, soluble solids and pH effect on *Alicyclobacillus acidoterrestris* viability in lemon juice concentrate. *Journal of Industrial Microbiology and Biotechnology* **35**, 141–144.

McDonald, C. J., Lloyd, S. W., Vitale, M. A., Petersson, K. & Inning, F. (2000) Effects of pulsed electric fields on microorganisms in orange juice using electric fields strengths of 30 and 50 kV/cm. *Journal of Food Science* **5**, 984–989.

Mcnamee, C., Noci, F., Cronin, D. A., Lyng, J. G., Morgan, D. J. & Scannell, A. G. M. (2010) PEF based hurdle strategy to control *Pichia fermentans*, *Listeria innocua* and *Escherichia coli* K12 in orange juice. *International Journal of Food Microbiology* **138**, 13–18.

Meléndez-Martínez, A. J., Vicario, I. M. & Heredia, F. J. (2007) Review: analysis of carotenoids in orange juice. *Journal of Food Composition and Analysis* **20**, 638–649.

Miller, L. G. & Kasper, C. W. (1994) *Escherichia coli* 0157:H7 acid tolerance and survival in apple cider. *Journal of Food Protection* **57**, 645–.

Min, S., Evrendilek, G.A. & Zhang, H.Q. (2007). Pulsed electric fields: processing system, microbial and enzyme inhibition, and shelf life extension of foods. *IEEE Plasma Science* **35**, 59–73.

Min, S., Jin, Z.T., Min, S.K., Yeom, H. & Zhang, Q.H. (2003) Commercial-scale pulsed electric field processing of orange juice. *Journal of Food Science* **68**, 1265–1271.

Monfort, S., Gayán, E., Saldaña, G., Puértolas, E., Condón, S., Raso, J. & Álvarez, I. (2010) Inactivation of *Salmonella typhimurium* and *Staphylococcus aureus* by pulsed electric fields in liquid whole egg. *Innovative Food Science and Emerging Technologies* **11**, 306–313.

Morales-De la Peña, M., Salvia-Trujillo, L., Rojas-Graü, M.A.A. & Martín-Belloso, O. (2010a) Impact of high intensity pulsed electric fields on antioxidant properties and quality parameters of a fruit juice-soymilk beverage in chilled storage. *LWT – Food Science and Technology* **43** (6), 872–881.

Morales-De la Peña, M., Salvia-Trujillo, L., Rojas-Graü, M.A.A. & Martín-Belloso, O. (2011) Impact of high intensity pulsed electric fields or heat treatments on the fatty acid and mineral profiles of a fruit juice–soymilk beverage during storage. *Food Control*, **22**, 1975–1983.

Moshonas, M. G. & Shaw, P. E. (2000) Changes in volatile flavor constituents in pasteurized orange juice during storage. *Journal of Food Quality* **23**, 61–71.

Nagy, S. & Smoot, J. M. (1977) Temperature and storage effects on percent retention and percent U.S. recommended dietary allowance of vitamin C in canned single-strength orange juice. *Journal of Agricultural and Food Chemistry* **25**, 135–138.

Nienaber, U. & Shellhammer, T. (2001a) High-pressure processing of orange juice: combination treatments and shelf life study. *Journal of Food Science* **66**, 332–336.

Nienaber, U. & Shellhammer, T. (2001b) High-pressure processing of orange juice: kinetics of pectinmethylesterase inactivation. *Journal of Food Science* **66**, 328–331.

Nikdel, S., Chen, C.S., Parish, M.E., Mackellar, D.G. & Friedrich, L.M. (1993). Pasteurization of citrus juice with microwave energy in a continuous-flow unit. *Journal of Agriculture and Food Chemistry* **41**, 2116–2119.

Odriozola-Serrano, I., Soliva-Fortuny, R., & Martín-Belloso, O. (2008) Phenolic acids, flavonoids, vitamin C and antioxidant capacity of strawberry juices processed by high-intensity pulsed electric fields or heat treatments. *European Food Research Technology* **228**, 239–248.

Oms-Oliu, G., Odriozola-Serrano, I., Soliva-Fortuny, R., Elez-Martínez, P. & Martín-Belloso, O. (2012) Stability of health-related compounds in plant foods through the application of non thermal processes. *Trends in Food Science & Technology* **23**, 111–123.

Palgan, I., Muñoz, A., Noci, F., Whyte, P., Morgan, D.J., Cronin, D.A. & Lyng, J.G. (2012) Effectiveness of combined pulsed electric field (PEF) and manothermosonication (MTS) for the control of *Listeria innocua* in a smoothie type beverage. *Food Control* **25**, 621–625.

Parish, M. (1998a) Orange juice quality after treatment by thermal pasteurization or isostatic high pressure. *Lebensmittel-Wissenschaft & Technologie* **31**, 439–442.

Parish, M.E. (1998b) Coliforms, *Escherichia coli* and *Salmonella serovars* associated with a citrus-processing facility implicated in a salmonellosis outbreak. *Journal of Food Protection* **61**, 280–284.

Pettipher, G., Osmundson, M. & Murphy, J. (1997) Methods for the detection and enumeration of *Alicyclobacillus acidoterrestris* and investigation of growth and production of taint in fruit juice and fruit juice-containing drinks. *Letters in Applied Microbiology* **24**, 185–189.

Pina-Pérez, M.C., Benlloch-Tinoco, M., Rodrigo, D. & Martínez, A. (2013a). *Cronobacter sakazakii* inactivation by microwave processing. *Food and Bioprocess Technology* **7**, 821–828.

Pina-Pérez, M., García-Fernández, M. M., Rodrigo, D. & Martínez-López, A. (2010) Monte Carlo simulation as a method to determine the critical factors affecting two strains of *Escherichia coli* inactivation kinetics by high hydrostatic pressure. *Foodborne Pathogens and Disease* **7**, 459–466.

Pina-Pérez, M.C., Martínez-López, A. & Rodrigo, D. (2012) Cinnamon antimicrobial effect against *Salmonella typhimurium* cells treated by pulsed electric fields (PEF) in pasteurized skim milk beverage. *Food Research International* **48**, 777–783.

Pina-Pérez, M.C., Martínez-López, A., & Rodrigo, D. (2013b). Cocoa powder as a natural ingredient revealing an enhancing effect to inactivate *Cronobacter sakazakii* cells treated by pulsed electric fields in infant milk formula. *Food Control* **32**, 87–92.

Plaza, L., Sanchez-Moreno, C., De Ancos, B., Elez-Martínez, P., Martín-Belloso, O. & Pilar Cano, M. (2011) Carotenoid and flavanone content during refrigerated storage of orange juice processed by high-pressure, pulsed electric fields and low pasteurization. *LWT – Food Science and Technology* **44**, 834–839.

Plaza, L., Sanchez-Moreno, C., Elez-Martínez, P., De Ancos, B., Martín-Belloso, O. & Pilar Cano, M. (2006) Effect of refrigerated storage on vitamin C and antioxidant activity of orange juice processed by high-pressure or pulsed electric fields with regard to low pasteurization. *European Food Research and Technology* **223**, 487–493.

Polydera, A., Galanou, E., Stoforos, N. & Taoukis, P. (2004) Inactivation kinetics of pectin methylesterase of Greek navel orange juice as a function of high hydrostatic pressure and temperature process conditions. *Journal of Food Engineering* **62**, 291–298.

Primo Yúfera, E. (1997) *Química De Los Alimentos*. Ed Síntesis, Madrid.

Puligundla, P., Abdullah, S.A., Choi, W., Jun, S., Oh, S.E. & Ko, S. (2013) Potentials of microwave heating technology for select food processing applications – a brief overview and update. *Journal of Food Processing & Technology* **4**, 1–9.

Raso, J., Calderón, M.L., Gongora, M., Barbosa-Cánovas, C. & Swanson, B.G. (1998a) Inactivation of mold ascospores and conidiospores suspended in fruit juices by pulsed electric fields. *Lebensmittel-Wissenschaft & Technologie* **31**, 668–672.

Raso, J., Calderón, M.L., Gongora, M., Barbosa-Cánovas, C. & Swanson, B.G. (1998b) Inactivation of Zygosaccharomyces bailii in fruit juices by heat, high hydrostatic pressure and pulsed electric fields. *Journal of food science*, **63**, 1042–1044.

Riener J., Noci, F., Cronin, D.A., Morgan, D.J. & Lyng, J.G. (2009) Combined effect of temperature and pulsed electric fields on pectin methyl esterase inactivation in red grapefruit juice (Citrus paradisi). *European Food Research and Technology* **228**, 373–379.

Rivas, A., Rodrigo, D., Company, B., Sampedro, F. & Rodrigo, M. (2007) Effects of pulsed electric fields on water-soluble vitamins and ACE inhibitory peptides added to a mixed orange juice and milk beverage. *Food Chemistry* **104**, 1550–1559.

Rivas, A., Rodrigo, D., Martínez, A., Barbosa-Cánovas, G.V. & Rodrigo, M. 2006. Effect of PEF and heat pasteurization on the physical–chemical characteristics of blended orange and carrot juice. *LWT – Food Science and Technology* **39**, 1163–1170.

Robertson, G. L. & Samaniego, C. M. L. (1986) Effect of initial dissolved oxygen levels on the degradation of ascorbic acid and the browning of lemon juice during storage. *Journal of Food Science* **51**, 184–187.

Rodrigo D, Barbosa-Cánovas, G.V., Martínez, A. & Rodrigo, M. (2003) Pectin methylesterase and natural microflora of fresh and mixed orange and carrot juice treated with pulsed electric fields. *Journal of Food Protection* **66**, 2336–2342.

Rupasinghe, H.P.V. & Yu, L.J. (2012) Emerging Preservation Methods for Fruit Juices and Beverages, Food Additive, Prof. Yehia El-Samragy (Ed.), ISBN: 978-953-51-0067-6, InTech, Available from: http://www.intechopen.com/books/food-additive/emerging -preservation-methods-3-for-fruit-juices-andbeverages.

Sagong, H.G., Park, S.H., Choi, Y.J., Ryu, S. & Kang, D.H. (2011) Inactivation of *Escherichia coli* O157:H7, *Salmonella typhimurium*, and *Listeria monocytogenes* in orange and tomato juice using ohmic heating. *Journal of Food Protection* **74**, 899–904.

Saldaña, G., Puertolas, E., Condón, S., Álvarez, I. & Raso, J. (2010) Modeling inactivation kinetics and occurrence of sublethal injury of a pulsed electric field-resistant strain of *Escherichia coli* and *Salmonella typhimurium* in media of different pH. *Innovative Food Science & Emerging Technologies* **11**, 290–298.

Sampedro, F., Geveke, D.J., Fan, X., Rodrigo, D. & Zhang, Q.H. (2009) Shelf-life study of an orange juice–milk based beverage after PEF and thermal processing. *Journal of Food Science* **74**, 107–112.

Sampedro, F., Rodrigo, D. & Hendrickx, M. (2008) Inactivation kinetics of pectin methyl esterase under combined thermal-high pressure treatment in an orange juice-milk beverage. *Journal of Food Engineering* **86**, 133–139.

Sampedro, F., Rodrigo, D. & Martínez, A. (2011) Modelling the effect of pH and pectin concentration on the PEF inactivation of *Salmonella enterica serovar Typhimurium* by using the Monte Carlo simulation. *Food Control* **22**, 420–425.

Sánchez-Moreno, C., Cano, M.P., De Ancos, B., Plaza, L., Olmedilla, B., Granado, F., Elez-Martínez, P., Martín-Belloso, O. & Martín, A. (2004). Pulsed electric fields-processed orange juice consumption increases plasma vitamin C and decreases F2-isoprostanes in healthy humans. *Journal of Nutritional Biochemistry.* **15**(10): 601–607.

Sánchez-Moreno, C., De Ancos, B., Plaza, L., Elez-Martínez, P. & Cano, M.P. (2009) Nutritional approaches and health-related properties of plant foods processed by high pressure and pulsed electric fields. *Critical Reviews in Food Science and Nutrition* **49**, 552–576.

Sánchez-Moreno, C., Plaza, L., De Ancos, B. & Cano, M. (2003) Vitamin C, provitamin A cartenoids, and other carotenoids in high-pressurized orange juice during refrigerated storage. *Journal of Agricultural and Food Chemistry* **51**, 647–653.

Sánchez-Moreno, C., Plaza, L., Elez-Martínez, P., De Ancos, B., Martín-Belloso, O. & Cano, M. (2005) Impact of high pressure and pulsed electric fields on bioactive compounds and antioxidant activity of orange juice in comparison with traditional thermal processing. *Journal of Agricultural and Food Chemistry* **53**, 4403–4409.

Schilling, S., Schmid, S., Jaeger, H., Ludwig, M., Dietrich, H., Toepfl, S., Knorr, D., Neidhart, S., Schieber, A. & Carle, R. (2008). Comparative study of pulsed electric field and thermal processing of apple juice with particular consideration of juice quality and enzyme deactivation. *Journal of Agricultural & Food Chemistry*, **56**, 4545–4554.

Sentandreu, E., Carbonell, D., Rodrigo, D. & Carbonell, J.V. (2006) Pulsed electric fields versus thermal treatment: equivalent processes to obtain equally acceptable citrus juices. *Journal of Food Protection* **69**, 2016–2018.

Sharma, S.K., Zhang, Q.H. & Chism, G.W. (1998) Development of a protein fortified fruit beverage and its quality when processed with pulsed electric field treatment. *Journal of Food Quality* **21**, 459–473.

Shaw, P. E., Moshonas, M. G., Hearn, C. J. & Goodner, K. L. (2000) Volatile constituents in fresh and processed juices from grapefruit and new grapefruit hybrids. *Journal of agricultural and food chemistry* **48**, 2425–2429.

Silva, F. M., Gibbs, P., Vieira, M. C. & Silva, C. L. M. (1999) Thermal inactivation of Alicyclobacillus acidoterrestris spores under different temperature, soluble solids and pH conditions for the design of fruit processes. *International Journal of Food Microbiology* **51**, 95–103.

Silva, F. V. M., Tan, E. K. & Farid, M. (2012) Bacterial spore inactivation at 45–65 degrees C using high pressure processing: study of Alicyclobacillus acidoterrestris in orange juice. *Food Microbiology* **32**, 206–211.

Soliva-Fortuny, R.C., Martín-Belloso, O., Escolà-Hernández, J. & Elez-Martínez, P. (2005) Inactivation of *Lactobacillus brevis* in orange juice by high-intensity pulsed electric fields. *Food Microbiology* **22**, 311–319.

Somolinos, M., García, D., Mañas, P., Condón, S. & Pagán, R. (2008) Effect of environmental factors and cell physiological state on pulsed electric fields resistance and repair capacity of various strains of *Escherichia coli*. *International Journal of Food Microbiology* **124**, 260–267.

Stanley, D.W. (1991). Biological membrane deterioration and associated quality losses in food tissues. In *Critical Reviews in Food Science and Nutrition*, (ed. F.M. Clydedale), 487–553. CRC Press, New York.

Szlavik, J., Paiva, D.S., MØrk, N., Van Den Berg, F., Verran, J., Whitehead, K., Knøchel, S. & Nielsen, D.S. (2012) Initial adhesion of *Listeria monocytogenes* to solid surfaces under liquid flow. *International Journal of Food Microbiology* **152**, 181–188.

Tajchakavit, S. & Ramaswamy, H. S. (1995) Continuous-flow microwave heating of orange juice: evidence of nonthermal effects. *International Microwave Power Institute* **30**, 141–148.

Tajchakavit, S. & Ramaswamy, H. S. (2007). Continuous-flow microwave inactivation kinetics of pectin methyl esterase in orange juice. *Journal of food processing and Preservation* **21**(5), 365–378.

Tajchakavit, S., Ramaswamy, H.S. & Fustier, P. (1998) Enhanced destruction of spoilage microorganisms in apple juice during continuous flow microwave heating. *Food Research International* **31**, 713–722.

Teo, A., Ravishankar, S. & Sizer, C. (2001) Effect of low-temperature, high-pressure treatment on the survival of *Escherichia coli* O157:H7 and *Salmonella* in unpasteurized fruit juices. *Journal of Food Protection* **64**, 1122–1127.

Terefe, N. S., Buckow, R. & Versteeg, C. (2014) Quality-related enzymes in fruit and vegetable products: effects of novel food processing technologies, Part 1: high-pressure processing. *Critical Reviews in Food Science and Nutrition* **54**, 24–63.

Terefe, N. S., Buckow, R. & Versteeg, C. (2015) Quality related enzymes in plant based products: effects of novel food processing technologies. Part 2: pulsed electric field processing. *Critical Reviews in Food Science and Nutrition* **55**, 1–15.

Tewari, G. & Juneja, V. K. (2007) Advances in thermal and non-thermal food preservation. *International Journal of Dairy Technology* **62**, 285–286.

Timmermans, R.A.H., Mastwijk, H.C., Knol, J.J., Quataert, M.C.J., Vervoort, L., Van Der Plancken, I., Hendrickx, M.E. & Matser, A.M. (2011) Comparing equivalent thermal, high pressure and pulsed electric field processes for mild pasteurization of orange juice. Part I: impact on overall quality attributes. *Innovative Food Science & Emerging Technologies* **12**, 235–243.

Toepfl, S., Heinz, V. & Knorr, D. (2007) High intensity pulsed electric fields applied for food preservation. *Chemical Engineering and Processing* **46**, 537–546.

Torregrosa, F., Cortés, C., Esteve, M.J. & Frígola, A. (2005) Effect of high intensity pulsed electric fields processing and conventional heat treatment on orange–carrot juice carotenoids. *Journal of Agricultural & Food Chemistry* **53**, 9519–9525.

Torregrosa, F., Esteve, M.J., Frígola, A. & Cortés, C. (2006) Ascorbic acid stability during refrigerated storage of orange-carrot juice treated by high pulsed electric field and comparison with pasteurized juice. *Journal of Food Engineering* **73**, 339–345.

Tribess, T. B. & Tadini, C. C. (2006) Inactivation kinetics of pectin methylesterase in orange juice as a function of pH and temperature/time process conditions. *Journal of the Science of Food and Agriculture* **86**, 1328–1335.

U.S. Food and Drug Administration (FDA) (USFDA), 2013, Methods to reduce/eliminate pathogens from produce and fresh-cut produce. Available at: http://www.fda.gov/Food/FoodScienceResearch/SafePracticesforFoodProcesses/ucm091363.htm (Accessed May, 2013).

Valko, M., Rhodes, C. J., Moncol, J., Izakovic, M. & Mazur, M. (2006) Free radicals, metals and antioxidants in oxidative stress-induced cancer. *Chemico-Biological Interactions* **160**, 1–40.

Van Loey, A.M., Verachtert, B. & Hendrickx, M.E. (2002) Effects of high electric field pulses on enzymes. *Trends in Food Science and Technology* **12**, 94–102.

Vega-Mercado, H., Martín-Belloso, O., Qin, B. L., Chang, F. J., Góngora-Nieto, M. M., Barbosa-Cánovas, G. V. & Swanson, B.G. (1997) Non-thermal food preservation: pulsed electric fields. *Trends in Food Science and Technology* **8**, 151– 157.

Venkatesh, M. S. & Raghavan, G. S. V. (2004) An overview of microwave processing and dielectric properties of agri-food materials. *Biosystems Engineering*, **88**, 1–18.

Vikram, V. B., Ramesh, M. N. & Prapulla, S. G. (2005) Thermal degradation kinetics of nutrients in orange juice heated by electromagnetic and conventional methods. *Journal of Food Engineering* **69**, 31–40.

Villamiel, M., Castillo, M. D. D., Martín, C. S. & Corzo, N. (1998) Assessment of the thermal treatment of orange juice during continuous microwave and conventional heating. *Journal of the Science of Food and Agriculture* **78**, 196–200.

Walking-Ribeiro, M., Noci, F., Cronin, D.A., Lyng, J.G. & Morgan, D.J. (2009) Shelf life and sensory evaluation of orange juice after exposure to thermosonication and pulsed electric fields. *Food and Bioproducts Processing* **87**, 102–107.

Walking-Ribeiro, M., Noci, F., Cronin, D.A., Lyng, J.G. & Morgan, D.J. (2010). Shelf life and sensory attributes of a fruit smoothie-type beverage processed with moderate heat and pulsed electric fields. *LWT – Food Science & Technology* **43**, 1067–1073.

Whitney, B., Williams, R., Eifert, J. & Marcy, J. (2007) High-pressure resistance variation of *Escherichia coli* O157:H7 strains and *Salmonella* in triptic soy broth, distilled water and fruit juice. *Journal of Food Protection* **70**, 2078–2083.

Wouters, P., Álvarez, I. & Raso, J. (2001) Critical factors determining inactivation kinetics by pulsed electric field food processing. *Trends in Food Science & Technology* **12**, 112–121.

Yamazaki, K., Teduka, H. & Shinano, H. (1996) Isolation and identification of *Alicyclobacillus acidoterrestris* from acidic beverages. *Bioscience, Biotechnology and Biochemistry* **60**, 543–545.

Yeom, H. W., Streaker, C. B., Zhang, Q. H. & Min, D. B. (2000) Effects of pulsed electric fields on the quality of orange juice and comparison with heat pasteurization. *Journal of Agricultural and Food Chemistry* **48**, 4597–4605.

Yeom, W.H., Zhang, Q.H. & Chism, G.W. (2002) Inactivation of pectin methyl esterase in orange juice by pulsed electric fields. *Journal of Food Science* **67**, 2154–2159.

Zadyraka, Y. (2012) Case study Advanced Microwave Technologies. Queen Margaret University Available at: http://www.qmu.ac.uk/marketing/press_releases/microwave_technology.htm.

Zhang, Q.H. (1997) Integrated pasteurization and aseptic packaging using high voltage pulsed electric field in Proceedings of ICEF7. Seventh International Congress of Engineering and Food, The Brighton Center, UK, pp. K13–K15.

Zhang, Q., Barbosa-Canovas, G.V. & Swanson, B.G. (1995) Engineering aspects of pulsed electric fields pasteurization. *Journal of Food Engineering* **25**, 268–281.

Zulueta, A., Barba, F.J., Esteve, M.J. & Frígola, A. (2010b) Effects on the carotenoid pattern and vitamin A of a pulsed electric field-treated orange juice–milk beverage and behavior during storage. *European Food Research and Technology* **231**, 525–534.

Zulueta, A., Esteve, M.J., Frasquet, I. & Frígola, A. (2007) Fatty acid profile changes during orange juice-milk beverage processing by high-pulsed electric field. *European Journal of Lipid Science and Technology* **109**, 25–31.

Zulueta, A., Esteve, M.J. & Frígola, A. (2010a) Ascorbic acid in orange juice–milk beverage treated by high intensity pulsed electric fields and its stability during storage. *Innovative Food Science & Emerging Technologies* **11**, 84–90.

3

Prunus Fruit Juices

Gamze Toydemir[1], Dilek Boyacioglu[2], Robert D. Hall[3], Jules Beekwilder[3], and Esra Capanoglu[2]*

[1]*Department of Food Engineering, Faculty of Engineering, Alanya Alaaddin Keykubat University, Kestel-Alanya, Antalya, Turkey*
[2]*Department of Food Engineering, Faculty of Chemical & Metallurgical Engineering, Istanbul Technical University, Maslak, Istanbul, Turkey*
[3]*Wageningen UR, Plant Research International, Bioscience, Wageningen UR, Wageningen, The Netherlands*

3.1   Introduction

Prunus fruit species, including apricot (*Prunus armeniaca*), cherry (sweet cherry (*Prunus avium*) and sour cherry (*Prunus cerasus*)), peach (*Prunus persica*), and plum (*Prunus domestica*), belong to the genus *Prunus* of Rosaceae (rose) family and are categorized as "stone fruits," having seeds enclosed in a hard, stone-like endocarp (Siddiq, 2006a). These stone fruits, which originated in Europe, Central Asia, and China, have prime economic importance on the fresh and processed fruits market throughout the temperate and the subtropical world (Fraignier *et al.*, 1995; Janick *et al.*, 2011). Turkey is the leading apricot and cherry producer in the world, whereas the world's leading peach- and plum-producing country is China (FAOSTAT, 2013) (Table 3.1).

The *Prunus* fruits are mostly processed as frozen, canned, dried, jam, puree, and juice, among which juice is one of the major processed products that is demanded and consumed throughout the year. Consumers generally accept fruit juice drinks as a convenient alternative to fresh fruits; thus, fruit juice production presents a high economical benefit (Etiévant *et al.*, 2010). On the other hand, fruit juice products have a short shelf life due to microbial and enzymatic degradation, thus, requiring the inactivation of these factors to give an extended shelf life (Espachs-Barroso *et al.*, 2003). Thermal pasteurization has traditionally been applied as the most popular technology for destroying microorganisms and inactivating enzymes, as well as for juice stabilization, but this treatment mostly results in some negative effects, including nonenzymatic

*Corresponding author: Esra Capanoglu, capanogl@itu.edu.tr

Innovative Technologies in Beverage Processing, First Edition.
Edited by Ingrid Aguiló-Aguayo and Lucía Plaza.
© 2017 John Wiley & Sons Ltd. Published 2017 by John Wiley & Sons Ltd.

Table 3.1   *Prunus* fruits: Production in leading countries

Fruit	Country (production in metric tons)				
Apricots	Turkey (811,609)	Iran (457,308)	Uzbekistan (430,000)	Algeria (319,784)	Italy (198,290)
Cherries	Turkey (494,325)	USA (301,205)	Iran (200,000)	Italy (131,175)	Uzbekistan (100,000)
Peaches and nectarines	China (11,924,085)	Italy (1,401,795)	Spain (1,329,800)	USA (964,890)	Greece (666,200)
Plums and sloes	China (6,100,000)	Serbia (738,278)	Romania (512,459)	Chile (306,354)	Turkey (305,393)

Source: FAOSTAT (2013).

browning, losses of essential nutrients, and changes in physicochemical and organoleptic properties (Espachs-Barroso *et al.*, 2003; Echavarria *et al.*, 2011). For the reason that today's consumers demand high quality, fresh-like, and microbiologically safe foods (Mittal & Griffiths, 2005), novel nonthermal technologies are gaining popularity in processing fruit juices as promising alternatives to avoid the negative effects of heat pasteurization (Espachs-Barroso *et al.*, 2003). These innovative processing methods, which are based on entirely new principles, enable us to enhance product safety and shelf life without altering the color, flavor, and nutrient contents of foods (Furukawa *et al.*, 2001; Ray *et al.*, 2001).

This chapter reviews the conventional processing techniques as well as the application of the alternative nonthermal treatments in *Prunus* fruit juice processing, by describing both the aim and the basic concepts of these technologies, and highlighting their effects on microbial quality, enzymatic activity, and physical, chemical, and sensory properties of *Prunus* fruit juices.

3.2   Conventional Processing Techniques

3.2.1   Cherry and Sour Cherry

Cherries (*Prunus* spp.) belong to the stone fruit family, Rosaceae, genus *Prunus*, with the most important species included in the subgenus Cerasus, those are known as sweet cherry (*P. avium* L.) and sour cherry or tart cherry (*P. cerasus* L.) (Ferretti *et al.*, 2010). The most important cherry-producing countries have been listed as Turkey, the United States, Iran, Italy, and Uzbekistan, mounting up to >1,000,000 MT in 2013 (Table 3.1).

Sweet cherries are mostly consumed as fresh fruit and less as processed products, whereas, with regard to sour cherries, more than 85% of production is utilized for processing into various products, including frozen fruits, canned products, marmalades, and fruit juices (USDA Foreign Agricultural Service, 2012) but is predominantly used in the fruit juice industry.

Conventional industrial-scale sour cherry juice (nectar) production includes a number of steps, represented in Fig. 3.1. Fresh sour cherry fruits, after washing and selection steps, are separated from their stalks. Subsequently, mash heating (80 °C, 90 s) and mash pressing (110 bar) steps take place, which provide enzyme inactivation and generate the first juice (73% juice yield), respectively. In the mash pressing unit, press cake is the waste material, which is subsequently subjected to three further press cake

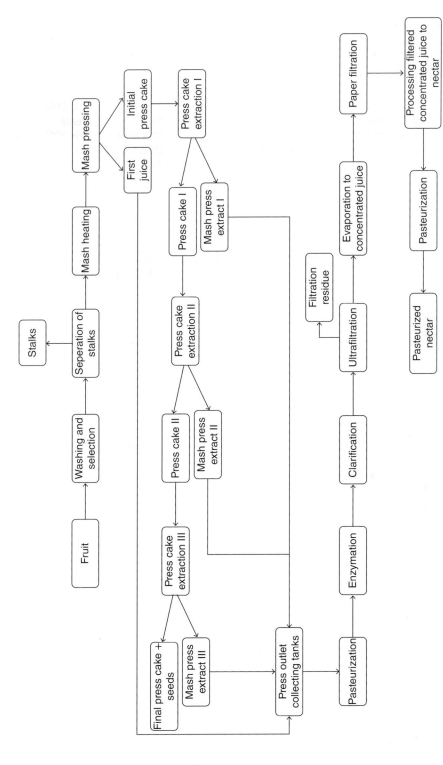

Figure 3.1 Flowchart of conventional industrial-scale sour cherry juice production. Toydemir *et al.* (2013a). Reproduced with permission of Elsevier.

extraction steps with certain amounts of water addition, in order to increase the juice yield (juice yield increased to 85%). The whole juice obtained from these mash pressing and press cake extraction steps is further pasteurized (95 °C, 90 s), enzyme treated (50 °C, 2 h), clarified (50 °C, 1 h), filtered, and evaporated (65–80 °C) to concentrated juice (65 °Brix). Paper filtration is applied to the concentrated juice to eliminate harmful bacteria (*Alycyclobacillus* spp.) that are resistant to pasteurization temperatures. The concentrated juice is subsequently processed with water (to bring the °Brix to 12.5), sucrose syrup (56% sucrose on dry-weight basis), and citric acid (to regulate the pH as 3.5) addition to generate the nectar. Pasteurization (95 °C, 45 s) of this nectar is the last step, after which the final pasteurized nectar sample is obtained (Toydemir, 2013; Toydemir *et al.*, 2013a, b).

3.2.2 Apricot, Peach, and Nectarine

Apricots (*P. armeniaca* L.), peaches (*P. persica* L.), and nectarines (*P. persica* L., var. nectarine) (smooth-skinned fuzz-less peach), like cherries, belong to the genus *Prunus* of the Rosaceae family (Lurie & Crisosto, 2005; Yigit *et al.*, 2009). Turkey is the leading apricot producer followed by Iran, Uzbekistan, Algeria, and Italy and mounts up to >2,000,000 MT in 2013 (Table 3.1); the top-ranked countries in peach and nectarine production are listed as China, Italy, Spain, the Unites States, and Greece, mounting up to a total of >17,000,000 MT in 2013 (Table 3.1) (FAOSTAT, 2013).

Approximately 15–20% of the apricots produced are consumed freshly and the rest are processed as canned, dried, frozen, jam, juice, and puree (Siddiq, 2006a). For peaches and nectarines, the freshly consumed part is more than 50% (almost 100% for nectarines), whereas the processed peach products include canned, frozen, and dried peaches as well as peach jam, jelly, and juice (Rieger, 2004).

A flowchart of conventional peach and apricot juice production is shown in Fig. 3.2. Peaches are generally extracted as a pulp or puree rather than juice. Peach juice drink is processed via dilution of peach pulp (pulp obtained after fruit mashing and after fiber and other coarse material is removed) or peach pulp base (pulp with added sugar, acid, color, and stabilizer). A heating process (93 °C, 2 min; blanching/softening step) applied before pulping enables the pulping process to be made easier, reduces

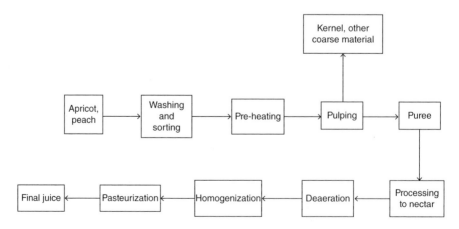

Figure 3.2 Flowchart of conventional peach and apricot juice production.

the oxidation, and stabilizes for cloudiness. Pulping the fully ripe peaches ends up with a puree having a yield of approximately 49% (v/w). For peach nectar, sucrose syrup (30°Brix) is added (39% v/v) to the puree, after which the pH is regulated to 3.7–3.9 by using citric acid. During these steps, the puree may have air mixed in with it which can lead to undesirable changes in nectar's color and flavor. Therefore, deaeration is needed before pasteurization of the nectar. Then the nectar is pasteurized at 88–93 °C to have the final pasteurized nectar product and quickly cooled to 1.7 °C for aseptic packaging. Apricot juice is processed in the same manner as peach juice (Bates *et al.*, 2001).

3.2.3 Plum

Plum is a common name for the tree of many species that belong to the genus *Prunus* of Rosaceae (rose) family and that have drupaceous (fleshy) fruit. Plum fruit is known to be cultivated since prehistoric times, longer perhaps than any other fruit except apple (Anon., 2004). China is the leading plum (and sloe)-producing country, which is followed by Serbia, Romania, Chile, and with a total production rate of >18,000,000 MT in 2013 (Table 3.1) (FAOSTAT, 2013).

Plums have a high potential in the fresh fruit market and/or as a processed product, both possessing the same ratios. Plums are mostly processed as dried plum (prune), prune juice, and whole canned plums. Other processed forms, including plum paste, plum sauce, and plum juice, are reported to have low production and consumption levels, in comparison to the equivalent products obtained from other fruits, such as apples, cherries, and apricots (Espie, 1992; Chang *et al.*, 1994).

For plum juice production, the sorted, washed, and drained fruits are subjected to steam heating for about 10 min in order to prepare the fruits for pulping. This pre-heating process also helps to inactivate the naturally occurring enzymes and prevent darkening. The puree obtained from the pulping step is first cooled to 50 °C, which is then mixed with an appropriate enzyme. The common crushing and pressing steps of juice production are not sufficient to obtain an actual juice from plum fruit; thus, an additional treatment with a macerating enzyme is necessary. The enzyme mixture is left to stand for 6–12 h until juice drains readily through a sample of cheesecloth. Subsequently, the juice is pressed, after which the obtained juice is filtered and the Brix is adjusted to around 23 °Brix. Then, the pasteurization step – either flash pasteurization to 88 °C and filled into bottles at 82–85 °C or first filled into bottles and heated to 85 °C for 30 min – is applied to generate the final pasteurized juice product (Bates *et al.*, 2001).

On the other hand, prunes, which are dried plums, have mostly been preferred for their juice (Bates *et al.*, 2001). Prune juice, a brownish-to-reddish brown liquid with the taste and flavor of prunes, is essentially a water extract of dried prunes with a Brix of 18.5°. It is not applicable to extract juice directly from prunes having a moisture content of approximately just 18%; thus, they are subjected to an aqueous extraction. This process involves heating the prunes in appropriate volumes of water (with a volume of four to five times the weight of the fruit), which enables the extraction of fruit solids without any effect on fruit flavor and color. Under atmospheric cooking, the juice extraction step can take several hours, including 1 h boiling followed by 10 h simmering, whereas, this time duration can significantly be shortened, to 10–15 min, by the use of pressure cooking. After extraction, filtration (pits and undissolved solids are removed) and pasteurization (88 °C for 1 min) steps take place to give the final product (Siddiq, 2006b). Figure 3.3 shows the major steps included in prune juice processing.

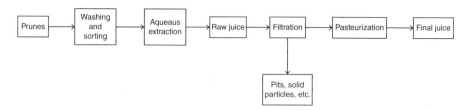

Figure 3.3   General flowchart for traditional prune juice processing.

3.3   Influence of Conventional Processing Techniques on Juice Quality

Food-processing procedures have long been used to prolong shelf life, to have certain products available out of season, to optimize products especially suited for home consumption, to develop strategies to design new or alternative food products with alternative/supplemented flavor and texture, to maintain or improve nutritional characteristics, and to increase quality and value. Food-processing strategies can range from being simple to being considerably complicated depending on the desired end product. For example, fresh produce requires nonthermal treatments, such as washing, selection, packaging, transportation, and storage at the point of retail. On the other hand, production of fully processed products generally includes multiple nonthermal as well as thermal treatments, such as washing, selection, cutting, removal of the seed and skin, blanching, evaporation, pasteurization, canning, and prolonged storage.

In the case of fruits and vegetables, they are commonly consumed as fresh produce but they are also processed into a variety of food products including juices, pastes, and canned foods. These processed products can also be valuable sources of nutrients. However, the various processing methods applied can have marked effects on fruit and vegetable quality, including organoleptic properties, physicochemical parameters, and the presence of bioactive compounds.

Among the processed products from *Prunus* fruit species, fruit juices are gaining more special interest, and hence, command higher economical value, depending on the fact that the fruit juice drinks, in general, are the most potent alternatives to their fresh fruit counterparts (Etiévant *et al.*, 2010). On the other hand, although fruit juice products are considered as safe because of their relatively low pH values due to the naturally existing organic acids (Parish, 1997), several foodborne outbreaks – caused by pathogenic microorganisms – associated with unpasteurized juices have been reported during recent decades (Mosqueda-Melgar *et al.*, 2008). Moreover, enzymatic degradation may occur in unpasteurized juices and often results in unacceptable changes in color and flavor (quality depreciation). Thermal pasteurization procedures have long been applied as the most efficient way of eliminating pathogenic microorganisms and spoilage bacteria, and to prevent enzymatic changes in fruit juices. However, this conventional technology may cause impairment of nutritive and sensory attributes as well as incurring loss in health-related bioactive compounds.

A study on the thermal degradation of the color of peach puree reported darkening of the color, which corresponded to a decrease in the *L*-value of the color scale. It was also indicated that peach puree lost its yellowness and became more red when heated, which was related with a decrease in the *b*-value and an increase in the *a*-value

(Avila & Silva, 1999). The color change in concentrated apple, peach, and plum pulp during heating at 56, 66, 80, and 94 °C for 700 min has also been investigated. The severity of damage was estimated using color differences (ΔE) and Hunter L parameters. The data analysis on the rate of color deterioration in concentrated fruit pulps as a function of temperature revealed the presence of more than one mechanism being involved (Lozano & Ibarz, 1997). Conventional sour cherry nectar processing resulted in 13% loss in fruit anthocyanins, whereas the loss for procyanidins was 38%. The final pasteurization (95 °C, 45 s) step applied to the processed nectar led to slight decreases in the levels of the sour cherry bioactives (Toydemir et al., 2013b). The total carotenoid content of fresh peach significantly decreased, by 65%, after pasteurization (90 °C, 5 min), while the antioxidant activity and the concentration of total phenolics were not significantly altered by this heat treatment (Oliveira et al., 2012). The antioxidant activity of unpeeled peach puree was determined to decrease by 20% after pasteurization at 100 °C for 30 min (Talcott et al., 2000). Similarly, a 20% reduction in total phenolic content of peach samples was observed as a result of pasteurization at 110 °C (Asami et al., 2003).

3.4 Novel Processing Techniques

High quality, fresh-like, and microbiologically safe foods are gaining increasing interest of today's consumers (Mittal & Griffiths, 2005). This is leading to the development of novel technologies that bring together the superiority of these techniques over the conventional techniques, which could be listed as the use of low processing temperatures, low energy consumption, and retention of nutritional and sensory attributes, while still providing the inactivation of pathogenic microorganisms and enzymes (Smith et al., 2002). These novel technologies include pulsed electric fields (PEFs) and hydrostatic high-pressure (HP) treatments as the most studied for fruit juice products. Besides these, there are also other innovative technologies being tested and applied for these products, which are all summarized in the following sections.

3.4.1 Pulsed Electric Fields

The PEF technology is a nonthermal food-processing technique that is used as an alternative preservation process for fruit juices. This novel technology aims to increase the shelf life of food products by inactivating the microorganisms and by decreasing the activity of enzymes, without any undesirable heat and chemical effects on flavor, color, and nutrients (Cserhalmi, 2006).

PEF treatment involves the application of high voltage pulses (typically 20–80 kV/cm) for a very short time (microseconds to milliseconds) to fluid foods placed between two electrodes in batch or continuous flow treatments (Mosqueda-Melgar et al., 2008). The typical components of a PEF system used for treatment of a pumpable fluid include a PEF generation unit (composed of a high voltage generator and a pulse generator), a treatment chamber, a suitable product handling system, and a set of monitoring and controlling devices. Although a number of theoretical models have been suggested to date, there is still no conclusive mechanistic model explaining the inactivation of microorganisms that are exposed to PEFs (Cserhalmi, 2006; Soliva-Fortuny et al., 2009). The most widely accepted theory comprises an electromechanical model introduced by Zimmermann *et al.* (1974), which is based on the dielectric breakdown of cell membranes, resulting in changes in membrane structure and permeability that occur at a critical breakdown voltage (Cserhalmi, 2006; Soliva-Fortuny *et al.*, 2009).

The potential use of PEF technology against *Penicillium expansum*, a fungal pathogen that causes fruit decay and produces carcinogenic mycotoxin patulin, was studied in different fruit juice/nectars, including sour cherry juice, peach, and apricot nectars. Inhibitory effects of PEF treatments, applied as a function of varying electric field strengths (0–34 kV/cm) and processing times (0–218 µs), on *P. expansum* spores, inoculated into juice/nectar samples at the level of 10^5–10^6 cfu/ml, were determined based on spore germination rate, germination tube elongation, and light and scanning electron microscopy (SEM) observations. During PEF treatment, the sample temperature was measured as 21 °C at the highest value of electric field strength (34 kV/cm), which enabled performing the whole treatment as nonthermal. The results obtained for germination tube elongation and spore germination rates indicated that a complete inhibitory effect was observed for all samples with an increasing electric field strength and processing time. In addition, after exposure to 30 kV/cm electric field strength, fungal spores showed degenerative changes in the morphology due to the light microscopic observations, whereas this degradation on morphology was visible after the PEF treatment of 17 and 30 kV/cm, when the measurements were performed using SEM (Akdemir Evrendilek *et al.*, 2008).

Altuntas *et al.* (2010) studied the effect of PEF treatment on sour cherry juice quality parameters, including pH, °Brix, titratable acidity, conductivity, color (L^*, a^*, and b^*), nonenzymatic browning index, metal ion concentration, total ascorbic acid, and total anthocyanin contents, as well as on microbial inactivation. Applied PEF treatment parameters (0 (control), 17, 20, 23, 27, and 30 kV/cm of electric field strengths at 131 µs treatment time and 0 (control), 66, 105, 131, 157, and 210 µs treatment times at 17 kV/cm) did not cause any significant changes on the measured physical and chemical properties of sour cherry juice ($P > 0.05$). On the other hand, increased electric field strength and longer treatment times were reported to lead to increased rates of microbial inactivation, based on the inactivation rates of *Escherichia coli* O157:H7, *Staphylococcus aureus*, *Listeria monocytogenes*, *Erwinia carotowora*, *Pseudomonas syringae* subs. *syringae*, *Botrytis cinerea*, and *P. expansum*. These results indicated that PEF is a promising alternative to process sour cherry juice, providing significant rates of microbial inactivation with the conservation of important physical and chemical attributes.

The inactivation of *Lactobacillus brevis* in peach juice and *Saccharomyces cerevisiae* in a model solution of peach juice was studied using high-intensity PEF treatment (Arantegui *et al.*, 1999, 2000). A destruction rate of 2 log reduction was obtained for *L. brevis* in peach juice by applying up to 16 pulses of 4 µs width and with an electric field strength between 20 and 32.5 kV/cm at a temperature of <20 °C (Arantegui *et al.*, 2000). For the inactivation of *S. cerevisiae* in a model solution of peach juice, 20 and 40 µs pulses at electric field strength of 5, 5.6, and 6.2 kV/cm were applied, and this resulted in 3 log reduction after 20 pulses at temperatures lower than 20 °C (Arantegui *et al.*, 1999).

Inactivation of enzymes in different fruit juices by the use of PEF has also been studied. Among the *Prunus* fruit juices, the activity of polyphenol oxidase (PPO) enzyme in peach juice was reported to be decreased by 70% in bipolar mode PEF treatment at 24.3 kV/cm electric field strength and for a total duration time of 5 ms using 20 µs pulse widths (Giner *et al.*, 2000).

3.4.2 High-Pressure Processing

High-pressure (HP) processing, which is also named as hydrostatic HP, has already been established as an alternative to classical thermal processing for the inactivation of microorganisms (Matser *et al.*, 2004) and several enzymes (Terefe *et al.*, 2014) without any requirement of chemical preservatives and/or additives at comparatively lower temperatures than thermal processes (Chakraborty *et al.*, 2014). Hydrostatic HP has the potential to produce high quality foods with "fresh-like" characteristics and improved functionalities (Patras *et al.*, 2009). As a result, this technique has been reported to fulfill consumer demand for healthy and natural food products together with the characteristics of high quality, greater safety, and increased shelf life (Rastogi *et al.*, 2007).

A commercial application of hydrostatic HP consists of subjecting the packaged or unpacked food in water at pressures of 100–900 MPa for 1–20 min at room temperature, which results in the inactivation of vegetative microorganisms and enzymes while maintaining physicochemical and nutritive properties, allowing the storage of the product at 4–6 °C (Cheftel, 1995). The applied pressure may affect different cellular targets involved in bacterial inactivation, which include inhibition of protein synthesis and reduction in the number of ribosomes at 50 MPa, partial protein denaturation at 100 MPa, damage to the cell membrane and internal cell structure at 200 MPa, and rupturing of the cell membrane and the excretion of internal substances as a result of irreversible denaturation of enzymes and proteins at 300 MPa and higher pressures (Abe, 2007). Unlike thermal processing, the moderate temperatures used in hydrostatic HP maintain the texture, flavors, nutrients, and other sensory quality attributes of the product (Balasubramaniam *et al.*, 2008). These benefits have led to the implementation of this technique in several fruit and vegetable processed products especially in fruit juices (Balasubramaniam & Farkas, 2008; Buzrul *et al.*, 2008; Barba *et al.*, 2013).

The efficiency of hydrostatic HP treatment on inactivation of pathogenic microorganisms in different *Prunus* fruit juices was studied in several studies (Table 3.2) (Alpas & Bozoglu, 2003; Bayindirli *et al.*, 2006). Among the various pressure (250 and 350 MPa)–time (5, 10, and 20 min)–temperature (30 and 40) combinations tried, a combined treatment of 5 min at 350 MPa, at 40 °C provided a significant viability loss ($P < 0.05$) of more than 8-log cycles for the selected relatively pressure-resistant pathogens, including *S. aureus*, *E. coli* O157:H7, *Salmonella enteretidis*, and *L. monocytogenes,* in apricot and cherry juices (Table 3.2) (Alpas & Bozoglu, 2003; Bayindirli *et al.*, 2006).

Huang *et al.* (2013) investigated the effects of hydrostatic HP at 300–500 MPa for 5–20 min and high temperature short time (HTST) at 110 °C for 8.6 s on enzymes, phenolics, carotenoids, and color of apricot nectars. According to the results, the enzymes of PPO and peroxidase (POD) in apricot nectar were significantly inactivated, whereas the activity of pectinmethylesterase (PME) was not changed by the applied hydrostatic HP treatments. The effects of hydrostatic HP treatments on total and individual phenolics, total carotenoids and individual carotenes, and color closely correlated to the pressure level and treatment time. Mostly, hydrostatic HP treatments increased total and individual phenolics in apricot nectars, although these were significantly lower than those in HTST-treated apricot nectars. Hydrostatic HP treatments showed no effect on total carotenoids and individual carotenes in apricot nectars except that the treatment at 500 MPa/20 min increased total carotenoids and β-carotene.

Table 3.2 Effect of hydrostatic HP treatment on survival of foodborne pathogens in selected *Prunus* fruit juices

Juice	Pressure (MPa)	Time (min)	Temperature (°C)	*S. aureus* C[a]	*S. aureus* P[b]	*E. coli* O157:H7 C[a]	*E. coli* O157:H7 P[b]	*S. enteretidis* C[a]	*S. enteretidis* P[b]	*L. monocytogenes* C[a]	*L. monocytogenes* P[b]	References
				log$_{10}$ cfu/ml								
Apricot	250	5	30	8.20	4.15	8.28	3.43	8.53	3.75	8.20	4.15	Alpas and Bozoglu (2003); Bayindirli et al. (2006)
		10		8.20	3.90	8.28	3.23	8.53	3.32	–	–	Bayindirli et al. (2006)
		20		8.20	3.20	8.28	2.38	8.53	2.50	–	–	
	350	5	30	8.20	1.26	8.28	1.00	8.53	0.53	8.20	1.26	Alpas and Bozoglu (2003); Bayindirli et al. (2006)
		5	40	8.20	ND	8.28	ND	8.53	ND	8.20	ND	Alpas and Bozoglu (2003); Bayindirli et al. (2006)
Cherry	250	5	30	8.20	3.86	8.28	3.00	8.53	3.06	8.34	4.00	Bayindirli et al. (2006)
		10		8.20	3.50	8.28	2.53	8.53	1.87	–	–	
		20		8.20	2.50	8.28	1.43	8.53	1.27	–	–	
	350	5	30	8.20	0.76	8.28	0.61	8.53	ND	8.34	0.90	Alpas and Bozoglu (2003); Bayindirli et al. (2006)
		5	40	8.20	ND	8.28	ND	8.53	ND	8.34	ND	

[a]Control sample.
[b]Pressurized sample.

In another study performed by González-Cebrino et al. (2013), physicochemical parameters (color, soluble solid content, titratable acidity, and pH), bioactive compounds (total phenols and anthocyanins), total antioxidant activity, and PPO activity were evaluated after the application of hydrostatic HP treatment on a plum puree (cv. Crimson Globe, with red flesh and peel). The inactivation of indigenous microorganisms (total aerobic mesophilic and psychrotrophic counts, and moulds and yeasts) of the plum purée after all HP treatments was sufficient to reduce spoilage microorganisms present in plum purée to undetectable levels in most of HP-treated purées. No significant changes were observed in pH and soluble solid content after processing. Processing induced a significant increase in L^*, a^*, hue, and chroma, which could enhance the redness perception of treated purées. All HP-treated purées showed a slight reduction in anthocyanins and antioxidant activity after processing. Nevertheless, total phenols were not significantly affected by HP treatments. These authors concluded that HP processing keep most quality parameters and nutritional compounds levels, although, it did not achieve the inhibition of browning-related enzymes, which could reduce the shelf life of plum products during storage.

The effect of hydrostatic HP processing at 400–600 MPa and 25 °C for 5–25 min on the inactivation of PPO and pectin methylesterase in peach juice has been investigated (Rao et al., 2014). Both enzymes were found to be inactivated significantly at 500 and 600 MPa, with increasing the pressure and time, and the inactivation kinetics was fitted by the first-order model. Moreover, ascorbic acid levels were better preserved and the color and sensory quality attributes were better maintained by the use of this novel processing technique compared to the use of conventional thermal treatment (Rao et al., 2014).

Some different forms of HP processing applied to fruit juice drinks, such as HP thermal and HP homogenization treatments, have been described. HP thermal treatment based on the combination of high pressures (500–900 MPa) with relatively high temperatures (70–120 °C) over a short treatment time. This technology aims to reduce the undesired effects of conventional thermal technologies by a rapid increase in temperature during compression, which is followed by a temperature decrease in the product upon decompression. In this way, HP thermal treatments ensure quick microbial elimination, a uniform and rapid increase in product temperature, a reduced thermal impact and processing time, and an extended shelf life (Subramanian et al., 2006). These listed advantages of HP thermal treatments would offer a better color, flavor, and aroma retention, and so, an increased overall quality in the processed product compared to those obtained via the traditional thermal treatments (Gupta et al., 2010).

Garcia-Parra et al. (2014) investigated the effect of HP thermal treatment on the color, the activity of the PPO enzyme, bioactive compounds (total polyphenols and anthocyanins) content, and the antioxidant activity of red fleshed and peel plum puree. HP thermal treatments, which included different pressure levels (300, 600, and 900 MPa) and different initial processing temperatures (60, 70, and 80 °C) applied for 1 min, were found to give a similar inactivation level of the PPO enzyme as the traditional thermal treatment (85 °C, 5 min).

In addition, the original color, the anthocyanin content, and the antioxidant activity of the processed purées were conserved more effectively by the use of HP thermal treatment compared to the use of traditional thermal treatments. The optimal reduction in PPO activity and the best maintained levels of bioactive compounds were achieved with the application of 600 MPa of pressure with the initial temperature of 70 °C.

Another promising technology that includes the use of high pressures is the HP homogenization (including the use of a continuous high-pressure homogenizer supplied with a homogenizing PS-type valve), which has recently been used for the stabilization of the liquid food mixtures, while providing the recovery of cell proteins and other biochemicals from microorganisms after cell disintegration (Patrignani *et al.*, 2009). The effectiveness of HP homogenization treatment in deactivating pathogenic and spoilage microorganisms in model systems as well as in real foods has been reported in several studies (Lanciotti *et al.*, 1994; Guerzoni *et al.*, 2002; Diels *et al.*, 2003; Brinez *et al.*, 2006; Taylor *et al.*, 2007; Pathanibul *et al.*, 2009). The HP homogenization treatment is also reported to have an effect on food constituents, including proteins, fat, and polysaccharides specifically, with a resulting modification of their functional properties and susceptibility to enzymatic attack as well as of food microstructure and rheological properties (Sandra & Dalgleish, 2005). HP homogenization has been indicated to have great potential in several fields of the food industry, among which the fruit juice products have been considered as an interesting and feasible field of application in order to achieve microbial inactivation while preserving the initial quality attributes of the fresh products (Donsi *et al.*, 2011). There are numerous literature studies that have focused on the application of HP homogenization treatment in fruit juices, which all showed that this treatment enabled the reduction of both naturally occurring and deliberated inoculated spoilage microbiota in fruit juices (Tahiri *et al.*, 2006; Briñez *et al.*, 2007; Campos & Cristianini, 2007; Betoret *et al.*, 2009; Patrignani *et al.*, 2009, 2010; Suarez-Jacobo *et al.*, 2010; Bevilacqua *et al.*, 2012). Several physical characteristics, such as the mean particle size of the suspended solids (Stipp & Tsai, 1988; Grant, 1989; Donsì *et al.*, 2009), juice viscosity (Campos & Cristianini, 2007; Betoret *et al.*, 2009; Donsì *et al.*, 2009; Patrignani *et al.*, 2010), the natural bioactive compounds of the juices, such as the flavonoids in citrus juices (Betoret *et al.*, 2009), and the freshness and texture attributes (Lacroix *et al.*, 2005), were reported to be better preserved by HP homogenization treatments compared to thermal treatments. Among the *Prunus* fruit juices, apricot juice has been subjected to HP homogenization treatments, at 100 MPa ranging from 1 to 8 passes, to evaluate the potential of this novel technology to inactivate *S. cerevisiae* yeast inoculated into the juice at levels of 3 and 6 $\log_{10}$ cfu/ml.

Under these applied parameters, the maximum temperature reached by the samples was recorded to be 40 °C. In juice samples inoculated with 6.0 $\log_{10}$ cfu/ml, a significant viability decrease (2.2 $\log_{10}$ cfu/ml) was obtained with four repeated passes at 100 MPa, while three passes at 100 MPa were adequate to reduce the cell viability under the detection limit in the same samples inoculated at a level of 3.0 $\log_{10}$ cfu/ml. This study showed an enhanced viscosity of apricot juices that were treated up to five passes at 100 MPa (Patrignani *et al.*, 2009).

3.4.3　Other Innovative Technologies

There are several other research studies that investigate the opportunity of new technologies to be used for juice processing. In a study performed by Cendres *et al.* (2011), the possibility of extracting juice from selected fruits (plum, apricot, and grape) using microwaves and gravity was evaluated. The process uses microwaves for hydrodiffusion of juice from the inside to the exterior of fruit material and earth gravity to collect the fruit juice outside the microwave cavity. According to the results, the obtained juices were brightly colored, as this method allows inactivation of endogenous enzymes such as PPO prior to fruit grinding. This technology permitted fast and efficient extraction even from "hard to press" fruits, such as plums and apricots, yielding an original, brightly colored product.

Another nonthermal technology is the application of nonthermal plasma that provides efficient inactivation of microorganisms with moderate heating of the treated sample (Moisan *et al.*, 2001). Exposure of a process gas to a strong electric field generates the nonthermal plasma, which presents a partially ionized gas. Besides these ionized gas molecules, this process also leads to the formation of other reactive chemical species, including free radicals, heat, and UV light, all of which are potentially jointly involved in different reactions leading to the inactivation of microorganisms (Keener, 2008). Garofulic *et al.* (2015) studied the effect of cold atmospheric pressure gas-phase plasma treatment on anthocyanins and phenolic acids in sour cherry Marasca juice. Plasma-treated sour cherry Marasca juice, by applying a short treatment time (3 min) to a certain volume of juice (3 ml), was found to have higher anthocyanin and phenolic acid concentrations compared to its pasteurized counterpart.

On the other hand, optimization of several fruit-juice-processing steps has also been investigated by researchers. One of these applications is cryoconcentration of fruit juices in which water is removed as ice instead of vapor compared to the classical evaporation step. It can be an alternative to thermal concentration, but the achievable concentration is lower than the values obtained by thermal concentration (Aider & de Halleux, 2008). Cherry, apricot, and orange juices were successfully cryoconcentrated (Aider & de Halleux, 2008; Omran *et al.*, 2013). In addition, novel clarification (electroflotation and ultrafiltration) was found to improve the efficiency of cryoconcentration. According to the results, at similar total soluble solid contents, cryoconcentrated samples showed significantly higher retention of aroma, ascorbic acid, and total antioxidant activity compared to those thermally concentrated. Thermal concentration induced formation of hydroxymethylfurfural more than the cryoconcentration process used for concentration of orange juice (Omran *et al.*, 2013).

3.5 Conclusion and Future Trends

The novel processing technologies, including particularly PEF and HP processing, have recently been extensively studied for their possible applications to pasteurize liquid foods as an alternative to conventional heat treatments. The use of these treatments looks promising as a route to obtaining safe and stable products while maintaining their fresh-like nutritional and organoleptic quality attributes as well as bioactive potential. Nevertheless, the application of these innovative technologies still needs additional systematic research that will focus on a better understanding of the process parameters in order to meet the industrial and regulatory demands for the commercialization of these methods. The studies that have been carried out in the related literature represent systems greatly differing both in processing parameters and in the studied food material, which makes it extremely difficult to compare results from different research groups. Therefore, the major points that need to be defined further, as have also been summarized by Cserhalmi (2006) and Barbosa-Canovas *et al.* (2000), are as follows:

- development of processing conditions that allow a more reliable comparison among different systems
- effects of the process on different food materials and food components
- treatment uniformity
- optimization and control of critical process factors
- standardization and development of effective methods for monitoring consistent delivery of a specified treatment
- process system design, evaluation, and cost reduction.

References

Abe, F. (2007) Exploration of the effects of high hydrostatic pressure on microbial growth, physiology, and survival: perspectives from piezophysiology. *Bioscience, Biotechnology and Biochemistry* **71,** 2347–2357.

Aider, M. & de Halleux, D. (2008) Production of concentrated cherry and apricot juices by cryoconcentrated technology. *LWT – Food Science and Technology* **17,** 1–8.

Akdemir Evrendilek, G., Tok, M.F., Soylu, E.M., & Soylu, S. (2008) Inactivation of *Penicillum expansum* in sour cherry juice, peach and apricot nectars by pulsed electric fields. *Food Microbiology* **25,** 662–667.

Alpas, H. & Bozoglu, F. (2003) Efficiency of high pressure treatment for destruction of *Listeria monocytogenes* in fruit juices. *FEMS Immunology and Medical Microbiology* **35,** 269–273.

Altuntas, J., Akdemir Evrendilek, G., Sangun, M.K., & Zhang, H.Q. (2010) Effects of pulsed electric field processing on the quality and microbial inactivation of sour cherry juice. *International Journal of Food Science & Technology* **45,** 899–905.

Anon. (2004) Plum. In: *The Columbia Electronic Encyclopedia,* 6th edition. Columbia University Press, New York. [Online], Available: http://www.columbia.edu/cu/cup/ [17 April 2015].

Arantegui, J., Feliu, M.J., Barbosa-Canovas, G.V., & Martin, O. (1999) Inactivation of *Saccharomyces cerevisiae* in a model solution of peach juice by electrical pulses of high field intensity. *Abstracts of papers and posters of European Conference of Emerging Food Science and Technology* Tampere, Finland, Nov. 22–24.

Arantegui, J., Marin, E., Rosell, J.R., & Martin, O. (2000) A mathematical model for the inactivation of *Lactobacillus brevis* by pulsed electric field. *In: Proceedings of the 3rd International Conference on Predictive Modeling in Food,* Leuven, Belgium, Sept. 12–15.

Asami, D.K., Hong, Y.J., Barrett, D.M., & Mitchell, A.E. (2003) Processing-induced changes in total phenolics and procyanidins in clingstone peaches. *Journal of the Science of Food and Agriculture* **83,** 56–63.

Avila, I.M.L.B. & Silva, C.L.M. (1999) Modelling kinetics of thermal degradation of colour in peach puree. *Journal of Food Engineering* **39,** 161–166.

Balasubramaniam, V.M. & Farkas, D. (2008) High-pressure food processing. *Food Science and Technology International* **14,** 413–418.

Balasubramaniam, V.M., Farkas, D., & Turek, E.J. (2008) Preserving foods through high-pressure processing. *Food Technology* **62,** 32–38.

Barba, F.J., Esteve, M.J., & Frigola, A. (2013) Physicochemical and nutritional characteristics of blueberry juice after high pressure processing. *Food Research International* **50**, 545–549.

Barbosa-Canovas, G.V., Pierson, M.D., Zhang, Q.H., & Schaffner, D.W. (2000) Pulsed electric fields. Special supplement: kinetics of microbial inactivation for alternative food processing technologies. *Journal of Food Science*, **65**, 65–79.

Bates, R.P., Morris, J.R., & Crandall, P.G. (2001) Tree fruit: apple, pear, peach, plum, apricot, and plums. In: *Principles and Practices of Small- and Medium-Scale Fruit Juice Processing* (eds R.P. Bates, J.R. Morris & P.G. Crandall), FAO Agricultural Services Bulletin 146, Chapter 13. FAO, USA.

Bayindirli, A., Alpas, H., Bozoglu, F., & Hizal, M. (2006) Efficiency of high pressure treatment on inactivation of pathogenic microorganisms and enzymes in apple, orange, apricot and sour cherry juices. *Food Control* **17**, 52–58.

Betoret, E., Betoret, N., Carbonell, J.V., & Fito, P. (2009) Effects of pressure homogenization on particle size and the functional properties of citrus juices. *Journal of Food Engineering* **92**, 18–23.

Bevilacqua, A., Corbo, M.R., & Sinigaglia, M. (2012) Use of natural antimicrobials and high pressure homogenization to control the growth of *Saccharomyces bayanus* in apple juice. *Food Control* **24**, 109–115.

Brinez, W.J., Roig-Sagues, A.X., Hernández-Herrero, M.M., & Guamis-Lopez, B. (2006) Inactivation of *Listeria innocua* inoculated into milk and orange juice using ultrahigh-pressure homogenization. *Journal of Food Protection* **69**, 86–92.

Briñez, W.J., Roig-Sagues, A.X., Herrero, M.M.H., & Lopez, B.G. (2007) Inactivation of *Staphylococcus* spp. strains in whole milk and orange juice using ultra high pressure homogenisation at inlet temperatures of 6 and 20 °C. *Food Control* **18**, 1282–1288.

Buzrul, S., Alpas, H., Largeteau, A., & Demazeau, G. (2008) Modeling high pressure inactivation of *Escherichia coli* and *Listeria innocua* in whole milk. *European Food Research International* **227**, 443–448.

Campos, F.P. & Cristianini, M. (2007) Inactivation of *Saccharomyces cerevisiae* and *Lactobacillus plantarum* in orange juice using ultra high-pressure homogenisation. *Innovative Food Science and Emerging Technologies* **8**, 226–229.

Cendres, A., Chemat, F., Maingonnat, J.F., & Renard, C.M.G.C. (2011) An innovative process for extraction of fruit juice using microwave heating. *LWT – Food Science and Technology* **44**, 1035–1041.

Chakraborty, S., Kaushik, N., Srinivasa Rao, P., & Mishra, H.N. (2014) High-pressure inactivation of enzymes: a review on its recent applications on fruit purees and juices. *Comprehensive Reviews in Food Science and Food Safety* **13**, 578–596.

Chang, T.S., Siddiq, M., Sinha, N.K., & Cash, J.N. (1994) Plum juice quality affected by enzyme treatment and fining. *Journal of Food Science* **59**, 1065–1069.

Cheftel, J.C. (1995) Review: high-pressure, microbial inactivation and food preservation/Revisión: alta-presión, inactivación microbiológica y conservación de alimentos. *Food Science and Technology International* **1**, 75–90.

Cserhalmi, Z. (2006) Non-thermal pasteurization of fruit juice using high voltage pulsed electric fields. In: *Handbook of Fruits and Fruit Processing* (ed Y.H. Hui), 1st edition, pp. 279–291. Blackwell Publishing, Iowa, USA.

Diels, A.M.J., Wuytack, E.Y., & Michiels, C.W. (2003) Modelling inactivation of *Staphylococcus aureus* and *Yersinia enterocolitica* by high-pressure homogenisation at different temperatures. *International Journal of Food Microbiology* **87**, 55–62.

Donsì, F., Esposito, L., Lenza, E., Senatore, B., & Ferrari, G. (2009) Production of shelf-stable Annurca apple juice with pulp by high pressure homogenization. *International Journal of Food Engineering* **5,** 1556–3758.

Donsi, F., Annunziata, M., Sessa, M., & Ferrari, G. (2011) Nanoencapsulation of essential oil to enhance their antimicrobial activity in foods. *LWT – Food Science and Technology* **44,** 1908–1914.

Echavarria, A.P., Torras, C., & Pagan, J. (2011) Fruit juice processing and membrane technology application. *Food Engineering Reviews* **3,** 136–158.

Espachs-Barroso, A., Barbosa-Canovas, G.V., & Martin-Belloso, O. (2003) Microbial and enzymatic changes in fruit juice induced by high-intensity pulsed electric fields. *Food Reviews International* **19,** 253–273.

Espie, M. (1992) *Michigan Agricultural Statistics.* pp. 29–30. Michigan Agricultural Statistics Service, Lansing, MI.

Etiévant, P., Bellisle, F., Dallongeville, J., Etilé, F., Guichard, E. & Padilla, M., *et al.* (eds) (2010) *Food Related Behavior: What are the Determinants? Which Actions for which Effects? (Les comportements alimentaires. Quels en sont les déterminants? Quelles actions, pour quels effets?)* Expertise Scientifique Collective, France: INRA, rapport (277 pp.).

Ferretti, G., Bacchetti, T., Belleggia, A., & Neri, D. (2010) Cherry antioxidants: from farm to table. *Molecules* **15,** 6993–7005.

Food and Agriculture Organization of the United Nations Statistics Division (FAOSTAT). (2013) *Food and Agricultural Commodities Production/Countries by Commodity,* [Online], Available: http://faostat3.fao.org/browse/rankings/countries_by_commodity /E [17 April 2015].

Fraignier, M.P., Marques, L., Fleuriet, A., & Macheix, J.J. (1995) Biochemical and immuno-chemical characteristics of polyphenol oxidases from different fruits of *Prunus. Journal of Agricultural and Food Chemistry* **43,** 2375–2380.

Furukawa, S., Noma, S., Yoshikawa, S., Furuya, H., Shimoda, M., & Hayakawa, I. (2001) Effect of filtration of bacterial suspension on the inactivation ratio in hydrostatic pressure treatment. *Journal of Food Engineering* **50,** 59–61.

Garcia-Parra, J., Gonzalez-Cebrino, F., Cava, R., & Ramirez, R. (2014) Effect of a different high pressure thermal processing compared to a traditional thermal treatment on a red flesh and peel plum purée. *Innovative Food Science and Emerging Technologies* **26,** 26–33.

Garofulic, I.E., Jambrak, A.R., Milosevic, S., Dragovic-Uzelac, V., Zoric, Z., & Herceg, Z. (2015) The effect of gas phase plasma treatment on the anthocyanin and phenolic acid content of sour cherry Marasca (*Prunus cerasus* var. Marasca) juice. *LWT – Food Science and Technology* **62,** 894–900.

Giner, J., Gimeno, V., Espachs, A., Elez, P., Barbosa-Canovas, G.V., & Martin, O. (2000) Inhibition of tomato (*Licopersicon esculentum* Mill.) pectin methylesterase by pulsed electric fields. *Innovative Food Science and Emerging Technologies,* **1,** 57–67.

González-Cebrino, F., Durán, R., Delgado-Adámez, J., Contador, R., & Ramírez, R. (2013) Changes after high-pressure processing on physicochemical parameters, bioactive compounds, and polyphenol oxidase activity of red flesh and peel plum purée. *Innovative Food Science and Emerging Technologies* **20,** 34–41.

Grant, P.M. (1989) Citrus juice concentrate processor. US Patent No. 4,886,574.

Guerzoni, M.E., Vannini, L., Lanciotti, R., & Gardini, F. (2002) Optimisation of the formulation and of the technological process of egg-based products for the prevention of

Salmonella enteritidis survival and growth. *International Journal of Food Microbiology* **73**, 367–374.

Gupta, R., Balasubramaniam, V.M., Schwartz, S.J., & Francis, D.M. (2010) Storage stability of lycopene in tomato juice subjected to combined pressure-heat treatments. *Journal of Agricultural and Food Chemistry* **58**, 8305–8313.

Huang, W., Bi, X., Zhang, X., Liao, X., Hu, X., & Wu, J. (2013) Comparative study of enzymes, phenolics, carotenoids and colour of apricot nectars treated by high hydrostatic pressure and high temperature short time. *Innovative Food Science and Emerging Technologies* **18**, 74–82.

Janick, J., Swietlik, D., & Zimmerman, R.H. (2011) Preface. In: *Origin and Dissemination of Prunus Crops.* (ed J. Janick), 1st edition, pp. 3–10. International Society for Holticultural Science (ISHS), Leuven, Belgium.

Keener, K.M. (2008) Atmospheric non-equilibrium plasma. *Encyclopedia of Agricultural, Food and Biological Engineering* **1**, 1–5.

Lacroix, N., Fliss, I., & Makhlouf, J. (2005) Inactivation of pectin methylesterase and stabilization of opalescence in orange juice by dynamic high pressure. *Food Research International* **38**, 569–576.

Lanciotti, R., Sinigaglia, M., Angelini, P., & Guerzoni, M.E. (1994) Effects of homogenization pressure on the survival and growth of some food spoilage and pathogenic microorganisms. *Letters in Applied Microbiology* **18**, 319–322.

Lozano, J.E. & Ibarz, A. (1997) Colour changes in concentrated fruit pulp during heating at high temperatures. *Journal of Food Engineering* **31**, 365–373.

Lurie, S. & Crisosto, C.H. (2005) Chilling injury in peach and nectarine. *Postharvest Biology and Technology* **37**, 195–208.

Matser, A.M., Krebbers, B., van den Berg, R.W., & Bartels, P.V. (2004) Advantages of high-pressure sterilisation on quality of food products. *Trends in Food Science & Technology* **15**, 79–85.

Mittal, G.S. & Griffiths, M.W. (2005) Pulsed electric field processing of liquid foods and beverages. In: *Emerging Technologies for Food Processing* (ed D.W. Sun), 1st edition, pp. 99–139. Academic Press, Boston, USA.

Moisan, M., Barbeau, J., Moreau, S., Pelletier, J., Tabrizian, M., & Yahia, L.H. (2001) Low-temperature sterilization using gas plasmas: a review of the experiments and an analysis of the inactivation mechanisms. *International Journal of Pharmaceutics* **226**, 1–21.

Mosqueda-Melgar, J., Raybaudi-Massilia, R.M., & Martin-Belloso, O. (2008) Non-thermal pasteurization of fruit juices by combining high-intensity pulsed electric fields with natural antimicrobials. *Innovative Food Science and Emerging Technologies* **9**, 328–340.

Oliveira, A., Pintado, M., & Almeida, D.P.F. (2012) Phytochemical composition and antioxidant activity of peach as affected by pasteurization and storage duration. *LWT – Food Science and Technology* **49**, 202–207.

Omran, M.N., Pirouzifard, M.K., Aryaey, P., & Nejad, M.H. (2013) Cryoconcentration of sour cherry and orange juices with novel clarification method; Comparison of thermal concentration with freeze concentration in liquid foods. *Journal of Agricultural Science Technology* **15**, 941–950.

Parish, M.E. (1997) Public health and non-pasteurized fruit juices. *Critical Reviews in Microbiology* **23**, 109–119.

Pathanibul, P., Taylor, T.M., Davidson, P.M., & Harte, F. (2009) Inactivation of *Escherichia coli* and *Listeria innocua* in apple and carrot juices using high pressure homogenization and nisin. *International Journal of Food Microbiology* **129**, 316–320.

Patras, A., Brunton, N.P., Da Pieve, S., & Butler, F. (2009) Impact of high pressure processing on total antioxidant activity, phenolic, ascorbic acid, anthocyanin content and colour of strawberry and blackberry purées. *Innovative Food Science & Emerging Technologies* **10,** 308–313.

Patrignani, F., Vannini, L., Kamdem, S.L.S., Lanciotti, R., & Guerzoni, M.E. (2009) Effect of high pressure homogenization on *Saccharomyces cerevisiae* inactivation and physico-chemical features in apricot and carrot juices. *International Journal of Food Microbiology* **136,** 26–31.

Patrignani, F., Vannini, L., Kamdem, S.L.S., Lanciotti, R., & Guerzoni, M.E. (2010) Potentialities of high-pressure homogenization to inactivate *Zygosaccharomyces bailii* in fruit juices. *Journal of Food Science* **75,** M116–M120.

Rao, L., Guo, X., Pang, X., Tan, X., Liao, X., & Wu, J. (2014) Enzyme activity and nutritional quality of peach (*Prunus persica*) juice: Effect of high hydrostatic pressure. *International Journal of Food Properties* **17,** 1406–1417.

Rastogi, N.K., Raghavarao, K.S.M.S., Balasubramaniam, V.M., Niranjan, K., & Knorr, D. (2007) Opportunities and challenges in high-pressure processing of foods. *Critical Reviews in Food Science and Nutrition* **47,** 69–112.

Ray, B., Kalchayanand, N., Dunne, P., & Sikes, A. (2001) Microbial destruction during high pressure processing of food. In: *Novel Processes and Control Technologies in the Food Industry* (eds F. Bozoglu, B. Ray, & T. Deak), NATO Science Series, Vol. 338, pp. 95–122. IOS Press, Amsterdam, The Netherlands.

Rieger, M. (2004) *Mark's Fruit Crops Homepage, University of Georgia*, [Online], Available: http://www.uga.edu/fruit [17 April 2015].

Sandra, S. & Dalgleish, D.G. (2005) Effects of ultra-high-pressure homogenization and heating on structural properties of casein micelles in reconstituted skim milk powder. *International Dairy Journal* **15,** 1095–1104.

Siddiq, M. (2006a) Apricots. In: *Handbook of Fruits and Fruit Processing.* (ed Y.H. Hui), 1st edition, pp. 279–291. Blackwell Publishing, Iowa, USA.

Siddiq, M. (2006b) Plums and prunes. In: *Handbook of Fruits and Fruit Processing.* (ed Y.H. Hui), 1st edition, pp. 279–291. Blackwell Publishing, Iowa, USA.

Smith, K., Mittal, G.S., & Griffith, M.W. (2002) Pasteurization of milk using pulsed electric field and antimicrobials. *Journal of Food Science* **67,** 2304–2308.

Soliva-Fortuny, R., Balasa, A., Knorr, D., & Martin-Belloso, O. (2009) Effects of pulsed electric fields on bioactive compounds in foods: a review. *Trends in Food Science & Technology* **20,** 544–556.

Stipp, G.K. & Tsai, C.H. (1988) Low viscosity orange juice concentrates useful for high brix products having lower pseudoplasticity and greater dispersibility. U.S. Patent No. 4,946,702, The Procter & Gamble Company.

Suarez-Jacobo, A., Gervilla, R., Guamis, B., Roig-Sagues, A.X., & Saido, J. (2010) Effect of UHPH on indigenous microbiota of apple juice: A preliminary study of microbial shelf-life. *International Journal of Food Microbiology* **136,** 261–267.

Subramanian, A., Ahn, J., Balasubramaniam, V.M., & Rodriguez-Saona, L. (2006) Determination of spore inactivation during thermal and pressure-assisted thermal processing using FT-IR spectroscopy. *Journal of Agricultural and Food Chemistry* **54,** 10300–10306.

Tahiri, I., Makhlouf, J., Paquin, P., & Fliss, I. (2006) Inactivation of food spoilage bacteria and *Escherichia coli* O157:H7 in phosphate buffer and orange juice using dynamic high pressure. *Food Research International* **39,** 98–105.

Talcott, S.T., Howard, L.R., & Brenes, C.H. (2000) Contribution of periderm material and blanching time to the quality of pasteurized peach puree. *Journal of Agricultural and Food Chemistry* **48,** 4590–4596.

Taylor, T.M., Roach, A., Black, D.G., Davidson, P.M., & Harte, F. (2007) Inactivation of *Escherichia coli* K-12 exposed to pressures in excess of 300 MPa in a high-pressure homogenizer. *Journal of Food Protection* **70,** 1007–1010.

Terefe, N.S., Buckow, R., & Versteeg, C. (2014) Quality-related enzymes in fruit and vegetable products: effects of novel food processing technologies Part 1: high-pressure processing. *Critical Reviews in Food Science & Nutrition* **54,** 24–63.

Toydemir, G. (2013) The effects of nectar processing on sour cherry antioxidant compounds: Changes in metabolite profile and bioavailability. *PhD thesis* Istanbul Technical University, Istanbul, Turkey.

Toydemir, G., Capanoglu, E., Kamiloglu, S., Boyacioglu, D., de Vos, R.C.H., Hall, R.D., & Beekwilder, J. (2013a) Changes in sour cherry (*Prunus cerasus* L.) antioxidants during nectar processing and *in vitro* gastrointestinal digestion. *Journal of Functional Foods* **5,** 1402–1413.

Toydemir, G., Capanoglu, E., Gomez Roldan, M.V., de Vos, R.C.H., Boyacioglu, D., Hall, R.D., & Beekwilder, J. (2013b) Industrial processing effects on phenolic compounds in sour cherry (*Prunus cerasus* L.) fruit. *Food Research International* **53,** 218–225.

USDA Foreign Agricultural Service. (2012) *Turkey Stone Fruit Annual Report 2011*, [Online], Available: http://static.globaltrade.net/files/pdf/2011091420410839.pdf [8 August 2012].

Yigit, D., Yigit, N., & Mavi, A. (2009) Antioxidant and antimicrobial activities of bitter and sweet apricot (*Prunus armeniaca* L.) kernels. *Brazilian Journal of Medical and Biological Research* **42,** 346–352.

Zimmermann, U., Pilwat, G., & Riemann, F. (1974) Dielectric breakdown on cell membranes. *Biophysical Journal* **14,** 881–889.

4

Vegetable Juices

Rogelio Sánchez-Vega[1], David Sepúlveda-Ahumada[1], and Ma. Janeth Rodríguez-Roque[2]*

[1]*Centro de Investigación en Alimentación y Desarrollo A.C. Avenida Río Conchos S/N, Parque Industrial, Cuauhtémoc-Chihuahua, Mexico*
[2]*Universidad Autónoma de Chihuahua, Facultad de Ciencias Agrotecnológicas Cd. Universitaria s/n, Chihuahua, Mexico*

4.1   Introduction

Vegetables are important sources of a great variety of substances with beneficial effects on health, such as essential minerals (i.e., potassium, calcium, and magnesium), vitamins (i.e., vitamin C, folate, and provitamin A), dietary fiber, carotenoids, phenolic compounds, and organosulfur compounds (Table 4.1) (Hui & Evranuz, 2016; Wootton-Beard *et al.*, 2011). These compounds provide protection against cancer (Boffetta *et al.*, 2010; Freedman *et al.*, 2008; van Duijnhoven *et al.*, 2009), cardiovascular desease (Heidemann *et al.*, 2008; Wang *et al.*, 2011), age-related cognitive function disease (Morris *et al.*, 2006), and type 2 diabetes mellitus (Villegas *et al.*, 2008). For this reason, the Food and Agricultural Organization of the United Nations (FAO) and World Health Organization (WHO) recommend intake of at least 400 g or five servings of fruit and vegetables per day for preventing chronic diseases and maintaining health (FAO/WHO, 2003).

On the other hand, vegetables are highly perishable commodities, susceptible to rapid senescence and rotting unless properly processed and preserved. One of the most important approaches for solving this problem is to process vegetables into juicing. Juices can be defined as the product obtained from healthy and ripe vegetables from one or more species, which keep the sensorial characteristics of the vegetable from which it is made (Rodríguez-Roque, 2014). Recently, the number of commercially available vegetable juices has augmented because these products provide the same nutritive and health benefits as whole vegetables. In addition, vegetable juice consumption is easier; more convenient (mainly for children and elderly) (Falguera & Ibarz, 2014); adds new nutrients or increases their concentration when different juices

*Corresponding author: Ma. Janeth Rodríguez-Roque, mjrodriguez@uach.mx

Innovative Technologies in Beverage Processing, First Edition.
Edited by Ingrid Aguiló-Aguayo and Lucía Plaza.
© 2017 John Wiley & Sons Ltd. Published 2017 by John Wiley & Sons Ltd.

Table 4.1   Main nutrients and bioactive compounds contained in vegetables

Nutrient/bioactive compounds	Vegetables
Carotenoids	
• α-Carotene	Carrots and green leafy vegetables
• β-Carotene	Broccoli, brussels sprouts, carrots, spinach, and tomatoes
• Lutein	Yellow/green vegetables
• Lycopene	Tomatoes
Fiber	All vegetables
Folates	Green leafy vegetables and potatoes
Furanocoumarins	Celery
Glycoalkaloids	Potato and aubergine
Glucosinolates	Brassicas
Minerals	
• Calcium	Green vegetables
• Iron	Green vegetables
• Magnesium	Green vegetables
• Potassium	All vegetables
Phenols	
• Flavonoids	Onions
• Flavonols (quercetin)	Broccoli, onions
• Hydroxycinnamates	Broccoli, potatoes
Vitamins	
• Vitamin C	Green vegetables (i.e., broccoli, Brussels sprouts, pea, and spinach), potatoes, and sweet red pepper
• Vitamin E	Avocado and sweet red pepper
• Vitamin K	Green leafy vegetables

are blended; enhances visual appeal; and offers a broad range of new flavors, aromas, and tastes (Rodríguez-Roque *et al.*, 2014).

Regarding vegetable juice preservation, it is important to highlight that consumers demand food products that provide variety, convenience, adequate shelf life, high nutritional quality, and fresh-like appearance. For this reason, food preservation processing methods are becoming more sophisticated and diverse to meet such demands (Barbosa-Cánovas *et al.*, 2005). Thermal processing is the commonest method for inactivating enzymes and destroying microorganisms from foods, beverages, and juices. However, this treatment may reduce the quality (sensorial and nutritional) and freshness of vegetable juices (Sánchez-Vega *et al.*, 2015). Nonthermal food preservation technologies, including high hydrostatic pressure (HHP), High-intensity pulsed electric fields (HIPEF), ultrasound treatment (UST), and ultraviolet treatment (UVT), have been developed as alternatives to heat treatments for obtaining safety and nutritious products with fresh-like appearance (Vasantha Rupasinghe & Yu, 2012).

This chapter reviews in detail the most relevant research about application of conventional and nonthermal processing technologies of vegetable juices and highlights the influence of these techniques on microbial quality, enzymatic activity, physicochemical, and sensorial properties. Some fruits, such as avocado, cucumber, pepper, pumpkin, squash, and tomato, are considered as vegetables as they are usually consumed with main course dishes (USDA, 2016); for this reason, in this chapter, some of these products are also considered.

4.2 Conventional Processing Technologies

Thermal processing is the conventional method used for preserving food, beverages, and juices as it destroys spoilage and pathogenic microorganisms. It also inactivates degradative enzymes (i.e., polyphenol oxidase which is associated with juice browning) and it minimizes flavor changes resulting from lipase and proteolytic activity. Scalding, pasteurization ($T < 100\,°C$), and sterilization ($T > 100\,°C$) are the commonest ways for thermally processing vegetable juices. Among them, juice pasteurization is the most widely used technique for reducing five logarithms of the most resistant microorganisms and it is accomplished using different combinations of time and temperature. In this sense, vegetable juices could be processed by low temperature (63–65 °C) and long time (minutes) or through high temperature (70–95 °C) and short time (seconds) (Vasantha Rupasinghe & Yu, 2012). Despite the effectivity of this treatment on food safety, deleterious effects on nutritional and organoleptic food attributes (Odriozola-Serrano *et al.*, 2013), mainly in those treatments using heat with long processing time, have been reported.

4.2.1 Influence of Conventional Processing on Microbial Quality

Thermal treatment ensures food safety by inactivating pathogenic and spoilage microorganism that can cause severe infections in humans. Pasteurization is a mild heat treatment that inactivates pathogenic microorganisms but does not destroy heat-resistant microorganism and enzymes. The Food and Drug Administration (2004) suggests that thermal process requires treatments of at least 71.1 °C during 3 s for achieving 5-log reductions of microorganisms of public health significance, such as *Escherichia coli* O157:H7, Salmonella, and *Listeria monocytogene*s. However, most vegetable juices had low pH (<4.6), requiring being processed with temperatures up to 100 °C for 15–20 s (Boz & Erdoğdu, 2016). Table 4.2 summarizes some chronological studies concerning microorganism inactivation in heat processed vegetable juices. In these studies, thermal processing was carried out using temperature between 80 and 121 °C during 8.6–600 s for reducing around 5-log reductions of molds, yeast, *Staphylococcus aureus*, *E. coli*, *Lactobacillus*, *Salmonella enteritidis*, mesophilic, and aerobic bacteria from asparagus, carrot, cucumber, and tomato juices.

4.2.2 Influence of Conventional Processing on Nutritional Attributes

Many compounds, such as vitamins, minerals, and dietary fiber, are present in vegetable juices. In addition to these compounds, a whole range of non-nutrients (secondary metabolites) have been associated with the health benefits of consuming vegetables and their juices. Secondary metabolites have biological activity that is related to their capability for neutralizing free radicals (antioxidant capacity), and

Table 4.2 Chronological studies on microorganism elimination from thermally and nonthermally processed vegetable juices

Vegetable juice		Processing conditions	Microorganism studied	Grade of elimination	References
Tomato juice	Thermal processing	92 °C for 90 s	Mold and yeast	<10 CFU/ml	Min *et al.* (2003a,b)
Cucumber juice		85 °C for 15 s	Mold and yeast	Inactivated	Zhao *et al.* (2013)
Green asparagus juice		121 °C for 3 min	Total mesophilic bacteria count	Not detected	Chen *et al.* (2015)
Carrot juice		110 °C for 8.6 s	Total plate count	4.88 log CFU/ml	Zhang *et al.* (2016)
			Mold and yeast	Not detected	
Cucumber juice		110 °C for 8.6 s	Total aerobic bacteria	3.58 log cycles	Liu *et al.* (2016)
			Yeast and mold	Not detected	
Carrot juice		80 °C for 10 min	*Staphylococcus aureus*	5.0–5.5 log CFU/ml	Khandpur and Gogate (2016)
			Escherichia coli		
			Lactobacillus		
			Salmonella enteritidis		
			yeasts and molds		
Spinach juice		80 °C for 10 min	*Staphylococcus aureus*	5.5–6.0 log CFU/ml	Khandpur and Gogate (2016)
			Escherichia coli		
			Lactobacillus		
			Salmonella enteritidis		
			yeasts and molds		

Product	Treatment	Microorganism	Result	Reference
Tomato juice	Nonthermal processing			
	HIPEF: 40 kV/cm. Pulse duration time: 2 μs. Treatment time: 57 μs	Mold and yeast	<10 CFU/ml	Min et al. (2003a,b)
Carrot juice	UST: 19.3 kHz, 700–800 W, 1 min, 60 °C	Escherichia coli K12	2.5 log CFU	Zenker et al. (2003)
Carrot juice	HHP: 350 MPa	Escherichia coli K12	5 log cycles	Pathanibul et al. (2009)
		Listeria innocua		
Formulated carrot juice	HIPEF: 27 kV/cm	Escherichia coli O157:H7	Around 4.5 log cycles	Akin and Evrendilek (2009)
Carrot juice	UVT: 1.5 kJ/l	Escherichia coli	3.0 log CFU	Koutchma et al. (2009)
Carrot juice	HPH: >250 MPa	Total plate count	3.0 log CFU	Pathanibul et al. (2009)
Carrot juice	HPH: 350 MPa	Escherichia coli	>5 log CFU	Pathanibul et al. (2009)
Cucumber juice	HHP: 400 MPa for 4 min 500 MPa for 2 min	Listeria innocua Mold and yeast	5 log CFU Inactivated	Zhao et al. (2013)
Carrot juice	HHP: 300–500 MPa for 15 min at 40–60 °C	Total plate count	No detected	Gong et al. (2015)
Green asparagus juice	HHP: 200–600 MPa for 10–20 min	Total mesophilic bacteria count	Not detected	Chen et al. (2015)
Carrot juice	HHP: 550 MPa for 2–10 min	Total plate count Mold and yeast	4.3 log CFU/ml Not detected	Zhang et al. (2016)
Cucumber juice	HHP: 500 MPa for 5 min	Total aerobic bacteria Yeast and mold	3.18 log cycles Not detected	Liu et al. (2016)

(continued overleaf)

Table 4.2 (*continued*)

Vegetable juice	Processing conditions	Microorganism studied	Grade of elimination	References
Carrot juice	UST: 20 kHz, 100 W, 15 min, $T < 30\,°C$	Staphylococcus aureus Escherichia coli Lactobacillus Salmonella enteritidis yeasts and molds	3.0–3.5 log CFU/ml	Khandpur and Gogate (2016)
Carrot juice	UST + UVT UST: 20 kHz, 100 W, 15 min, $T < 30\,°C$ UVT: 2 UV lamps with 8 W power dissipation	Staphylococcus aureus Escherichia coli Lactobacillus Salmonella enteritidis yeasts and molds	5.0–5.5 log CFU/ml	Khandpur and Gogate (2016)
Spinach juice	UST: 20 kHz, 100 W, 15 min, $T < 30\,°C$	Staphylococcus aureus Escherichia coli Lactobacillus Salmonella enteritidis yeasts and molds	3.5–4.0 log CFU/ml	Khandpur and Gogate (2016)

CFU, colony forming units; HHP, high hydrostatic pressure; HIPEF, high-intensity-pulsed electric fields; HPH, high-pressure homogenization; UST, ultrasound treatment; and UVT, ultraviolet treatment.

they are also responsible for quality features (i.e., color, appearance, flavor, and aroma, among others) (Tomás-Barberán & Gil, 2008). Antioxidants from vegetable juices can easily undergo thermal degradation and/or consumption in the Maillard reaction pathway during conventional processing. For instance, vitamin C and carotenoids are thermolabile compounds susceptible to chemical and enzymatic oxidation during thermal processing (Ball, 2006; Rodríguez-Roque, 2014). Regarding phenolic compounds, processing is known to change the physicochemical features of these constituents, such as changes in their structure (hydroxylation, methylation, isoprenylation, dimerization, and glycosylation, among others) and/or the formation of phenolic derivatives (by partial degradation of the combined forms or by losing the moieties between phenols and sugars) (Rodríguez-Roque *et al.*, 2015). In addition, pasteurized products may require additional preservation methodologies (i.e., refrigeration) for maintaining the nutritional and sensory features of thermally treated products (Martín-Belloso *et al.*, 2014). As stated in Table 4.3, several authors observed significant reductions in the concentration of a wide range of bioactive compounds from thermally treated vegetable juices, including asparagus, broccoli, carrot, cucumber, tomato, spinach, and a blend of orange–carrot juice. Thermal processing was carried out at 70–121 °C during 8.6 s and up to 50 h. The general trend is that increasing the temperature and time processing parameters, the reductions/losses of bioactive compounds also augmented. Vitamin C and chlorophylls were the most affected compounds by thermal processing, being reduced up to 90%. Other compounds, such as lycopene, α-carotene, β-carotene, and lutein, showed losses up to 66%, 56%, 70%, and 48.8%, respectively, after thermal treatment. Controversial results were found in the concentration of lycopene, β-carotene, and vitamin A, which was up to 12% higher in thermally treated vegetable juices than in untreated products. The concentration of total phenolic compounds and the antioxidant capacity from vegetable juices was also documented and diminished in the range 0–46.7% after thermal processing.

4.2.3 Influence of Conventional Processing on Organoleptic Attributes

Similar to nutritional properties, the organoleptic features of vegetable juices change after thermal processing. As temperature and processing time increase, products deterioration take place due to onset of different biochemical reactions occurring in the juices (Ortega-Rivas & Salmerón-Ochoa, 2014). Heat processing also induces physical and chemical changes, resulting in the degradation of flavor, color, volatile compounds (aroma), and nutritional value of foods to varying degrees (Hjelmqwist, 2005). At the same time, the high temperature increases the rate of nonenzymatic browning (Millard reaction and sugar caramelization), the starch gelatinization, and protein denaturation, thus decreasing the sensorial features of beverages. Unpleasant flavors, odors, and product deterioration are also commonly caused by microbial growth and the presence of yeasts in vegetable juices.

Table 4.4 shows the chronological studies related with the physicochemical characteristics of thermally treated vegetable juices. According to Table 4.4, thermal processing consisted on applying temperatures between 85 and 121 °C during 8.6–180 s. The analyzed vegetable juices were that coming from asparagus, cucumber, and tomato. Overall, no significant changes in the color (Hue, ΔE, and L^*), pH, soluble solids, titratable acid, and volatile compounds (*trans*-2-hexanal and 2-isobutylthiazole) were

Table 4.3 Chronological studies on the vegetable juice bioactive compound and antioxidant capacity retention through thermal and nonthermal processing

Vegetable juice	Processing conditions	Bioactive compound	Outcomes	References
Broccoli juice	80–120 °C	Chlorophyll	Retention of 10% at 100 °C during 37 min	Van Loey et al. (1998)
Tomato juice	95 or 70 °C for up to 50 h	Ascorbic acid	Reduction of 90% after 4 h Formation of Maillard reaction products	Anese et al. (1999)
Tomato juice	92 °C for 90 s	Lycopene	Retention of 33.6% after 112 days of storage	Min et al. (2003a,b)
Orange–carrot juice	98 °C for 21 s	β-Carotene and vitamin A	Retention of 105.8% and 107.8%, respectively	Torregrosa et al. (2005)
Blend of orange–carrot juice	98 °C for 21 s	Ascorbic acid	Retention of 83%	Torregrosa et al. (2006)
Tomato juice	90 °C for 30 or 60 s	Lycopene and Vitamin C	Retention of 104.7% and 79.2–80.4%, respectively	Odriozola-Serrano et al. (2008)
Carrot juice	90 °C for 30 or 60 s	Vitamin C, β-carotene Total phenolic and antioxidant capacity	Retention of 86.6–89.0%, 110–113%, 91.6–96.4% and 94.3%, respectively	Quitao-Teixeira et al. (2009)
Tomato juice	90 °C for 30 or 60 s	Lycopene and phenolic compounds	Retention of 104–107% and 100–101%, respectively	Odriozola-Serrano et al. (2009)
Cucumber juice	85 °C for 15 s	Chlorophyll a and chlorophyll b	Retention of 72.82% and 82.76%, respectively	Zhao et al. (2013)
Broccoli juice	90 °C for 60 s	Chlorophyll a and chlorophyll b	Retention of 24.23%, and 18.28%, respectively	Sánchez-Vega et al. (2014)

Product	Treatment	Compounds	Results	Reference
Green asparagus juice	121 °C for 3 min	Ascorbic acid, total phenolics, and antioxidant activity	Retention of 77%, 90%, and 168%, respectively	Chen et al. (2015)
Carrot juice	80 °C for 10 min	Vitamin C, total phenol content, and antioxidant capacity	Retention of 42.8%, 64.7%, and 101–96.9%, respectively	Khandpur and Gogate (2015)
Spinach juice	80 °C for 10 min	Vitamin C, total phenol content, and antioxidant capacity	Retention of 32%, 54.7%, and 55.3%, respectively	Khandpur and Gogate (2015)
Broccoli juice	90 °C for 60 s	β-Carotene, lutein, vitamin C, total phenolics, and antioxidant capacity	Retention of 30.6%, 51.2%, 68%, 71.8%, and 72.4%, respectively	Sánchez-Vega et al. (2015)
Carrot juice	110 °C for 8.6 s	Lutein, α-carotene, β-carotene, total phenol content, and antioxidant capacity (DPPH)	Retention of 94.87%, 43.77%, 56.98%, 88.50%, and 90.42%, respectively	Zhang et al. (2016)
Broccoli juice	HHP: 200–800 MPa/30–80 °C	Chlorophyll	No significant changes at 800 MPa and 40 °C	Van Loey et al. (1998)
Tomato juice	HIPEF: 40 kV/cm	Lycopene	Retention of 47.2% after 112 days of storage	Min et al. (2003a,b)
Orange–carrot juice	HIPEF: 25–40 kV/cm for 30–100 μs	β-Carotene and vitamin A	Retention of 98.6% and 101%, respectively	Torregrosa et al. (2005)

(continued overleaf)

Table 4.3 *(continued)*

Vegetable juice	Processing conditions	Bioactive compound	Outcomes	References
Blend of orange–carrot juice	HIPEF: 25–40 kV/cm for 30–340 µs stored during 50 days	Ascorbic acid	Retention of 90%	Torregrosa *et al.* (2006)
Broccoli juice	HHP: 500 MPa – 10 min	Sulforaphane and indole-3-carbinol	Inhibition of mutagenicity	Mandelová and Totušek (2007)
White cabbage juice, red cabbage juice, broccoli juice, cauliflower juice, and brussels sprouts juice	HHP: 500–600 MPa, 10 min	Total isothiocyanates	4.21, 3.85, 11.27, 10.92, and 21.46 μM/L, respectively	Tříska *et al.* (2007)
Tomato juice	HIPEF: 35 kV/cm for 1000 µs at 50–250 Hz in bipolar or monopolar mode at 1–7 µs.	Lycopene, vitamin C, and antioxidant capacity	Retention of 146%, 99%, and 92.3%, respectively	Odriozola-Serrano *et al.* (2007)
Tomato juice	HIPEF: 35 kV/cm for 1500 µs in bipolar 4 µs pulses at 100 Hz.	Lycopene, vitamin C, and phenolic compounds	Retention of 107.6–110%, 86.5%, and 98.89%, respectively	Odriozola-Serrano *et al.* (2008, 2009)
Carrot juice	HIPEF: 35 kV/cm for 1500 µs of 6 µs bipolar pulses	Vitamin C, β-carotene, total phenolic, and antioxidant capacity	Retention of 95.1%, 123.3%, 102%, and 93.4%, respectively	Quitão-Teixeira *et al.* (2009)

Carrot juice	UVT: seven passes at flow rate of 0.5 l/s and total accumulative dose of 1.5 kJ/l	Vitamin C, retinol, vitamin A, and β-carotene	Retention of 100–85%, 100%, 100%, and 100%, respectively	Koutchma et al. (2009)
Cucumber juice	HHP: 400 MPa for 4 min; 500 MPa for 2 min	Chlorophyll a and chlorophyll b	No significant changes respect to untreated juice	Zhao et al. (2013)
Broccoli juice	HIPEF: 15–35 kV/cm for 500–2000 μs in bipolar or monopolar mode. 100 Hz and 4 μs	Chlorophyll a and chlorophyll b	Retention of 116% and 120.7%, respectively	Sánchez-Vega et al. (2014)
Carrot juice	HHP: 300–500 MPa for 15 min at 40–60 °C	Vitamin C and carotenoid	Retention of 94% and 103%, respectively	Gong et al. (2015)
Carrot juice	UST: 100 W, 20 kHz, 15 min, 30 °C; UST + UVT: 2 lamps of 8 W	Vitamin C, total phenol content, and antioxidant capacity	Retention of 60.4–42.5%, 114.7–104.7%, 101.0%, and −98.9%, respectively	Khandpur and Gogate (2015)

Table 4.3 (*continued*)

Vegetable juice	Processing conditions	Bioactive compound	Outcomes	References
Spinach juice	UST: 100 W, 20 kHz, 15 min, 30 °C UST + UVT: 2 lamps of 8 W	Vitamin C, total phenol content, and antioxidant capacity	Retention of 74.7–60%, 107.9–101.0%, and 102.1–99.4%, respectively	Khandpur and Gogate (2015)
Green asparagus juice	HHP: 200–600 MPa for 10–20 min	Ascorbic acid, total phenolics, and antioxidant activity	Retention of 86.5%, 92.5%, and 138%, respectively	Chen *et al.* (2015)
Broccoli juice	HIPEF: 15–35 kV/cm for 500–2000 μs in bipolar or monopolar mode at 100 Hz and 4 μs	β-Carotene, lutein, vitamin C, total phenolics, and antioxidant capacity	Retention of 130.5%, 121.2%, 90.1%, 96.1%, and 95.9%, respectively	Sánchez-Vega *et al.* (2015)
Carrot juice	HHP: 550 MPa for 2–10 min	Lutein, α-carotene, β-carotene, total phenol content, and antioxidant capacity (DPPH)	Retention of 92.31%, 75.55%, 80.68%, 100.21%, and 97%, respectively	Zhang *et al.* (2016)

HHP, high hydrostatic pressure; HIPEF, high-intensity-pulsed electric fields; UST, ultrasound treatment; and UVT, ultraviolet treatment.

Table 4.4 Chronological studies on the physicochemical characteristics of vegetable juice through thermal and nonthermal processing

Vegetable juice	Processing conditions	Parameter	Outcomes	References
Tomato juice	92 °C for 90 s	trans-2-Hexenal, 2-isobutylthiazole, brown color, and 5-hydroxymethyl furfural	Retention of 98%, 99%, 0.34%, and 4.50%, respectively	Min and Zhang (2003)
Tomato juice	90 °C for 30 or 60 s	Color: parameter hue	No significant changes with respect to untreated juice	Aguiló-Aguayo et al. (2007)
Tomato juice	90 °C for 60 s	5-Hydroxymethyl furfural	2.9 mg/l	Aguiló-Aguayo et al. (2009a)
Cucumber juice	85 °C for 15 s	Color difference (ΔE)	1.4	Zhao et al. (2013)
Green asparagus juice	121 °C for 3 min	Color difference (ΔE)	1.74	Chen et al. (2015)
		Color: parameter L^*	Significant increment	
		pH	No significant changes with respect to untreated juice	
		Soluble solids		
Cucumber juice	110 °C for 8.6 s	Color difference (ΔE), pH, titratable acid, and soluble solids	No significant changes respect with to untreated juice	Liu et al. (2016)
Broccoli juice	HHP: 0.1–850 MPa at 30–90 °C	Color: parameter a^*	10% of greenness loss at 800 MPa and 50 °C	Weemaes et al. (1999)
Tomato juice	HIPEF: 40 kV/cm for 57 µs	trans-2-Hexenal, 2-isobutylthiazole, brown color, and 5-hydroxymethyl furfural	Retention of 110%, 108%, 0.29%, and 4.02%, respectively	Min and Zhang (2003)
Tomato juice	HIPEF: 35 kV/cm for 1500 µs. 4 µs bipolar pulses at 100 Hz	Color: parameter hue	No significant changes with respect to untreated juice	Aguiló-Aguayo et al. (2007)

(continued overleaf)

Table 4.4 (*continued*)

Vegetable juice	Processing conditions	Parameter	Outcomes	References
Carrot juice	HIPEF: 35 kV/cm for 1000 µs at 1–7 µs in bipolar or monopolar mode and 50–250 Hz	Color: parameters L^*, a^*, and b^* and browning index	No significant changes with respect to untreated juice	Quitão-Teixeira *et al.* (2007)
Formulated carrot juice	HIPEF: 27 kV/cm	pH, acidity, grades brix, conductivity, nonenzymatic browning, and color	No significant changes with respect to control were observed	Akin and Evrendilek (2009)
Tomato juice	HIPEF: 35 kV/cm for 1000 µs. 50–250 Hz and 1–7 µs	5-Hydroxymethyl furfural	2.1–2.8 mg/l	Aguiló-Aguayo *et al.* (2009a)
		Color difference (ΔE)	0.5–1.0	
Cucumber juice	HHP: 400 MPa for 4 min	Color difference (ΔE)	0.49 and 0.57	Zhao *et al.* (2013)
	500 MPa for 2 min			
Green asparagus juice	HHP: 200–600 MPa for 10–20 min	Color: parameter L^*	Significant diminution	Chen *et al.* (2015)
		pH	No significant changes with respect to untreated juice	
		Soluble solids		
Carrot juice	HHP: 300–500 MPa for 15 min at 40–60 °C	Color difference (ΔE)	Up to 5.29	Gong *et al.* (2015)
Cucumber juice	HHP: 500 MPa for 5 min	Color difference (ΔE)	0.75	Liu *et al.* (2016)
		pH	No significant changes with respect to untreated juice	
		Titratable acid		
		Soluble solids		

HHP, high hydrostatic pressure; HIPEF, high-intensity-pulsed electric fields.

Table 4.5 Chronological studies on the vegetable juice enzyme inactivation through thermal and nonthermal processing

Vegetable juice	Processing conditions	Enzyme studied	Grade of inactivation (%)	References
Tomato juice	92 °C for 90 s	Lipoxygenase	100	Min et al. (2003a,b)
Tomato juice	90 °C for 60 or 30 s	Peroxidase, polygalacturonase, and pectin methylesterase	90 or 79, 44 or 22, and 96 or 71, respectively	Aguiló-Aguayo et al. (2007)
Broccoli juice	60 °C for 10 min	Myrosinase	90	Van Eylen et al. (2007)
Carrot juice	90 °C for 60 s	Peroxidase	94.5	Quitao-Teixeira et al. (2009)
Cucumber juice	85 °C for 15 s	Lipoxygenase	41.77	Zhao et al. (2013)
Broccoli juice	90 °C fo- 60 s	Chlorophyllase	99.5	Sánchez-Vega et al. (2014)
Carrot juice	HHP: 6C0 MPa for 10 min	Polyphenol oxidase	89.9	Kim et al. (2001)
		Pectin methylesterase	Residual activity: 129	
Carrot juice	HHP: 600 MPa for 10 min	Peroxidase	63.4	Park et al. (2002)
		Polyphenol oxidase	87	
		Pectin methylesterase	56	
		Lipoxygenase	79	
Tomato juice	HIPEF: 40 kV/cm. Pulse duration time: 2 μs. Treatment time: 57 μs	Lipoxygenase	53	Min et al. (2003a,b)
Tomato juice	HIPEF: 10–35 kV/cm for 20–70 μs and 3 μs of pulse width	Lipoxygenase	80	Min et al. (2003a,b)

(continued overleaf)

Table 4.5 (*continued*)

Vegetable juice	Processing conditions	Enzyme studied	Grade of inactivation (%)	References
Carrot juice	HHP: 800 MPa for 36 min	Pectin methylesterase	90	Balogh et al. (2004)
Tomato juice	HIPEF: 35 kV/cm for 1500 μs. 4 μs bipolar pulses at 100 Hz	Peroxidase	97	Aguiló-Aguayo et al. (2007)
		Polygalacturonase	12	
		Pectin methylesterase	82	
Broccoli juice	HHP: 100–600 MPa at 10–60 °C	Myrosinase	86.8	Van Eylen et al. (2007)
Broccoli juice	HHP: 100–600 MPa at 10–60 °C	Myrosinase	86.8	Van Eylen et al. (2007)
Carrot juice	HIPEF: 35 kV/cm for 1000 μs at 1–7 μs in bipolar or monopolar mode and 50–250 Hz	Peroxidase	74.6	Quitão-Teixeira et al. (2007)
Tomato juice	HIPEF: 35 kV/cm for 1000–2000 μs. 1–7 μs bipolar or monopolar pulses at 50–250 Hz	Peroxidase	100	Aguiló-Aguayo et al. (2008)
Carrot juice	HIPEF: 35 kV/cm for 1500 μs. 6 μs bipolar pulses	Peroxidase	93	Quitao-Teixeira et al. (2009)

	Treatment	Enzyme	Residual activity:	Reference
Tomato juice	HIPEF: 35 kV/cm for 1000 μs. 1–7 μs bipolar or monopolar pulses at 50–250 Hz	Lipoxygenase Hydroperoxide lyase	20–125 6.7–105.2	Aguiló-Aguayo et al. (2009b)
Carrot–orange juice blend	UVT: 24.5 kJ/l	Pectin methylesterase	18	Caminiti et al. (2012)
Cucumber juice	HHP: 400 MPa for 4 min 500 MPa for 2 min	Lipoxygenase	20 14	Zhao et al. (2013)
Broccoli juice	HIPEF: 15–35 kV/cm for 500–2000 μs in bipolar or monopolar mode. 100 Hz and 4 μs	Chlorophyllase	73.7	Sánchez-Vega et al. (2014)
Carrot juice	HHP: 300–500 MPa for 15 min at 40–60 °C	Peroxidase	Up to 83.5	Gong et al. (2015)

HHP, high hydrostatic pressure; HIPEF, high-intensity-pulsed electric fields; and UVT, ultraviolet treatment.

observed in vegetable juices treated by heat. Similarly, thermal processing showed a great efficiency for inactivating oxidative enzymes (Table 4.5): lipoxygenase was inactivated 100%, peroxidase between 79% and 94%, poligalacturonase from 22% to 44%, pectin methylesterase 71–96%, myrosinase 90%, and chlorophyllase 99.5% from broccoli, carrot, cucumber, and tomato juices.

4.3　Nonthermal Processing Technologies

Currently, nonthermal technologies are undergoing extensive research with the purpose of obtaining harmless vegetable juices, with better fresh-like appearance and higher nutritional and organoleptic features, than those thermally treated (Aguiló-Aguayo et al., 2009a; Rodríguez-Roque, 2014). In addition, nonthermal processing technologies allow decreasing the processing cost (they utilize less energy than thermal technology) or adding value to the product (Barbosa-Cánovas et al., 2005).

Food can be nonthermally treated through different techniques such as irradiation, HHP, HIPEFs, light pulses and oscillating magnetic fields, ultrasound, and micro and ultrafiltration, among others (Zhang et al., 2011). The temperature is not the main factor for inactivating microorganism and enzymes in these treatments (only a slight increase in the temperature during processing has been documented) (Raso & Barbosa-Cánovas, 2003).

In vegetable juices, the most applied nonthermal technologies are HHP, HIPEF, high-pressure homogenization (HPH), UST, and UVT due to their capability for inactivating microorganisms and enzymes without compromising the fresh-like appearance, the nutritional, and sensorial qualities of these products (Odriozola-Serrano et al., 2013; Sánchez-Moreno et al., 2009). For all these reasons, nonthermal food preservation technologies are considered as an alternative to thermal treatments for obtaining highly nutritional and functional juices and beverages.

4.3.1　Influence of Nonthermal Processing on Microbial Quality

Nonthermal processing methods have demonstrated to meet the requirement of the FDA for reducing 5-log of microorganisms in vegetable juices. In this sense, HHP processing uses pressures in the range 100–600 MPa, the pressure transmitting medium usually being water, and it allows obtaining safe products with reduced processing time (Welti-Chanes et al., 2005). Pathogenic microorganisms, including bacteria, yeast, molds, and virus, can be inactivated by HHP. The inactivation of these microorganisms depends on processing parameters such as pressure, temperature, rate of pressure change, treatment time, and pressurization mode (single or in repeated cycles) (Cullen et al., 2012). Particularly, HHP has been applied for inactivating E. coli, Listeria innocua, mold, yeast, and mesophilic and aerobic bacteria from carrot, cucumber, and green asparagus juices processed between 200 and 600 MPa for 2–20 min (Table 4.2).

During HIPEF processing, a high-voltage electrical field (between 15 and 70 kV/cm) is applied across the food for a few microseconds. HIPEF processing induces microbial inactivation by applying an external electric field on cells, thus causing an irreversible electroporation (disruption of cell membrane) (Odriozola-Serrano et al., 2013). Microorganism inactivation depends on processing parameters such as electric field strength, temperature, flow rate, treatment time, pulse shape, pulse width, pulse polarity, and frequency. In addition, the composition of food, pH, and electrical conductivity of food also influence microorganism inactivation (Park *et al.*, 2014). This technology has been applied for inactivating mold, yeast, and *E. coli* from tomato and formulated carrot juice (Table 4.2), but no more information has been found on this regard.

UST produces breakdown of cell walls, disruption of membranes and damage of DNA of microorganisms due to a high localized temperature and pressure induced by cavitation (when ultrasonic waves propagate in liquid, small bubbles are formed and collapse thousands of times in a second) (O'Donnell *et al.*, 2010). However, a combination of sonication with mild heat or pressure is essential for reducing 5-log of microorganisms (Salleh-Mack & Roberts, 2007). Although US is considered as a promising technology for obtaining safe products, only few reports concerning vegetable juices (carrot and spinach juices) have been made for inactivating spoilage microorganisms such as *E. coli*, *S. aureus*, *Lactobacillus*, *S. enteritidis*, yeasts, and molds (Table 4.2).

Other nonthermal processing techniques used for inactivating microorganisms in vegetable juices are UVT and HPH.

UVT includes radiation from 100 to 400 nm and is divided into UV-A (320–400 nm), UV-B (280–320 nm), and UV-C (200–280 nm). In 2001, the U.S. FDA approved UV-C as treatment for juice products with the purpose to reduce human pathogens and other microorganisms. UV-C light has germicidal effect against bacteria, viruses, yeast, molds, protozoa, and algae by damaging their cell walls and DNA (Koutchma *et al.*, 2016). The degree of cell damage depends on UV dose applied, type of both microorganisms and medium (Ngadi *et al.*, 2003). The radiation source must consist of LPM lamps, which emit 90% of their UV-light at a wavelength of 253.7 nm for pasteurizing juices (Koutchma *et al.*, 2016). Reduction of 3-log CFU of total plate count and *E. coli* from carrot juice has been achieved using UVT (1.5 kJ/l), while reductions up to 5.5 logarithms were obtained when combining this technology with UST (Table 4.2). *S. aureus*, *Lactobacillus*, *S. enteritidis*, yeasts, and molds were reduced between 2.5 and 5.5 log when UVT and UST were combined for processing carrot and spinach juices.

In HPH, a liquid food is forced to pass through a narrow gap at high pressure (150–200 MPa) or at ultra-high-pressure homogenization (UHPH) (350–400 MPa). UHPH, also called dynamic high pressure, is a novel technology recently studied in foods for achieving inactivation of bacteria, yeast, fungi, endogenous flora, and even some viruses at a level equivalent at least to pasteurization (Dumay *et al.*, 2013). HPH destroys vegetative bacteria through mechanical disruption of the cell integrity, mainly caused by pressure and velocity gradients, turbulence, impingement, and cavitation (Donsì *et al.*, 2009). Common factors influencing microbial inactivation are microbial strains (i.e., Gram-negative bacteria are more sensitive to HPH than Gram-positive bacteria), pressure and temperature of homogenization, equipment, and medium of suspension, among others (Dielsa & Michiels, 2006). Regarding vegetable juices, 5 or more logarithmic reductions of *E. coli* and *L. innocua* have been achieved using HPH (250–300 MPa) in carrot juices (Table 4.2).

4.3.2 Influence of Nonthermal Processing on Nutritional Attributes

Nonthermal technologies have shown promising results for obtaining vegetable juices with high nutritional and healthy properties. The nutritional quality of vegetable juices is provided by the high levels of antioxidants such as lycopene (tomato juice), α- and β-carotenes (carrot, tomato, and green vegetables), and glucosinolates (Brassica), among others (Table 4.1).

In this sense, vegetable juices processed by HHP maintain a high concentration of diverse compounds with antioxidant properties (Table 4.3). For instance, no changes on chlorophyll concentration were observed in broccoli and cucumber juices processed by HHP. A high concentration of isothiocyanates was recovered from white and red cabbage juices, broccoli juice, cauliflower juice, and Brussels sprouts juice, all of them HHP processed. The vitamin C retention ranged from 86.5% to 94% in pressurized carrot and green asparagus juices. Carotenoids, such as lutein and α- and β-carotenes, recovered from carrot juice processed by HHP, were between 75% and 103%. The retention of total phenolic compounds and the antioxidant capacity ranged from 92.5% to 138% (processing details are shown in Table 4.3).

HIPEF processing preserves the sensory and nutritional characteristics of vegetable juices due to the very short processing time and low processing temperatures. Moreover, HIPEF facilitates the extraction of certain constituents by disrupting cell membranes and induces stress reactions in plant systems or cell cultures, resulting in an enhancement in the bioproduction of certain compounds (Soliva-Fortuny et al., 2009). For this reason, the retention of some vegetable juice antioxidants was higher than 100% (Table 4.3): 101% of vitamin A, up to 102% of phenolic compounds, 116% of chlorophyll a, 120.7% of chlorophyll b, 121.2% of lutein, β-carotene up to 130.5%, and lycopene up to 146%. On the other hand, the recovery of vitamin C ranged from 86.5% to 99%, while the antioxidant capacity was between 92.3% and 95.9% (see Table 4.3 for processing details).

Similar to HIPEF, UHPH also facilitates metabolite extraction. Emulsions formed during UHPH processing could entrap and protect poorly water-soluble bioactive components from unfavorable environmental conditions (i.e., pH, light, temperature, and oxygen) until consumption and digestion. Moreover, such delivery systems could release the entrapped/bound biomolecules to specific sites at controlled rates (Dumay et al., 2013). Unfortunately, no studies analyzing the influence of this technology on the nutritional quality of vegetable juices were found.

US has been considered as an effective, inexpensive, simple, and reproducible technology for improving the extraction of diverse food constituents. This technique may also increase the extent of hydroxylation, thus enhancing the antioxidant activity of compounds, such as flavonoids (Ashokkumar et al., 2008). At the same time, hydroxyl radicals can be produced during cavitation by oxidizing vegetable juice antioxidants (Soria & Villamiel, 2010). Beneficial or detrimental cavitation effect depends on both the processing conditions and the food matrix. Low frequency ultrasound is preferred when OH radicals adversely affect the vegetable juice antioxidants. As can be seen in Table 4.3, the retention of vitamin C was 60.4% and 74.7% in carrot and spinach juices treated by US, respectively. US-treated carrot juice showed a retention of 114.7% of total phenolic compounds, while in spinach juice, the recovery was 107.9%. The antioxidant capacity retention of spinach juice was 102.1% after US treatment.

According to Table 4.3, exposure to carrot juice to UV light resulted in 0–25% of degradation of vitamin C, while vitamin A, retinol, and β-carotene did not show significant differences before and after treatment. However, this treatment can adversely affect food quality by generating free radicals, thus oxidizing vitamins, proteins, antioxidants, and lipids. Adverse effects of this treatment depend on both food matrix and treatment conditions. For instance, vitamin A and β-carotene were oxidized only in the presence of visible light (Koutchma *et al.*, 2009). Closer examination of UV-light treatment on nutritional quality of vegetable juices is needed.

4.3.3 Influence of Nonthermal Processing on Organoleptic Attributes

The soluble solid content, pH, color, viscosity, flavor, and appearance, as well as the low concentration of oxidative enzymes, are factors that determine the organoleptic quality of vegetable juices. Given expectations for fresh-like appearance vegetable juices with similar organoleptic features than fresh and entire vegetables, the use of nonthermal processing technologies is in use.

HHP has low influence on color, aroma, flavor, and similar quality attributes because the uniform compression heating and the expansion cooling on decompression help to reduce the severity of processing on these parameters (Park *et al.*, 2014). Regarding the effect of HHP technology on vegetable juices, little changes in the color of broccoli and cucumber juices were observed; while the pH, titratable acid, and soluble solids of green asparagus and cucumber juices did not change (Table 4.4). HHP enables inactivation of oxidative enzymes, thus avoids undesirable changes (catalyzed by these enzymes) in taste and appearance of vegetable juices. According to Table 4.5, the grade of inactivation of vegetable juice enzymes processed by HHP was as follows: myrosinase from broccoli juice 86.8%, peroxidase from carrot juice between 63.4% and 89.9%, and lipoxygenase from carrot and cucumber juices 79% and 14–20%, respectively. Inactivation up to 90% of pectin methylesterase from carrot juice was obtained, but this enzyme also showed a residual activity of 129% after HHP treatment.

Overall, HIPEF processing has shown to maintain the aroma, color, and physicochemical features of vegetable juices, as well as inactivates oxidative enzymes (such as peroxidase, pectin methylesterase, and polygalacturonase), and avoids nonenzymatic browning (low 5-hydroxymethyl furfural) (Tables 4.4 and 4.5). HIPEF maintained the pH, acidity, grades brix, conductivity, color (L^*, a^*, b^*, Hue, and Chroma), and browning index at similar levels than untreated tomato and carrot juices. Similarly, no changes in the initial amount of 5-hydroxymethyl furfural were obtained in HIPEF-treated tomato juice. The concentration of volatile compounds *trans*-2-hexenal and 2-isobutylthiazole was up to 10% higher in HIPEF-treated tomato juice as compared with untreated juice. On the other hand, a low inactivation of some oxidative enzymes, such as lipoxygenase (residual activity between 20% and 124%), hydroperoxide lyase (residual activity ranging from 6.7% to 105.2%), and polygalacturonase (12% of inactivation), all from tomato juice, was observed. Despite the low inactivation of these enzymes, HIPEF processing satisfactorily inactivated the most important oxidative enzymes: peroxidase from tomato and carrot juices (inactivation between 74.6% and 100%), pectin methylesterase from tomato juice (inactivated 82%), and chlorophyllase from broccoli juice (73.7% of inactivation) (processing details are shown in Table 4.5).

Although no reports about the influence of UVT on the organoleptic quality of vegetable juices were found, it has been established that low doses may maintain the aroma, taste, and color at similar level than fresh products. On the contrary, high doses may induce adverse physicochemical changes by a wide range of photochemical reactions, such as changes in visual quality, texture, color, and enzyme activity, as well as the formation of off-flavors and aromas (Koutchma *et al.*, 2009). For instance, UV-C treatments (2.36, 4.74, and 14.22 kJ/m^2 at 254 nm) had little or no effects on the content of sugar and organic acids of lettuce (Allende *et al.*, 2006). Low UV-C doses (2.36 and 4.74 kJ/m^2 at 254 nm) did not damage the lettuce tissue, but softening and increased browning were found when applying 14.22 kJ/m^2. Similarly, only slight differences in overall appearance of "Lollo Rosso" lettuce treated with UV-C (between 0.81 and 4.06 kJ/m^2) were observed (Allende *et al.*, 2003). Nevertheless, the tissue of "Lollo Rosso" lettuce became brighter at 8.14 kJ/m^2 possibly due to the segregation of wax by the lettuce tissue as a protection against the UV-C stress.

Regarding oxidative enzymes, only one report was found to be focused on evaluating the influence of UVT on inactivation of enzymes from vegetable juices (Table 4.5). In this study, the residual activity of pectin methylesterase (catalyzes the hydrolysis of methyl ester groups from pectin and leads to the formation of a calcium pectate gel that causes cloud loss) from an orange–carrot juice blend was 82%. It has been established that enzyme inactivation depends on UVT dose, wavelength, and juice composition (Koutchma *et al.*, 2016).

4.4   Conclusion and Future Trends

Vegetable juices are becoming more popular since they are natural, healthy, easy to consume, and they have an appealing color and flavor. Moreover, they can be an outstanding way for incorporating nutrients, phytochemicals, antioxidants, fiber, and minerals into the diet. Thus, vegetable juices can be used as ingredient of new food products that fit into categories of functional foods and nutraceuticals.

The challenge of obtaining harmless vegetable juices has been overcome through both thermal and nonthermal processing technologies, which are efficient for reducing five logarithms of the most resistant and pathogenic microorganisms of vegetable juices.

The other main challenge has been the maintenance of high nutritional and organoleptic quality of vegetable juices: fresh and natural appearance, color, flavor, and taste, as much as freshly squeezed juice. Thermal treatment produces deleterious changes on nutritional and organoleptic features of processed vegetable juices, resulting in lower consumer preference. However, nonthermal processing technologies, mainly HIPEF and HHP, are promising approaches for obtaining vegetable juices with high nutritional and organoleptic quality.

Despite the nutritional value of vegetable juices, only few nonthermal processing techniques have been used by the beverage industry and food technologists for processing these products. Much more research is therefore needed for analyzing the influence of other nonthermal processing methods on bioactive constituents of vegetable juices. Furthermore, additional research is required for transferring the knowledge generated of these technologies at pilot sale to industry scale.

References

Aguiló-Aguayo, I., Soliva-Fortuny, R. & Martín-Belloso, O. (2007) Comparative study on color, viscosity and related enzymes of tomato juice treated by high-intensity pulsed electric fields or heat. *European Food Research and Technology* **227**, 599–606.

Aguiló-Aguayo, I., Soliva-Fortuny, R. & Martín-Belloso, O. (2009a) Avoiding non-enzymatic browning by high-intensity pulsed electric fields in strawberry, tomato and watermelon juices. *Journal of Food Engineering* **92**, 37–43.

Aguiló-Aguayo, I., Soliva-Fortuny, R. & Martín-Belloso, O. (2009b) Effects of high-intensity pulsed electric fields on lipoxygenase and hydroperoxide lyase activities in tomato juice. *Journal of Food Science* **74**, C595–C601.

Aguiló-Aguayo, I. *et al.* (2008) Inactivation of tomato juice peroxidase by high-intensity pulsed electric fields as affected by process conditions. *Food Chemistry* **107**, 949–955.

Akin, E. & Evrendilek, G.A. (2009) Effect of pulsed electric fields on physical, chemical, and microbiological properties of formulated carrot juice. *Food Science and Technology International* **15**, 275–282.

Allende, A., Padilla, E. & Artés, F. (2003) Changes in microbial and sensory quality of fresh processed UV-C treated "Lollo Rosso" lettuce. *Acta Horticulturae* **682**, 753–760.

Allende, A. *et al.* (2006) Effectiveness of two-sided UV-C treatments in inhibiting natural microflora and extending the shelf-life of minimally processed "Red Oak Leaf" lettuce. *Food Microbiology* **23**, 241–249.

Anese, M. *et al.* (1999) Antioxidant properties of tomato juice as affected by heating. *Journal of the Science of Food and Agriculture* **79**, 750–754.

Ashokkumar, M. *et al.* (2008) Modification of food ingredients by ultrasound to improve functionality: a preliminary study on a model system. *Innovative Food Science & Emerging Technologies* **9**, 155–160.

Ball, G.F.M. (ed) (2006) *Vitamins in Foods. Analysis, Bioavailability, and Stability*. Taylor & Francis Group, Boca Raton, FL, USA.

Balogh, T. *et al.* (2004) Thermal and high-pressure inactivation kinetics of carrot pectin-methylesterase: from model system to real foods. *Innovative Food Science and Emerging Technologies* **5**, 429–436.

Barbosa-Cánovas, G.V. *et al.* (2005) Use of magnetic fields as a nonthermal technology. In: *Novel Food Processing Technologies* (eds G.V. Barbosa-Cánovas *et al.*), pp. 443–452. Taylor & Francis Group, Boca Raton, FL, USA.

Boffetta, P. *et al.* (2010) Fruit and vegetable intake and overall cancer risk in the european prospective investigation into cancer and nutrition (EPIC). *Journal of the National Cancer Institute* **102**, 529–537.

Boz, Z. & Erdoğdu, F. (2016) Thermal processing: canning and aseptic processing. In: *Handbook of Vegetable Preservation and Processing* (eds Y.H. Hui & E.Ö. Evranuz), pp. 157–174. Taylor & Francis Group, Boca Raton, FL, USA.

Caminiti, I.M. *et al.* (2012) The effect of pulsed electric fields, ultraviolet light or high intensity light pulses in combination with manothermosonication on selected physico-chemical and sensory attributes of an orange and carrot juice blend. *Food and Bioproducts Processing* **90**, 442–448.

Chen, X. *et al.* (2015) Effect of high pressure processing and thermal treatment on physicochemical parameters, antioxidant activity and volatile compounds of green asparagus juice. *LWT – Food Science and Technology* **62**, 927–933.

Cullen, P.J.K., Tiwari, B.K. & Valdramidis, V.P. (2012) Status and trends of novel thermal and non-thermal technologies for fluid foods. In: *Novel Thermal and Non-Thermal Technologies for Fluid Foods* (eds P.J. Cullen, B.K. Tiwari, & V.P. Valdramidis), pp. 1–6. Elsevier Academic Press, San Diego, CA, USA.

Dielsa, A.M.J. & Michiels, C.W. (2006) High-pressure homogenization as a non-thermal technique for the inactivation of microorganisms. *Critical Reviews in Microbiology* **32**, 201–216.

Donsì, F. *et al.* (2009) Main factors regulating microbial inactivation by high-pressure homogenization: operating parameters and scale of operation. *Chemical Engineering Science* **64**, 520–532.

Dumay, E. *et al.* (2013) Technological aspects and potential applications of (ultra) high-pressure homogenisation. *Trends in Food Science & Technology* **31**, 13–26.

Falguera, V. & Ibarz, A. (eds) (2014) *Juice processing. Quality, Safety, and Value-Added Opportunities.* Taylor & Francis Group.

FAO/WHO. (2003) Diet, nutrition and the prevention of chronic diseases. Report of a Joint FAO/WHO Expert Consultation. Geneva, World Health Organization, WHO Technical Report Series.

Food and Drug Administration (2004). Validated pasteurization treatments for juice. In: Guidance for industry: Juice HACCP Hazards and Controls Guidance. First Edition. Retrieved from: https://www.fda.gov/food/guidanceregulation/ucm072557.htm#v.

Freedman, N.D. *et al.* (2008) Fruit and vegetable intake and head and neck cancer risk in a large United States prospective cohort study. *International Journal of Cancer* **122**, 2330–2336.

Gong, Y. *et al.* (2015) Comparative study of the microbial stability and quality of carrot juice treated by high-pressure processing combined with mild temperature and conventional heat treatment. *Journal of Food Process Engineering* **38**, 395–404.

Heidemann, C. *et al.* (2008) Dietary patterns and risk of mortality from cardiovascular disease, cancer, and all causes in a prospective cohort of women. *Circulation* **118**, 230–237.

Hjelmqwist, J. (2005) Commercial high-pressure equipment. In: *Novel Food Processing Technologies* (eds G.V. Barbosa-Cánovas, M.S. Tapia & M.P. Cano), CRC Press, Boca Raton, FL, USA, pp. 361–373.

Hui, Y.H. & Evranuz, Ö. (eds) (2016) *Handbook of Vegetable Preservation and Processing,* 2nd edition, Taylor & Francis Group, Boca Raton, FL, USA.

Khandpur, P. & Gogate, P.R. (2015) Effect of novel ultrasound based processing on the nutrition quality of different fruit and vegetable juices. *Ultrasonics Sonochemistry* **27**, 125–136.

Khandpur, P. & Gogate, P.R. (2016) Evaluation of ultrasound based sterilization approaches in terms of shelf life and quality parameters of fruit and vegetable juices. *Ultrasonics Sonochemistry* **29**, 337–353.

Kim, Y.S. *et al.* (2001) Effects of combined treatment of high hydrostatic pressure and mild heat on the quality of carrot juice. *Journal of Food Science* **66**,1355–1360.

Koutchma, T., Forney, L.J. & Moraru, C.I. (eds) (2009) *Ultraviolet Light in Food Technology: Principles and Applications.* Taylor & Francis Group, Boca Raton, FL, USA.

Koutchma, T. *et al.* (2016) Effects of ultraviolet light and high-pressure processing on quality and health-related constituents of fresh juice products. *Comprehensive Reviews in Food Science and Food Safety* **00**, 1–24.

Liu, F. *et al.* (2016) Potential of high-pressure processing and high-temperature/short-time thermal processing on microbial, physicochemical and sensory assurance of clear cucumber juice. *Innovative Food Science & Emerging Technologies* **34**, 51–58.

Mandelová, L. & Totušek, J. (2007) Broccoli juice treated by high pressure: chemoprotective effects of sulforaphane and indole-3-carbinol. *High Pressure Research* **27**, 151–156.

Martín-Belloso, O. *et al.* (2014) Non-thermal processing technologies. In: *Food Safety Management. A Practical Guide for the Food Industry.* (eds Y. Motarjemi & H. Lelieveld), pp. 443–465. Academic Press Inc..

Min, S., Jin, Z.T. & Zhang, Q.H. (2003a) Commercial scale pulsed electric field processing of tomato juice. *Journal of Agricultural and Food Chemistry* **51**(11), 3338–44.

Min, S., Min, S.K. & Zhang, Q.H. (2003b) Inactivation kinetics of tomato juice lipoxygenase by pulsed electric fields. *Journal of Food Science* **68**, 1995–2001.

Min, S. & Zhang, Q.H. (2003) Effects of commercial-scale pulsed electric field processing on flavor and color of tomato juice. *Journal of Food Science* **68**, 1600–1606.

Morris, M.C. *et al.* (2006) Associations of vegetable and fruit consumption with age-related cognitive change. *Neurology* **67**, 1370–1376.

Ngadi, M., Smith, J.P. & Cayouette, B. (2003) Kinetics of ultraviolate light inactivation of *Escherichia coli* O157:H7 in liquid foods. *Journal of the Science of Food and Agriculture* **83**, 1551–1555.

O'Donnell, C.P. *et al.* (2010) Effect of ultrasonic processing on food enzymes of industrial importance. *Trends in Food Science & Technology* **21**, 358–367.

Odriozola-Serrano, I., Soliva-Fortuny, R. & Martín-Belloso, O. (2008) Changes of health-related compounds throughout cold storage of tomato juice stabilized by thermal or high intensity pulsed electric field treatments. *Innovative Food Science & Emerging Technologies* **9**, 272–279.

Odriozola-Serrano, I. *et al.* (2007) Lycopene, vitamin C, and antioxidant capacity of tomato juice as affected by high-intensity pulsed electric fields critical parameters. *Journal of Agricultural and Food Chemistry* **55**, 9036–9042.

Odriozola-Serrano, I. *et al.* (2009) Carotenoid and phenolic profile of tomato juices processed by high intensity pulsed electric fields compared with conventional thermal treatments. *Food Chemistry* **112**, 258–266.

Odriozola-Serrano, I. *et al.* (2013) Pulsed electric fields processing effects on quality and health-related constituents of plant-based foods. *Trends in Food Science & Technology* **29**, 98–107.

Ortega-Rivas, E. & Salmerón-Ochoa, I. (2014) Nonthermal food preservation alternatives and their effects on taste and flavor compounds of beverages. *Critical Reviews in Food Science and Nutrition* **54**, 190–207.

Park, S.H., Lamsal, B.P. & Balasubramaniam, V.M. (2014) Principles of food processing. In: *Food Processing: Principles and Applications* (eds S. Clark, S. Jung, & B. Lamsal), pp. 1–15. Wiley-Blackwell.

Park, S.-J., Lee, J.I. & Parj, J. (2002) Effects of a combined process of high-pressure carbon dioxide and high hydrostatic pressure on the quality of carrot juice. *Journal of Food Science* **67**, 1827–1834.

Pathanibul, P. *et al.* (2009) Inactivation of *Escherichia coli* and *Listeria innocua* in apple and carrot juices using high pressure homogenization and nisin. *International Journal of Food Microbiology* **129**, 316–20.

Quitão-Teixeira, L.J. *et al.* (2007) Inactivation of oxidative enzymes by high-intensity pulsed electric field for retention of color in carrot juice. *Food and Bioprocess Technology* **1**, 364–373.

Quitao-Teixeira, L.J. *et al.* (2009) Comparative study on antioxidant properties of carrot juice stabilised by high-intensity pulsed electric fields or heat treatments. *Journal of the Science of Food and Agriculture* **89**, 2636–2642.

Raso, J. & Barbosa-Cánovas, G.V. (2003) Nonthermal preservation of foods using combined processing techniques. *Critical Reviews in Food Science and Nutrition* **43**, 265–285.

Rodríguez-Roque, M.J. (2014) *In vitro bioaccessibility of health-related compounds from beverages based on fruit juice, milk or soymilk: Influence of food matrix and processing.* Doctoral thesis, University of Lleida.

Rodríguez-Roque, M.J., Morales-de la Peña, M. & Martín-Belloso, O. (2014) Bebidas funcionales. In: *Los Alimentos Funcionales: Un Nuevo Reto Para la Industria de Alimentos.* (eds G.A. González-Aguilar, *et al.*), pp. 687. AGT Editor, Mexico.

Rodríguez-Roque, M.J. *et al.* (2015) Impact of food matrix and processing on the in vitro bioaccessibility of vitamin C, phenolic compounds, and hydrophilic antioxidant activity from fruit juice-based beverages. *Journal of Functional Foods* **14**, 33–43.

Salleh-Mack, S.Z. & Roberts, J.S. (2007) Ultrasound pasteurization: the effects of temperature, soluble solids, organic acids and pH on the inactivation of *Escherichia coli* ATCC 25922. *Ultrasonics Sonochemistry* **14**, 323–329.

Sánchez-Moreno, C. *et al.* (2009) Nutritional approaches and health-related properties of plant foods processed by high pressure and pulsed electric fields. *Critical Reviews in Food Science and Nutrition* **49**, 552–76.

Sánchez-Vega, R., Elez-Martínez, P. & Martín-Belloso, O. (2014) Effects of high-intensity pulsed electric fields processing parameters on the chlorophyll content and its degradation compounds in broccoli juice. *Food and Bioprocess Technology* **7**, 1137–1148.

Sánchez-Vega, R., Elez-Martínez, P. & Martín-Belloso, O. (2015) Influence of high-intensity pulsed electric field processing parameters on antioxidant compounds of broccoli juice. *Innovative Food Science & Emerging Technologies* **29**, 70–77.

Soliva-Fortuny, R. *et al.* (2009) Effects of pulsed electric fields on bioactive compounds in foods: a review. *Trends in Food Science & Technology* **20**, 544–556.

Soria, A.C. & Villamiel, M. (2010) Effect of ultrasound on the technological properties and bioactivity of food: a review. *Trends in Food Science & Technology* **21**, 323–331.

Tomás-Barberán, F.A. & Gil, M.I. (eds) (2008) *Improving the Health-Promoting Properties of Fruit and Vegetable Products*, Woodhead Publishing Limited, Cambridge, England.

Torregrosa, F. *et al.* (2005) Effect of high-intensity pulsed electric fields processing and conventional heat treatment on orange–carrot juice carotenoids. *Journal of Agriculture and Food Chemistry* **53**, 9519–9525.

Torregrosa, F. *et al.* (2006) Ascorbic acid stability during refrigerated storage of orange–carrot juice treated by high pulsed electric field and comparison with pasteurized juice. *Journal of Food Engineering* **73**(4), 339–345.

Tříska, J. *et al.* (2007) Comparison of total isothiocyanates content in vegetable juices during high pressure treatment, pasteurization and freezing. *High Pressure Research* **27**, 147–149.

USDA (2016) *USDA Definition of Specialty Crop*, Available: http://www.ams.usda.gov /AMSv1.0/getfile?dDocName=STELPRDC5082113 [30 Mar 2016].

van Duijnhoven, F.J.B. *et al.* (2009) Fruit, vegetables, and colorectal cancer risk: the European prospective investigation into cancer and nutrition. *The American Journal of Clinical Nutrition* **89**, 1441–52.

Van Eylen, D. *et al.* (2007) Kinetics of the stability of broccoli (*Brassica oleracea* cv. italica) myrosinase and isothiocyanates in broccoli juice during pressure/temperature treatments. *Journal of Agricultural and Food Chemistry* **55**, 2163–2170.

Van Loey, A. *et al.* (1998) Thermal and pressure-temperature degradation of chlorophyll in broccoli (*Brassica oleracea* L. italica) juice: a kinetic study. *Journal of Agricultural and Food Chemistry* **46**, 5289–5294.

Vasantha Rupasinghe, H.P. & Yu, L.J. (2012) Emerging preservation methods for fruit juices and beverages. In: *Food Additive* (ed Y. El-Samragy), pp. 65–82. InTech, Janeza Trdine, Croatia.

Villegas, R. *et al.* (2008) Vegetable but not fruit consumption reduces the risk of type 2 diabetes in Chinese women. *Journal of Nutrition* **138**, 574–580.

Wang, S. *et al.* (2011) How natural dietary antioxidants in fruits, vegetables and legumes promote vascular health. *Food Research International* **44**, 14–22.

Weemaes, C. *et al.* (1999) Pressure–temperature degradation of green color in broccoli juice. *Journal of Food Science* **64**, 504–508.

Welti-Chanes, J. *et al.* (2005) Fundamentals and applications of high pressure processing to foods. In: *Novel Food Processing Technologies* (eds G.V. Barbosa-Cánovas *et al.*), pp. 157–181. Taylor & Francis Group, Boca Raton, FL, USA.

Wootton-Beard, P.C., Moran, A. & Ryan, L. (2011) Stability of the total antioxidant capacity and total polyphenol content of 23 commercially available vegetable juices before and after in vitro digestion measured by FRAP, DPPH, ABTS and Folin-Ciocalteu methods. *Food Research International* **44**, 217–224.

Zenker, M., Hienz, V., & Knorr, D. (2003) Application of ultrasound-assisted thermal processing for preservation and quality retention of liquid foods. *Journal of food Protection* **66**, 1642–1649.

Zhang, H.Q. *et al.* (eds) (2011) *Nonthermal Processing Technologies for Food.* Wiley-Blackwell.

Zhang, Y. *et al.* (2016) Quality comparison of carrot juices processed by high-pressure processing and high-temperature short-time processing. *Innovative Food Science and Emerging Technologies* **33**, 135–144.

Zhao, L. *n* (2013) Comparing the effects of high hydrostatic pressure and thermal pasteurization combined with nisin on the quality of cucumber juice drinks. *Innovative Food Science* **17**, 27–36.

5

Exotic Fruit Juices

Zamantha Escobedo-Avellaneda, Rebeca García-García, and Jorge Welti-Chanes

[1]*Tecnológico de Monterrey, Escuela de Ingeniería y Ciencias, Centro de Biotecnología FEMSA, Tecnológico, Monterrey, Nuevo León, Mexico*

5.1 Introduction

Fruits are important to the human diet, they contribute greatly to the vitamin, mineral, and dietary fiber intake. Recently, the interest in fruits has increased due to their association with chronic disease treatment and prevention. The health-promoting and disease-preventing effects of fruits are related with their content of functional compounds also known as nutraceuticals, bioactive compounds, or phytochemicals. Despite their limited regional presence, exotic fruits are a very interesting group of fruits because of their pleasant sensory characteristics and their healthy properties (Devalaraj *et al.*, 2011). Acai, banana, guava, guanabana, jackfruit, kiwi, litchi, mamey, mango, mangosteen, papaya, passion fruit, pineapple, pomegranate, and prickly pear are among these fruits. Exotic fruits are normally classified as tropical (e.g., banana, mango, papaya, pineapple, guava, litchi, mangosteen, and passion fruit) and subtropical fruits (e.g., kiwi and pomegranate) due to the geographical region in which they are grown (Kader and Barrett, 2005).

Acai, native to northern South America, is a fruit with a high antioxidant activity (AOA). It is normally used to prepare smoothies, pulps, and a variety of beverages. Banana, thought to be originated from Malaysia, is a good source of carbohydrates, potassium, calcium, and β-carotene. It is typically consumed fresh, as puree or frozen to formulate other products (Mohapatra *et al.*, 2011; Nakasone & Paull, 1998). Guava, native to the American tropics, is an important subtropical fruit with high AOA and phenolic content (Nakasone & Paull, 1998). In the ancient Chinese Medicine, guava was used to treat diabetes and other chronic diseases (Luximon-Ramma *et al.*, 2003), and it is mainly consumed as nectar, juice, jam and canned halves, and slices (Chan, 1993). Guanabana originated from American tropics and Caribbean, and it is extensively grown in Mexico (Nakasone & Paull, 1998). It is commonly consumed as ice creams, pulp, and nectar (adjusted to pH 3.7 and 15 °Brix), and it is known as

*Corresponding author: Jorge Welti-Chanes, jwelti@itesm.mx

Innovative Technologies in Beverage Processing, First Edition.
Edited by Ingrid Aguiló-Aguayo and Lucía Plaza.
© 2017 John Wiley & Sons Ltd. Published 2017 by John Wiley & Sons Ltd.

a good source of vitamin K, riboflavin, and niacin. Jackfruit is indigenous to India, and it is widely grown in Bangladesh, Burma, Malaysia, Philippines, and Brazil. It is consumed both as a vegetable when unripe and as a fruit when ripe. It is also prepared as canned frozen fruit, jam puree, dehydrated bulbs, candies, concentrates, and powder (Narasimham, 1990). Kiwi, native to Northern China, is a good source of vitamin C, dietary fiber, and flavonoids. Kiwi seed contains about 62% of α-linolenic acid. The main products obtained from kiwi are puree, juice, dried products, and oils with high content of omega-3 (http://www.kiwifruitz.co.nz/). Litchi, indigenous to China, is a source of anthocyanins and other functional compounds (Nakasone & Paull, 1998). Mamey, native to Central America, grows from southern Mexico to southern Costa Rica. The fruit is eaten raw or used to prepare smoothies, ice cream, fruit bars, and jellies. Mango, a source of tannins and carotenoids native to South and Southeast Asia, is one of the most important tropical fruits. It is processed in nectars, purees, jams, and jellies, and it is also used in dairy and bakery products (Swi-Bea et al., 1993). Mongosteen, a fruit with high content of phenolic acids, tannins, xanthones, and anthocyanins, can be found in India, Myanmar, Sri Lanka, and Thailand (Xie et al., 2015). Passion fruit is native to southern Brazil and distributed to South America, Caribbean, Asia, Africa, India, and Australia. It is a good source of provitamin A, niacin, and ascorbic acid, and it is used to obtain juice and puree (Nakasone & Paull, 1998). Papaya, native to Central America, is widely used to obtain the enzyme papain. It contributes to the vitamin requirements and is also a good source of carotenoids. Puree and nectar are the main products elaborated from papaya, but jams, jellies, syrups, toppings, and dried products are also found (Chan, 1993). Pineapple, a fruit probably originated in South America, is a good source of vitamin C and carotenoids. Pineapple is highly consumed as canned slices, but it is also used to obtain single strength juice, concentrated juice, juice beverages, wine, wine coolers, drink mixes, frozen foodstuffs, and dairy products. It is also used in backed goods and as a flavoring agent in syrups (Hodgson and Hodgson, 1993). Pomegranate is a fruit native to southern Asia and/or the Middle East. It is eaten fresh, as juice, grenadine, wine, and desserts. Prickly pear is a fruit native to Mexico normally consumed as raw fruit of used to obtain juices. Prickly pear is a good source of vitamin C, fiber, and flavonoids (Jiménez-Aguilar et al., 2014).

Industrial production of fruit juices has increased due to their convenience and high content of functional compounds. The shelf life of fruits and derived products is limited due to the microbiological, biochemical, and enzymatic reactions constantly taking place. So, the application of the most adequate preservation method to increase stability and also to make fruits available in off-seasons and in regions outside from their origin, is necessary. Thermal treatments, such as pasteurization, sterilization, and concentration, are traditionally used to increase shelf life of juices, but at certain processing conditions, they could cause detrimental changes in overall quality. As a result of the increased consumer demand for minimally processed juices with fresh-like characteristics (sensory and nutritional quality), nonthermal methods have arisen. High hydrostatic pressure (HHP), pulsed electric fields (PEF), ultrasound, and UV-C irradiation are among these technologies. In this chapter, the traditional thermal processing techniques and new nonthermal preservation technologies with recent or potential application in exotic fruit juices are reviewed.

5.2   Exotic Fruits: Relevance in Human Nutrition and Health

Exotic fruits are not easily accessible to international markets, and they are sold at higher cost than most common fruits. Nevertheless, the interest to reach international markets has increased due to their appreciable sensory and nutritional attributes. Antioxidants of fruits such as carotenoids, phenolics, and flavonoids, among others have been associated with health benefits including lowering risk of chronic diseases such as cancer, heart disease, stroke, and others (De Souza *et al.*, 2010; Kader & Barrett, 2005). The color of fruits is an indicative of the type of phytochemicals that they content. Red/blue/purple fruits like pomegranate are good sources of flavonoids such as anthocyanins that contribute to their high AOA. Yellow and orange fruits such as mango, papaya, pineapple, and some red fruits are good sources of carotenoids. Exotic fruits such as guava, kiwi, litchi, and papaya are better sources of vitamin C than orange juice, and, in general, they have high content of phenolic compounds and provide high AOA. They also have high contents of fiber also with health benefices. Some fruits such as guava, mango, papaya, and passion fruit contribute greatly to the β-carotene content of the human diet, important because of its role as a vitamin A precursor (Table 5.1).

Many beneficial effects have been associated with the consumption of acai. Among these are included antidiarrheal and hypocholesterolemic activities (De Souza *et al.*, 2010; Kang *et al.*, 2010; Mertens-Talcott *et al.*, 2008). De Souza *et al.* (2010) suggested that the consumption of acai pulp during 6 weeks improves the antioxidant status in rats causing a hypocholesterolemic effect. They noted a reduction of the total and non-high-density-lipoprotein cholesterol and also in the superoxide dismutase activity. Phenolics (5–40 mg GAE/l) from acai protect cells against radical production induced by hyperglycemic conditions (25 mM glucose) and reversed glucose-induced intracellular and secretory inflammatory biomarkers (Noratto *et al.*, 2011). The xanthones, α-mangostin and γ-mangostin, are the major bioactive compounds in mangosteen fruit extracts, and they have AOA (Xie *et al.*, 2015). Antidepressive effects possibly induced by some alkaloids (annonaine, nornuciferine, and asimilobine) have been reported for guanabana pulp (Hasrat *et al.*, 1997). Many medical properties have been attributed to pomegranate (Crane & Campbell, 1990; Kumar, 1990). It has been shown that flavonoids from pomegranate inhibit low-density-lipoprotein oxidation and cardiovascular diseases through direct interaction of the flavonoids with lipoproteins and/or indirectly through accumulation of flavonoid in arterial macrophages (Aviram *et al.*, 2002). Antioxidant, anticarcinogenic, and antinflammatory properties of pomegranate compounds have also been published (Jurenka, 2008). It was shown that the consumption of prickly pear positively influence the redox balance of the body, decreasing oxidative damage of lipids, and reducing oxidative stress in healthy humans (Tesoriere *et al.*, 2004). Pectin presented in prickly pear fruits decreases plasma LDL concentrations by increasing hepatic apolipoprotein B/E receptor expression in guinea pigs fed with a hypercholesterolemic diet (Fernandez *et al.*, 1994). It decreases the total cholesterol (12%), low-density-lipoprotein cholesterol (15%), apolipoprotein B (9%), triglycerides (12%), fibrinogen (11%), blood glucose (11%), insulin (11%), and uric acid

Table 5.1 Water content and phytochemical composition of raw exotic fruits per each 100 g edible pulp (USDA, https://ndb.nal.usda.gov/ndb/nutrients/index, www.ars.usda.gov/nutrientdata/orac)

	Water (g)	Total fiber (g)	Total vitamin C (mg)	Vitamin A (µg)	β-Carotene (µg)	AOA (µmol TE)	TP (mg GAE)
Acai (*Euterpe oleracea*)	–	–	–	–	–	102,700	1390
Banana (*Musa paradisiaca*)	75	2.6	8.7	3	26	795	155
Common guava (*Psidium guajava*)	81	5.4	228.3	31	374	1,422	136
Guanabana[a] (*Annona muricata*)	85	0.1	11.1	0	0		
Jackfruit[a] (*Artocarpus heterophyllus*)	73	1.5	13.7	0	61		
Kiwi (*Actinidia deliciosa*)	83	3.0	92.7	4	52	862	211
Litchi (*Litchi chinensis*)	82	1.3	71.5	0	0		
Mamey (*Pouteria sapota*)	86	3.0	14.0	12	–		
Mango (*Mangifera indica*)	84	1.6	36.4	54	640	1,300	101
Mangosteen[a] (*Garcinia mangostana*)	81	1.8	2.9	2	16	2,510	85
Papaya (*Carica papaya*)	88	1.7	60.9	47	274	300	54
Passion fruit (*Passiflora edulis*)	84	0.2	18.2	47	525		
Pineapple (*Ananas comosus*)	86	1.4	47.8	3	35	562	122
Pomegranate (*Punica granatum*)	78	4.0	10.2	0	0	4,479	338
Prickly pear (*Opuntia ficus-indica*)	88	3.6	14.0	2	25		
Red guava	–	–	–	–	–	1,990	247
White guava	–	–	–	–	–	2,550	345

Source: USDA, https://ndb.nal.usda.gov/ndb/nutrients/index, www.ars.usda.gov/nutrientdata/orac.
AOA, antioxidant activity (ORAC method); TE, trolox equivalents; GAE, gallic acid equivalents; and TP, total phenolic.
[a]Canned nectar or syrup.

(10%), while body weight, high-density-lipoprotein cholesterol, apolipoprotein A-I, and lipoprotein(a) remained unchanged in nondiabetics with hyperlipidemia patients (Wolfram *et al.*, 2002).

In order to provide health benefices, functional compounds of fruits should be bioavailable; this implicates that the preservation process must not destroy them. The degree of bioavailability of each phytochemical is determined by its specific chemical properties and also for its ability to be released from foods during the digestion process. The release of functional compounds and nutrients from foods can be facilitated by the technological processes at which they are submitted. Among these operations are found cutting, chopping, and heating. However, technologies such as HHP, PE, and ultrasound can also contribute to their release from food matrices probably due to cellular disruption and/or due to a direct effect on the interactions between these compounds and macronutrients making them more bioavailable (Corrales *et al.*, 2009; de Ancos *et al.*, 2002; Escobedo-Avellaneda *et al.*, 2015; Hyun-Sun *et al.*, 2011; Kuo-Chiang *et al.*, 2008; Qiu *et al.*, 2006; Sánchez-Moreno *et al.*, 2005).

There are some studies showing the potential of HHP and PEF to increase the bioavailability of some nutraceuticals (Sánchez-Moreno *et al.*, 2003, 2006), but there is no information about exotic fruit juices.

5.3   Deterioration of Exotic Fruit Juices

Exotic fruits as well as other food systems are constantly exposed to biochemical and microbiological reactions that cause deterioration and limit their shelf life. Respiration, transpiration or water loss, and ethylene production are biological processes implicated in the deterioration of fruits. During respiration, carbohydrates and other nutrient are oxidized producing energy, water, and CO_2. Water loss is regulated by the plant dermal system and it not only causes loss of weight with economic implications but also softening, loss of crispness, juiciness, and nutritional quality. Morphological and anatomical characteristics of plants and factors to which they are exposed like temperature and relative humidity influence the rate of water loss (Kader & Barrett, 2005). Ethylene regulates many aspects of growth, development and senescence of plants, so the rate of their production influences the rate of deterioration of fruits. Based on their ethylene production and respiration rate, fruits are classified as climacteric and nonclimacteric fruits. In climacteric fruits, the respiration rate and ethylene production increase during storage, and therefore, they continue ripening after harvesting, while nonclimacteric fruits do not continue ripening once harvested and produce very small amounts of ethylene. Banana, guava, mango, papaya, and passion fruit are considered as climacteric fruits, while litchi, mangosteen, and pineapple are nonclimacteric. The respiration rate of fruits is a very important factor because it determines the postharvest life. As respiration rate increases, the postharvest life of fruits decreases. Pineapple and carambola have very low respiration rates (<35 mg/kg-h); green banana, litchi, papaya, jackfruit, passion fruit, and mangosteen present a low respiration rate (35–70); mango and guava have moderate rates (70–150); and ripe banana has a high respiration rate (150–300) (Nakasone and Paull, 1998). The control of these processes by adequate postharvest practices is very important to preserve the quality of fruits to be consumed as fresh products or to obtain juices.

Microbiological growth and enzyme activity are implicated in the deterioration of fruit juices, so they should be controlled by adequate preservation methods trying to preserve the sensory, nutritional, and functional quality of foods. An important factor affecting microbial spoilage of juices is the pH, so the design of preservation processes should take into account the pH of the juice in order to determine the most adequate treatment. The acids present in juices represent an advantage in processing because it reduces the intensity of the treatment required (Ramaswamy, 2005). Contamination with yeasts is the main concern in the fruit juice industry, but due to the high acidity of juices, lactic and acetic acid bacteria can also cause contamination problems. Molds such as *Byssochlamys*, *Talaromyces*, and *Neosartorya* also can grow in juices, and they produce ascospores that can tolerate some processing conditions; nevertheless, the level of ascospore contamination is usually low because they are eliminated during the washing process or they do not survive to the high acidity of most fruits. At least a 5-$\log_{10}$ reduction in the counts of pathogenic microorganisms such as *Escherichia coli* O157:H7, *Cryptosporidium parvum*, and *Salmonella* spp. must be achieved to assure safety of juices (McLellan & Padilla-Zakour, 2005).

Enzymes such as polyphenoloxidase (PPO), pectin methylesterase (PME), peroxidase (POD), lipoxygenase, and catalase are the main responsible of juice quality changes (color, texture, and flavor) during storage (Höhn *et al.*, 2005; Ramaswamy, 2005). In the treatment of high acidity juices, the inactivation of enzymes is often used as the design criteria for pasteurization and sterilization processes because they are usually more heat resistant than microorganisms. The peroxidase activity is frequently used as indicator for the destruction of heat-resistant enzymes (Ramaswamy, 2005).

5.4   Thermal and Nonthermal Technologies Used to Preserve Juices

Pasteurization and sterilization are two of the main methods used to preserve foods. Juice pasteurization is usually done by heat exchangers at temperature from 70 to 90 °C during 15 s to 3 min (hot filling) or by heating juices previously bottled, in continuous or in batch pasteurizer systems. Most pasteurized juices require refrigeration for storage and distribution Juice sterilization typically performed with aseptic systems, in which the juice and the container are sterilized separately and finally the containers are filled aseptically. The use of high temperature can produce sterile products with lower changes in quality as compared with more extreme conditions (McLellan & Padilla-Zakour, 2005).

Juice concentration is another important operation in the fruit processing industry, and it reduces storage volume requirements and transportation costs. Due to a_w reduction, concentration diminishes the probability of yeast growth. Single and multiple effect evaporators can be used to concentrate juices (McLellan & Padilla-Zakour, 2005). Usually, concentration must be combined with other preservation factors such as sugar addition, preservatives, or refrigeration to ensure product safety.

New preservation methods based on nonthermal principles have also been applied to preserve fruit juices with the aim of maintaining fresh-like characteristics. Some of these technologies include HHP, PEF, ultrasound, and UV-C irradiation (McLellan & Padilla-Zakour, 2005).

5.4.1 Thermal Processing

Under extreme processing conditions, thermal treatment of juices negatively affects their nutritional and sensory quality. Pasteurization is a process by which pathogenic microorganisms, spoilage bacteria, yeasts, and molds are destroyed due to the exposure of juices to temperature (Lozano, 2006). Low temperature–long time (63 °C, 30 min), high temperature–short time (HTST) (75 °C, 15 s), and ultrahigh temperature pasteurization (UHT) (>121 °C, 1–2 s) are pasteurization methods typically used (US Department of Health and Human Services, 2004). These techniques could cause some negative changes in flavor and aroma if not adequately applied, resulting in the well-known 'cooked' taste. In addition, inadequate processing conditions can cause loss of nutritional quality (Lozano, 2006).

In some sterilization process, juices are placed in cans or glass jars and subjected to temperatures up to 121 °C for 15 min in horizontal or vertical retorts (Körmendy, 2006). In this process, not only are pathogenic microorganism and spoilage bacteria inactivated but also their heat resistant spores; however, these severe processing conditions induce chemical and physical changes that modify the organoleptic and nutritional quality of juices reducing the bioavailability of certain bioactive compounds (Rawson *et al.*, 2011). In order to determine the heat processing conditions, factors such as type and heat resistance of the target microorganism, spore, or enzyme, pH of the product, storage conditions, and thermophysical properties of the food must be known.

Concentration of juice consists in reducing the a_w level by eliminating part of its water content. Consequently, the storage volume, transportation cost, and grown of yeast and other microorganisms are diminished. Concentrated juices are normally used to formulate beverages or to obtain reconstituted juices. During concentration, water evaporation is controlled to avoid loss of aroma constituents. Striping of volatiles or steam-stripping process can be used to recovery aroma compounds (McLellan & Padilla-Zakour, 2005).

5.4.1.1 Influence on Microbial Quality
The production of microbiologically safe products is of paramount importance in the food industry. Microorganism growth in fruit products is pH dependent. Foods can be classified according to their acidity as low acidity (pH > 4.6) or high acidity (pH < 4.6). Most of the exotic juices are considered to be of mid to high acidity (pH < 4), which implicates that *Salmonella*, *Staphylococcus*, lactic-acid-producing bacteria (*Lactobacillus* and *Leuconostoc*), molds, and yeast are the microorganisms of concern in juices (Körmendy, 2006). Norovirus and *Salmonella* are the leading viral and bacterial pathogens, respectively, documented to have caused outbreaks of infections associated with consumption of tropical fruits (Strawn *et al.*, 2011).

Among the mechanisms of microorganism inactivation by thermal processing are cell membranes damage, DNA injury, and denaturation of enzymes that control metabolic functions resulting in inhibition or death (Fellows, 2009; Hancock, 2013). Vegetative cells of yeast are inactivated with wet heat at 50–60 °C for 5 min, while spores require higher temperatures (70–80 °C) to be destroyed during the same time. Molds are destroyed when high acidity exotic fruits are subjected to wet heat at 62 °C for 30 min, although *Actinomycetes* only requires 15 min at 60 °C; however, thermoresistent molds (*Byssochlamys, Monascus, Phialophora, Talaromyces*, and *Neosartorya*) are capable of surviving even at the pasteurization temperatures mentioned (Holdsworth, 1988). Studies about the inactivation of *Neosartorya fisher* ascospores showed that this microorganism is more resistant in papaya than in pineapple juice. *D* (decimal reduction time) values obtained from papaya were 129.9, 19.0, and 1.9 min at 80, 85, and 90 °C, respectively, whereas in pineapple juice, they were 73.5, 13.2, and 1.5 at the same temperatures (Salomao *et al.*, 2007). The thermal treatment at 90 °C for 30 and 60 s significantly reduced microbial counts (aerobic bacteria, coliforms, yeasts, and molds) in Chokanan mango juice (Santhirasegaram *et al.*, 2013, 2015). Yeasts, molds, or aerobic bacteria were not found in mango pulp treated at 110 °C/8.6 s (HTST) (Liu *et al.*, 2013). Benlloch-Tinoco *et al.* (2014) reported 5-$\log_{10}$ reduction or more for *Listeria monocytogenes* in kiwi purees thermal treated at 97 °C for 19.3 min and 84 °C for 5 min.

5.4.1.2   Influence on Nutritional Quality

Thermal treatment severity to which exotic fruits are submitted determines the level of loss of bioactive compounds. Pasteurization of canned mango puree, pineapple juice, mango, and papaya decreases their anthocyanins, carotenoids, and ascorbic acid contents (Djioua *et al.*, 2009; Rattanathanalerk *et al.*, 2005; Beyers & Thomas, 1979). These compounds are sensible to oxygen, heat, and light, so their stability is related to the type and conditions of the pasteurization process applied. Heat or oxygen exposure may induce carotenoids isomerization, so the *trans*-carotenoids decrease while the *cis* configuration increases causing loss of color. Cano and De Ancos (1994) showed that the thermal treatment causes degradation of some important carotenoids as xanthophylls, producing changes in the nutritional value of foods. Similarly, thermal processing degrades anthocyanins and originates the formation of a variety of compounds (quinonas, 4-hydroxibenzoic acid, phloroglucinaldehyde, pelargonidin, and others) depending on the severity of the treatment (Rawson *et al.*, 2011). Other studies about the effect of thermal processing of phytochemical compounds of exotic fruits are shown in Table 5.2.

5.4.1.3   Influence on Organoleptic Attributes

An important challenge in exotic fruits processing is to preserve their organoleptic attributes such as color, aroma, flavor, and texture. Variations in these parameters due to processing can be indicative of both, negative organoleptic and nutritional changes (Fellows, 2009).

Temperature is the main factor affecting organoleptic attributes of exotic fruits. Pigments responsible for the exotic fruits color such as carotenoids, chlorophylls, and anthocyanins tend to degrade when the fruits are thermally processed. Chlorophylls are degraded to pheophytin providing the product with a brown coloration; carotenoids are converted into epoxides, and the degradation of anthocyanins causes a notable change in color (Holdsworth, 1988). Lespinard *et al.* (2012) compare the effects of thermal treatment on some quality parameters of kiwi, showing that the kiwi suffers a change of color (yellow brownish coloring) due to the formation of pheophytin, pyropheophytin, and pyropheophorbides as a result of chlorophyll degradation, as suggested also by Cano and Marin (1992). Maillard reaction affects exotic fruit product coloration. Rattanathanalerk *et al.* (2005) measured the hydroxymethylfurfural (HMF) and pigment formation in pineapple juice thermally treated at 55–95 °C, HMF and the pigment formation increased linearly with time following zero-order kinetic reactions, the reaction constant rate increased from 3.8×10^{-3} to

Table 5.2 Effect of thermal processing on the nutritional quality of exotic fruit products

Product	Processing conditions	Main findings	References
Pineapple juice	99 °C/17 min and 90 °C/3 min	Vitamin C decreased by 94% at 99 °C/17 min, while at 90 °C/3 min, it diminished 38%.	Hounhouigan *et al.* (2014); Akinyele *et al.* (1990)
Guava juice	88 °C/10 min	Vitamin C decreases by 3.3%.	Jawaheer *et al.* (2003)
Kiwi puree	84 °C/5 min	Vitamins A, C, and E, total phenolic, total flavonoids, and AOA decreased by 100%, 26.7%, 83.3%, 29.5%, 50.9%, and 77.6%, respectively. Total tannin content was not affected.	Benlloch-Tinoco *et al.* (2014)
Mango puree	85–93 °C/up to 16 min	Significant *trans–cis* isomerization of β-carotene was observed. The maximum retention of this carotenoid was 93% with 15.4% of vitamin A loss.	Vásquez-Caicedo *et al.* (2007)
Mango juice	90 °C/1 min	Carotenoids and ascorbic acid content decreased 40% and 65%, respectively.	Santhirasegaram *et al.* (2015)
Mango nectar	HTST at 110 °C/8.6 s	No significant change in phytochemical contents was observed immediately after treatment, nevertheless during the storage at 4 and 25 °C/16 weeks, L-ascorbic acid decreased by 10.4% and 18.4%, total phenolic 17% and 25.2%, total carotenoids 69.2% and 62.1%, and AOA 27.4% and 28.6% at 4 and 25 °C, respectively.	Liu *et al.* (2014)

12.3×10^{-3} min^{-1} when temperature changed from 55 to 95 °C. In addition, the rate constant for the melanoidins formation (compounds related with the brown coloration during the Maillard reaction) increased from 0.10×10^{-3} to 0.56×10^{-3} min^{-1} when temperature increased from 55 to 95 °C, resulting in pineapple juice with a darker color.

Ledeker *et al.* (2012, 2014) showed that mango puree pasteurized at 85 °C for 15 s presented a different flavor (cooked taste) to fresh mango puree and a decrease in sweetness. The reduction of the intense flavor is the result of the degradation and loss of aromatic components during the thermal process. Similarly, Yen and Lin (1999) showed that guava juice heat treated (95 °C, 5 min) had a decrease in the content of volatile compounds compared with the fresh product.

Concentration of pineapple juice by evaporation changes the flavor and the aroma profile of the juice, thermal treatment produced an intensity cooked and artificial flavor, as well as sweet aroma due to caramelization (De Vasconcelos *et al.*, 2009).

5.4.1.4   Influence on Enzymatic Activity   The control of enzymatic activity is a way to preserve some of the organoleptic and nutritional attributes of exotic fruit juice during storage, so the inactivation of enzymes is important. PPO and POD activity is related with changes in color and deterioration of nutritional value, and PME is associated with alterations of texture and cloud loss in juices. Sims *et al.* (1994) observed that the heat treatment at 90 °C/30 s inactivated PPO of banana puree and reduced browning. The thermal treatment of kiwi puree at 84 °C/5 min reduced POD activity by 90%; nevertheless, the color was deteriorated. The treatment also increased the consistency and viscosity of the puree (Benlloch-Tinoco *et al.*, 2014). No residual activity of POD was obtained in mango puree pasteurized at 80 and 85 °C, while at 65, 70, and 75 °C, residual enzymatic activities varying from 78% to 4.3% were obtained (Sugai and Tadini, 2006). In mango puree treated at temperatures between 85 and 93 °C, PPO was readily inactivated after 1 min processing, while residual POD activities of 4–6.3% were detected even after 16 min at all pasteurization temperatures (Vásquez-Caicedo *et al.*, 2007). PME activity was reduced by 97% in unclenched mango puree and was completely inactivated in the blanched product after the HTST treatment (Liu *et al.*, 2013).

5.4.2   Nonthermal Processing

The effect of nonthermal technologies on nutritional quality, enzymatic activity, and microbial safety of various foods has been extensively studied, but the information regarding exotic fruit juices is limited. In this section, the fundamentals and suitability of some of the most promising nonthermal technologies that have been applied in exotic fruit juices will be reviewed.

5.4.2.1   Fundaments of Nonthermal Technologies

High Hydrostatic Pressures   HHP is a nonthermal preservation method that can extend the shelf life of foods with minimal alterations of the nutritional and sensory properties (Bermúdez-Aguirre & Barbosa-Cánovas, 2011; Campus, 2010; Mújica-Paz *et al.*, 2011; Torres & Velazquez, 2005; Torres *et al.*, 2009; Welti-Chanes *et al.*, 2006); thus, HHP is a promising method to preserve exotic fruit juices. In HHP, previously packed products are placed on vessels filled with water (other pressuring liquids can be used) that exerts uniform pressure independently of the size and shape of the food (isostatic principle) causing microbiological death and enzymatic inactivation. Pressure can be applied in combination with moderate or high temperatures (pressure-assisted thermal processing) in order to decrease time and pressure levels required to reach the microbial or enzyme inactivation targets (Escobedo-Avellaneda *et al.*, 2011). The mechanisms for HHP inactivation without compromising nutritional quality are related with their prejudicial effects on weak bonds without altering the covalent bonds (Masson *et al.*, 2001; Welti-Chanes *et al.*, 2006). HHP inactivates microorganisms by causing changes in membrane permeability, denaturation of proteins, and inactivation of key enzymes, among other effects. Pressure, rate of depressurization, holding time, temperature, and CUT (*come up time*, time required to achieve the desired pressure level) are parameters affecting the effectiveness of HHP. Product-related parameters, such as pH, a_w, and composition, and microorganism factors, such as initial load, type of microorganism, and phase of growth, also affect results (Welti-Chanes *et al.*, 2005).

Pulsed Electric Fields PEFs involve the application of short pulses of high electric field intensity with duration of microseconds. In this method, the food (mainly liquids) is placed inside or passed through a treatment chamber with two electrodes in where high voltage pulses are applied (Barbosa-Cánovas & Sepúlveda, 2005). As well as with HHP, the effect of PEF on microorganism and enzymes can be reversible or irreversible depending on the degree of damage caused to cellular membranes and intensity of conformational changes caused to enzymes. The mechanisms of microorganism inactivation by PEF are related with membrane electroporation, DNA damage, and generation of toxic compounds (Knorr, 1999). The mechanisms for enzyme inactivation with PEF implicate protein unfolding, denaturation, breakdown of covalent bonds, and oxidation–reduction reactions (Martín-Belloso *et al.*, 2005).

Parameters such as electric field strength, processing time (pulse width multiplied by the number of pulses), pulse shape (exponential or square pulses that can be applied in a mono or bipolar way), and treatment chamber design (parallel plate, coaxial, or colinear configuration) affect the effectiveness of PEF (Knorr, 1999). In PEF the ionic strength, dielectric strength and conductivity of the product have a great influence on process effectiveness. The maximum level of electric field that a product can resist without presenting dielectric breakdown (in which an uncontrolled electric discharge occurs damaging the equipment and the product by heating) is defined by the dielectric strength. It is generally recognized that increasing the treatment time, applying square pulses (because there is a constant intensity for the total duration of the pulse), using bipolar pulses, lowering the conductivity (the temperature and applied power decreases causing an increase in the electric field intensity and overall effectiveness), lowering the temperature, and low contents of fat and proteins, favors the PEF treatment (Barbosa-Cánovas & Sepúlveda, 2005; Pagán *et al.*, 2005; Rodrigo *et al.*, 2005).

Ultrasound Ultrasound involves the application of sound waves of high frequency (18 kHz to 500 MHz), generating cycles of compression and expansion with subsequent cavitation when, in a liquid food, the sound is propagated as longitudinal waves. During the expansion cycle, some bubbles can be formed, and during the compression, they can implode generating spots with very high pressure and temperature that disrupt cells. The increment in pressure and temperature depends on the intensity of the treatment applied that can generate either stable or transient cavitation. In stable cavitation, the bubbles formed are small and their size tends to be uniform due to the low intensity of the ultrasound, and in transient cavitation, the size of bubbles varies strongly due to high intensity of the treatment. In this last type, the surface area of the bubbles is greater during the expansion step and it increases during each cycle until it reaches a critical size causing that bubbles implode originating a higher increase in temperature and pressure peaks. High-intensity ultrasound can generate emulsions, disrupt cells, promote chemical reactions, inhibit enzymes, tenderize meat, modify crystallization process, and enhance extraction yield of bioactive compounds. To increase the effectiveness of ultrasound, it must be combined with other factors such as pressure and temperature (Cheng *et al.*, 2007; Knorr *et al.*, 2011). Frequency and amplitude of the ultrasonic waves, hydrostatic pressure, and temperature are factors involved in the cavitation phenomena. The minimum oscillating pressure for cavitation to occur depends on the physical and chemical characteristics of the liquid media such as vapor pressure, tensile strength, solid concentration and dissolved gas. With regard to the sensitivity of microorganisms to ultrasound, it is generally recognized that larger cells are more sensitive than smaller cells and that Gram-negative bacteria are more sensitive than the Gram-positive one.

Ultraviolet Light   Treatment of foods with ultraviolet light radiation (UV-C) involves the application of nonionizing light at wavelength from 200 to 280 nm to decontaminate the surface of fruits. This technology does not leave residues, is easy to use requiring inexpensive equipment, and is lethal to most microorganisms (Keyser *et al.*, 2008). UV-C is considered to be germicidal against microorganisms such as bacteria, viruses, protozoa, yeast, molds, and algae (Bintsis *et al.*, 2000), and the highest germicidal effect is obtained at 254 nm (Guerrero-Beltrán & Barbosa-Cánovas, 2005). UV-C has also been considered for the decontamination of fruit juices. The mechanism of UV-C decontamination is related with DNA damage. The resistance of microorganisms to UV-C is determined by their ability to repair DNA.

Microorganisms in the logarithmic growth phase are more sensitive than in the stationary phase (López-Malo and Palou, 2005). The absorption characteristics and the amount of soluble solids in the liquid food influence the effectiveness of the UV-C treatment (Guerrero-Beltrán & Barbosa-Cánovas, 2005). Geometric configuration of the reactor, radiation power, wavelength, treatment time, dose and physical arrangements of the lamps, product flow profile, and radiation direction are the main process parameters to be considered when designing UV-C treatments. The FDA has approved the use of UV at 14 mJ/cm^2 to pasteurize apple juice (McLellan & Padilla-Zakour, 2005).

5.4.2.2   Application of Nonthermal Technologies in Exotic Fruit Juices   There are many studies about the application of nonthermal technologies in fruit juices, but the information concerning exotic fruit juices is limited. Only techniques such as HHP, PEF, ultrasound, and UV-C light have been applied in exotic fruit juices.

High Hydrostatic Pressures   The effect of HHP on different quality aspects of some exotic fruit juices, nectars, and purees has been studied. The treatment of mango nectar at 600 MPa/1 min decreased the counts of yeasts and molds to less than 1-$\log_{10}$ CFU/ml and total aerobic bacteria to less than 2-$\log_{10}$ CFU/ml, while PPO and POD were completely inactivated and not activity was detected during storage at 4 and 25 °C during 16 weeks. No significant changes in L-ascorbic acid, total phenolics, total carotenoids, and AOA measured with the FRAP method were obtained immediately after processing. But during 16 weeks, storage decrements of 16.3% and 29.6% in L-ascorbic acid, 19.1% and 27.9% in total phenolics, 67.7% and 65.3% in total carotenoids, and 20.8% and 21.1% in AOA at 4 and 25 °C, respectively, were obtained. The color of the product was not affected by the HHP process. Despite the changes obtained with the HHP treatment in the functional compounds of mango nectar during storage, the changes were lower when compared with the thermally treated product at 100 °C/8.6 s (Liu *et al.*, 2014).

Jiménez-Aguilar *et al.* (2015) studied the effects of HHP (400 or 550 MPa/room temperature/0–16 min) on the content of functional compounds and AOA of prickly pear beverages prepared with 10% peel and 90% pulp (Cristal and Rojo San Martín, Mexican varieties) either with (A) or without (N) the incorporation of acids and antimicrobials. These authors obtained that prickly pear beverages prepared from Cristal (A) and Rojo San Martin (A and N) varieties processed at 550 MPa/≥2 min showed significant increase in total phenolics (16–35%) and AOA (8–17%), with no significant changes in kaempferol and isorhamnetin contents, and 3–15% losses of vitamin C. Beverages formulated from the Rojo San Martin variety (A) treated at 550 MPa/$\geq$2 min showed significant increase in betaxanthins (6–8%) and betacyanin (4–7%). On the other hand, heat sterilization caused significant losses ($p < 0.05$) of vitamin C (46–76%), TP (27–52%), flavonoids (0–52%), betalains (7–45%), and AOA (16–45%). These results show that HHP treatments retain, and can even increase, the content of most phytochemical compounds of prickly pear beverages when compared to untreated samples, thus yielding products with higher nutraceutical quality than fresh or heat-treated beverages.

Other important studies about the application of HHP on exotic fruit juices or purees are presented in Table 5.3.

Pulsed Electric Fields Several studies using PEF have been effective to inhibit micro-organisms and enzymes, so for example for orange juice at least 5-$\log_{10}$ reductions were obtained when applied 40 kV/cm–150 µs (*Staphylococcus aureus*), 30 kV/cm–12 µs (*Listeria innocua, E. coli*), 12.5 kV/cm–800 µs (*Saccharomyces cerevisiae*), 30 kV/cm–15 µs (*Leuconostoc mesenteroides*), and 40 kV/cm–97 µs (molds and yeast) (Buckow *et al.*, 2013). In other fruit juices, residual PPO, POD, and PME activities lower than 10% were reached by applying the processing conditions presented in Table 5.4 (Martín-Belloso *et al.*, 2014).

Studies about PEF in exotic fruits products have shown that there is required the application of other preservation factors (hurdles) to increase the effectiveness of the PEF process. García-García *et al.* (2015) obtained promising results for *S. cerevisiae* inactivation when applying PEF in combination with preservatives and acids in functional beverages obtained by mixing peel and pulp of prickly pear. These authors found that the use of preservatives and acids in combination with PEF is necessary to obtain microbiologically safe products. The application of 50 exponential decay pulses in mango juice caused less than 1-$\log_{10}$ cell-cycle reduction of *E. coli* O157:H7 (2×10^7 cells/ml), while the combination with carvacrol (1.3 mM) reduced 5-$\log_{10}$ cell cycles with 20 pulses. The reduction of the initial microbial contamination to 2×10^4 cells/ml diminishes the addition of carvacrol from 1.3 to 0.31 mM while maintaining the synergistic effect in combination with PEF (Ait-Ouazzou *et al.*, 2013). When mango juice is treated either with mild heat ($\leq$52 °C) or PEF, there were only 1.3-$\log_{10}$ microbial reductions, but with the addition of preservatives, a significantly large microbial reduction is observed. The microbial counts were reduced in 4.4-$\log_{10}$ CFU/ml with the application of 20 pulses of 87 kV/cm at 52 °C in the presence of nisin and lysozyme (27.5 IU of nisin and 690 IU of lysozyme per milliliter of juice) (Zhang and Mittal, 2005). As the field strength increased (24 and 34 kV/cm), the inactivation of *E. coli* K12 in a tropical fruit smoothie (pineapple, banana, and coconut milk) increased from 2.8- to 4.2-$\log_{10}$ CFU/ml, and when PEF was combined with temperature (45–55 °C, 1 min), higher inactivation levels are observed (5.1- and 6.9-$\log_{10}$ CFU/ml) (Walkling-Ribeiro *et al.*, 2008).

Ultrasound Some studies indicate that the treatment of exotic fruit juices with ultrasound could be an alternative to thermal pasteurization. Garcia *et al.* (2011) showed that after the ultrasound treatment of pineapple juice at 376 W/cm^2 and 10 min, the PPO activity and viscosity decreased by 20% and 75%, respectively, while the phenolic

Table 5.3   Effect of HHP on different quality parameters of exotic fruit juices and purees

Product	Processing conditions	Main results	References
Banana puree (pH 3.4 and a_w 0.97)	517 and 689 MPa/ 10 min/21 °C	The color was retained during 15 days storage at 25 °C. Total microorganisms yeast and mold counts were <10 CFU/g through storage. The use of only HHP at both pressures was not enough to decrease PPO activity but with blanching during 7 min followed by 689 MPa/10 min, residual PPO activity was <5%.	Palou et al. (1999)
Guava puree	600 MPa/15 min/25 °C	The microorganisms were inactivated to <10 CFU/ml and the product exhibited no change in color, and the 'cloud' and ascorbic acid content were stable as compared with fresh samples. When compared with the thermally treated product (88–90 °C for 24 s), the HHP guava showed higher residual enzymatic activity, but the thermally treated one had marked changes in viscosity, turbidity, and color. HHP-treated puree had a final quality similar to the freshly extracted puree after storage at 4 °C/40 days.	Gow-Chin and Hsin-Tang (1996)
Guava juice	600 MPa/15 min/25 °C	The methanol, ethanol, and 2-ethylfuran contents increase due to the treatment. Nevertheless, the volatile distribution of the HHP treated juice was similar to the fresh juice when stored at 4 °C/30 days, while most of the volatile components in the guava juice were lost during pasteurization at 95 °C/5 min.	Gow-Chin and Hsin-Tang (1999)
Kiwi pulp	500 MPa/3 min/room temperature	L-Ascorbic acid content decreased by ≈20% immediately after processing, nevertheless in untreated sample the vitamin C content decreased during storage (4 °C/40 days) while for the HHP treated puree it remained constant.	Fernández-Sestelo et al. (2013)
Blanched and unblanched mango pulp	300 MPa/15 min, 400 MPa/5 min, 500 MPa/1 min and 600 MPa/1 min	No yeasts, molds, or aerobic bacteria were detected in mango pulps after HHP treatments. PME activity was significantly reduced at 600 MPa/1 min in unblanched puree and significantly activated in the blanched one. Treatments from 300 to 500 MPa activated PME regardless of blanching, but in unblanched puree PME was inactivated in 97%.	Liu et al. (2013)
Mango puree	100–600 MPa/ 1 s–20 min/30 °C	HHP treatments caused significant changes in color. Maximum retentions of 85%, 92%, and 90% of L-ascorbic acid, total phenolic and AOA, respectively were obtained after treatments.	Kaushik et al. (2014)
Yellow passion fruit pulp (pH 3.1 and 14.5 °Brix)	300 MPa/5 min/25 °C	No detectable levels of Salmonella, coliforms, yeast and molds were observed.	Laboissière et al. (2007)

Table 5.4 PEF processing conditions to obtain fruit juices with residual enzymatic activities lower than 10% by applying squared pulse shape

Enzyme	Pulse polarity	Pulse width (μs)	Field strength (kV/cm)	Frequency (Hz)	Time (μs)	Temperature (°C)
PPO	Monopolar	5	35	16	75	<42
	Bipolar	4	35	229	2000	<40
	Bipolar	4	35	600	5000	<40
POD	Monopolar	5	35	16	75	<42
	Bipolar	6	35	200	1500	<40
	Bipolar	4	35	188	1727	<40
PME	Bipolar	7	35	250	1000	<40
	Monopolar	7	35	150	1000	<40
	Bipolar	2.5	30		50	65–80

Source: Martín-Belloso *et al.* (2014). Reproduced with permission of Elsevier.

compounds were not affected. In addition, the ultrasound enhanced the juice color and it was more stable during 42 days of storage than the untreated product. Cheng *et al.* (2007) reported that ultrasound did not significantly affect the overall quality of guava juices preserving the L-ascorbic acid content.

Mango juice was subjected to 40 kHz frequency and 130 W during 15, 30, and 60 min resulting in no significant changes in pH, total soluble solids, and acidity. A significant increase in extractability of carotenoids (4–9%) and polyphenols (30–35%) was observed in the juice subjected to ultrasound treatment for 15 and 30 min when compared with the control, while the color and vitamin C content were not affected. The AOA increased in all sonicated juices regardless of the treatment time (Santhirasegaram *et al.*, 2013).

Ultraviolet Light Ultraviolet light has been applied in fruit juices with the aim of obtaining microbiologically safe products of higher quality than with thermal processing. Keyser *et al.* (2008) used a turbulent flow system to treat guava-pineapple juice (with energy dosages of 0, 230, 459, 919, and 1377 J/l) and mango nectar (0, 689, 1388, and 2066 J/l). The taste and color of treated juices was not altered. Mango nectar showed 1.4-$\log_{10}$ reductions in aerobic bacteria and 2.8-$\log_{10}$ in yeast and molds. At a UV dosage of 1377 and 2066 J/l, no viable aerobic bacteria, yeast, and molds were observed. Santhirasegaram *et al.* (2015) showed that the UV-C light treatments for 15, 30, and 60 min at 25 °C did not change the physicochemical properties of mango juice, while they increased the extractability of carotenoids (6%), polyphenols (31%), and flavonoids (3%) in juices treated for 15 and 30 min. In addition, UV-C light significantly reduced the microbial loads of the juices. The combined UV-C light treatment (5.6, 7.6, and 11.2 mJ/cm^2) with moderate heat (50, 55, and 60 °C during 10, 20, and 30 min) was used to treat pineapple juice. The MPE and bromelain activities were not reduced by the UV treatment. Increasing holding time and UV dosage lead to higher reduction in total phenolics. The treatment at

55 °C/10 min–5.6 mJ/cm^2 decreased PME activity by 60.5% while retained the 61.5% and 72.8% of the total phenolic content and bromelain activity, respectively (Sew *et al.*, 2014). Guava–pineapple juice treated with UV showed 3.3-log$_{10}$ reductions in aerobic bacteria and 4.5-log$_{10}$ in yeast and molds after 1377 J/l.

5.5   Conclusions and Future Trends

Despite their limited region of growth, exotic fruits are a very interesting group of fruits used to produce juices. The recognition of their health benefits due to its composition and agreeable sensorial characteristics has increased the interest to reach international markets. Nevertheless, there is still much research to develop in this field, especially concerning the effects of thermal and nonthermal technologies to preserve the overall quality of these juices. Both thermal and nonthermal treatments could cause important changes in the content of some functional compounds and organoleptic characteristics of exotic fruit juices, and especially the color and flavor are affected, but the degree of damage will depend greatly on the method and processing conditions used. Despite this, nonthermal technologies have demonstrated to cause lower damage to the overall quality of these juices, and in addition, the stability during storage is higher. Some nonthermal processing conditions can originate increments in the content of some functional compounds, suggesting an increase in their bioavailability, this has led to rise the interest in apply nonthermal technologies in exotic fruit juices. Thermal methods have the potential to inactivate microorganisms and enzymes, while achieving the same levels of inactivation with nonthermal technologies could be a major challenge that could be possible if the adequate processing conditions are applied. The information about the application of nonthermal technologies to exotic fruit juices is limited. Only techniques such as HHP, PEF, ultrasound, and UV-C light have been applied in fruits such as kiwi, mango, passion fruit, banana, pineapple, guava, and prickly pear.

References

Ait-Ouazzou, A., Espina, L., García-Gonzalo, D. & Pagán, R. (2013) Synergistic combination of physical treatments and carvacrol for *Escherichia coli* O157: H7 inactivation in apple, mango, orange, and tomato juices. *Food Control* **32**, 159–167.

Akinyele, I.O., Keshinro, O.O. & Akinnawo, O.O. (1990) Nutrient losses during and after processing of pineapples and oranges. *Food Chemistry* **37**, 181–188.

Aviram, M., Dornfeld, L., Kaplan, M., Coleman, R., Gaitini, D., Nitecki, S., Hofman, A., Rosenblat, M., Volkova, N., Presser, D., Attias, J., Hayek, T. & Fuhrman, B. (2002) Pomegranate juice flavonoids inhibit low-density lipoprotein oxidation and cardiovascular diseases: studies in atherosclerotic mice and in humans. *Drugs under Experimental and Clinical Research* **28**, 49–62.

Barbosa-Cánovas, G.V., and Sepúlveda, D. (2005) Present status and the future of FEF technology. In *Novel Food Processing Technologies* (eds G.V. Barbosa-Cánovas, M.S. Tapia & M.P. Cano), pp. 1–44. CRC Press, USA.

Benlloch-Tinoco, M., Martínez-Navarrete, N. & Rodrigo, D. (2014) Impact of temperature on lethality of kiwifruit puree pasteurization by thermal and microwave processing. *Food Control* **35**, 22–25.

Bermúdez-Aguirre, D. & Barbosa-Cánovas, G.V. (2011) An update on high hydrostatic pressure, from the laboratory to industrial applications. *Food Engineering Reviews* **3**, 44–61.

Beyers, M. & Thomas, A.C. (1979) γ Irradiation of subtropical fruits. Changes in certain nutrients present in mangoes, papayas, and litchis during canning, freezing, and γ irradiation. *Journal of Agricultural and Food Chemistry* **27**, 48–51.

Bintsis, T., Litopoulou-Tzanetaki, E. & Robinson, R. (2000) Existing and potential applications of ultraviolet light in the food industry-a critical review. *Journal of the Science of Food and Agriculture* **80**, 637–645.

Buckow, R., Ng, S. & Toepfl, S. (2013) Pulsed electric field processing of orange juice: a review on microbial, enzymatic, nutritional, and sensory quality and stability. *Comprehensive Reviews in Food Science and Food Safety* **12**, 455–467.

Campus, M. (2010) High pressure processing of meat, meat products and seafood. *Food Engineering Reviews* **2**, 256–273.

Cano, M. & De Ancos, B. (1994) Carotenoid and carotenoid ester composition in mango fruits as influenced by processing method. *Journal of Agricultural and Food Chemistry* **42**, 2737–2742.

Cano, M. & Marin, M.A. (1992) Pigment composition and color of frozen and canned kiwi fruit slices. *Journal Agricultural Food Chemistry* **40**, 2141–2146.

Chan, H.T. (1993) Passuion fruit, papaya, and guava juices. In *Fruit Juice Processing Technology*. (eds S. Nagy, C.S. Chen & P.E. Shaw), pp. 334–377. Agscience, Inc., USA.

Cheng, L.H., Soh, C.Y., Liew, S.C. & The, F.F. (2007) Effects of sonication and carbonation on guava juice quality. *Food Chemistry* **104**, 1396–1401.

Corrales, M., Fernández-García, A., Butz, P. & Tauscher, B. (2009) Extraction of anthocyanins from grape skins assisted by high hydrostatic pressure. *Journal of Food Engineering* **90**, 415–421.

Crane, J.H. & Campbell, C.W. (1990) Origin and distribution of tropical and subtropical fruits. In *Fruits of Tropical and Subtropical Origin: Composition, Properties, Uses* (eds S. Nagy, P.E. Swaw & W.F. Wardowski), pp. 1–65. Florida Science Source.

de Ancos, B., Sgroppo, S., Plaza, L. & Cano, M.P. (2002) Possible nutritional and health-related value promotion in orange juice preserved by high-pressure treatment. *Journal of the Science of Food and Agriculture* **82**, 790–796.

de Souza, M.O., Silva, M., Silva, M.E., Oliveira, R.P. & Pedrosa, M.L. (2010) Diet supplementation with acai (Euterpe oleracea Mart.) pulp improves biomarkers of oxidative stress and the serum lipid profile in rats. *Nutrition* **26**, 804–810

De Vasconcelos, F.H., De Souza, N.M., Maia, G. *et al.* (2009) Changes in flavor quality of pineapple juice during processing. *Journal of Food Processing and Preservation* **34**, 508–519.

Devalaraj, S., Jain, S. & Yadav, H. (2011) Exotic fruits as therapeutic complements for diabetes, obesity and metabolic syndrome. *Food Reseach International* **44**, 1856–1865.

Djioua, T., Charles, F., Lopez-Lauri, F. *et al.* (2009) Improving the storage of minimally processed mangoes (*Magnifera indica* L.) by hot water treatments. *Postharvest Biology and Technology* **52**, 221–226.

Escobedo-Avellaneda, Z., Gutiérrez-Uribe, J., Valdez-Fragoso, A., Torres, J.A. & Welti-Chanes, J. (2015) High hydrostatic pressure combined with mild temperature for the preservation of comminuted orange: effects on functional compounds and AOA. *Food and Bioprocess Technology and International Journal* **8**, 1032–1044.

Escobedo-Avellaneda, Z., Pateiro-Moure, M., Chotyakul, N., Torres, J.A., Welti-Chanes, J. & Pérez-Lamela, C. (2011) Benefits and limitations of food processing by high-pressure technologies: effects on functional compounds and nonbiotic contaminants. *CyTA – Journal of Food* **9**, 352–365.

Fellows, P.J. (2009) *Processing Technology. Principles and Practice.* 3ʳᵈ edition. CRC Press, FL, USA.

Fernandez, M.L., Lin, E.C., Trejo, A. & McNamara, D.J. (1994) Prickly pear (*Opuntia* sp.) pectin alters hepatic cholesterol metabolism without affecting cholesterol absorption in guinea pigs fed a hypercholesterolemic diet. *Journal of Nutrition* **124**, 817–824.

Fernández-Sestelo, A., Sendra de Saá, R., Pérez-Lamela, C., Torrado-Agrasa, A., Rúa, M.L. & Pastrana-Castro, L. (2013) Overall quality properties in pressurized kiwi purée: microbial, physicochemical, nutritive and sensory tests during refrigerated storage. *Innovative Food Science and Emerging Technologies* **20**, 64–72.

Garcia, M.C.M., Vidal-Fonteles, T., Tibério de Jesus, A.L., Lima-Almeida, F.D., Alcântara de Miranda, M.R., Narciso-Fernandes, F.A. & Rodrigues, S. (2011) High-intensity ultrasound processing of pineapple juice. *Food and Bioprocess Technology* **6**, 997–1006.

García-García, R., Escobedo-Avellaneda, Z., Tejada-Ortigoza, V., Martín-Belloso, O., Valdez-Fragoso, A. & Welti-Chanes, J. (2015) Hurdle technology applied to prickly pear beverages to inhibit *Saccharomyces cerevisiae* and *Escherichia coli*. *Letters in Applied Microbiology* **60**, 558–564.

Gow-Chin, Y. & Hsin-Tang, L. (1996) Comparison of high pressure treatment and thermal pasteurization effects on the quality and shelf life of guava puree. *International Journal of Food Science and Technology* **31**, 205–213.

Gow-Chin, Y. & Hsin-Tang, L. (1999) Changes in volatile flavor components of guava juice with high-pressure treatment and heat processing and during storage. *Journal of Agriculture and Food Chemistry* **47**, 2082–2087.

Guerrero-Beltrán, J.A. & Barbosa-Cánovas, G.V. (2005) Reduction of *Saccharomyces cerevisiae, Escherichia coli* and *Listeria innocua* in apple juice by ultraviolet light. *Journal of Food Process Engineering* **28**, 437–452.

Hancock, C.O. (2013) Heat sterilization. In *Russell, Hugo & Ayliffe's: Principles and Practice of Disinfection, Preservation and Sterilization* (eds A.P. Fraise, J.Y. Maillard, S.A. Sattar), 5ᵗʰ edition, pp. 277–293. Blackwell Publishing Ltd., UK.

Hasrat, J.A., De Bruyne, T., De Backer, J.P., Vauquelin, G. & Vlietinck, A.J. (1997) Isoquinoline derivatives isolated from the fruit of *Annona muricata* as 5-HTergic 5-HT$_{1A}$ receptor agonists in rats: unexploited antidepressive (lead) products. *Journal of Pharmacy and Pharmacology* **49**, 1145–1149.

Hodgson, A.S. & Hodgson, L.R. (1993) Pinneaple juice. In *Fruit Juice Processing Technology* (eds S. Nagy, C.S. Chen & P.E. Shaw), pp. 378–435. Agscience, Inc., USA.

Höhn, A., Sun, D., and Nolle, F. (2005) Enzymes in the fruit juice and wine industry. In *Processing Fruits, Science and Technology* (ed D.M. Barrett), 2ⁿᵈ edition, pp. 97–112. CRC Press, Boca Raton, FL.

Holdsworth, S.D. (1988) *Conservación de Frutas y Hortalizas.* Editorial Acribia, Zaragoza, España.

Hounhouigan, M.H., Linnemann, A.R., Soumanou, M. *et al.* (2014) Effect of processing on the quality of pineapple juice. *Food Reviews International* **30**, 112–133.

Hyun-Sun, L., Lee, H., Yu, H.J., Ju, D.W., Kim, Y., Kim, C.T., Kim, C.J., Cho, Y.J., Kim, N., Choi, S.Y. & Suh, H.J. (2011) A comparison between high hydrostatic pressure extraction and heat extraction of ginsenosides from ginseng (Panax ginseng CA Meyer). *Journal of the Science of Food and Agriculture* **91**, 1466–1473.

Jawaheer, B., Goburdhun, D. & Ruggoo, A. (2003) Effect of processing and storage of guava into jam and juice on the ascorbic acid content. *Plant Foods for Human Nutrition* **58**, 1–12.

Jiménez-Aguilar, D.M., Escobedo-Avellaneda, Z., Martín-Belloso, O., Gutiérrez-Uribe, J., Valdez-Fragoso, A., García-García, R., Torres, J.A. & Welti-Chanes, J. (2015) Effect of high hydrostatic pressure on the content of phytochemical compounds and AOA of prickly pears (*Opuntia ficus-indica*) beverages. *Food Engineering Reviews*. doi: 10.1007/s12393-015-9111-5

Jiménez-Aguilar, D.M., Mújica-Paz, H. & Welti-Chanes, J. (2014) Phytochemical characterization of prickly pear (Opuntia spp.) and of its nutritional and functional properties: a review. *Current Nutrition and Food Science* **10**, 57–69

Jurenka, J.S. (2008) Therapeutic applications of pomegranate (*Punica granatum* L.): a review. *Alternative Medicine Review: A Journal of Clinical Therapeutic* **13**, 128–144.

Kader, A.A. & Barrett, D.M. (2005) Classification, composition of fruits, and postharvest maintenance of quality. In *Processing Fruits, Science and Technology* (eds D.M. Barret, L. Somogyi, & H. Ramaswamy), 2nd edition, pp. 3–22. CRC Press, Boca Raton, Florida, USA.

Kang, J., Li, Z., Wu, T., Jensen, G.S., Schauss, A.G. & Wu, X. (2010) Anti-oxidant capacities of flavonoid compounds isolated from acai pulp (*Euterpe oleracea* Mart.). *Food Chemistry* **122**, 610–617.

Kaushik, N., Kaur, B.P., Rao, P.S. & Mishra, H.N. (2014) Effect of high pressure processing on color, biochemical and microbiological characteristics of mango pulp (*Mangifera indica* cv. Amrapali). *Innovative Food Science & Emerging Technologies* **22**, 40–50.

Keyser, M., Muller, I.A., Cilliers, F.P., Nel, W. & Gouws, P.A. (2008) Ultraviolet radiation as a non-thermal treatment for the inactivation of microorganism in fruit juice. *Innovative Food Science and Emerging Technologies* **9**, 348–354.

Knorr, D. (1999) Novel approaches in food-processing technology: new technologies for preserving foods and modifying function. *Current Opinion in Biotechnology* **10**, 485–491.

Knorr, D., Froehling A., Jaeger, H., Reineke K., Schlueter O. & Schoessler K. (2011) Emerging technologies in food processing. *Annual Review of Food Science and Technology* **2**, 203–235.

Körmendy, I. (2006) Fruit processing: principles of heat treatment. In *Handbook of Fruits and Fruit Processing: Science and Technology*. (eds Y.M. Hui, M.P. Cano & J. Barta). Blackwell Publishing, Iowa, USA.

Kumar, G.N.M. (1990) Pomegranate. In *Fruits of Tropical and Subtropical Origin: Composition, Properties, Uses*. (eds S. Nagy, P.E. Swaw & W.F. Wardowski). pp. 328–347. Florida Science Source.

Kuo-Chiang, H., Fa-Jui, T. & Yi, C. (2008) Evaluation of microbial inactivation and physicochemical properties of pressurized tomato juice during refrigerated storage. *LWT – Food Science and Technology* **41**(3), 367–375.

Laboissière, L.H.E.S., Deliza, R., Barros-Marcellini, A.M., Rosenthal, A., Camargo, L.M.A.Q. & Junqueira, R.G. (2007) Effects of high hydrostatic pressure (HHP) on sensory characteristics of yellow passion fruit juice. *Innovative Food Science and Emerging Technologies* **8**, 469–477.

Ledeker, C.N., Chambers, D.H., Chambers, IV E. *et al.* (2012) Changes in the sensory characteristics of mango cultivars during the production of mango puree and sorbet. *Journal of Food Science* **77**, S348–S355.

Ledeker, C.N., Suwonsichon, S., Chambers, D. *et al.* (2014) Comparison of sensory attributes in fresh mangoes and heat-treated mango puree prepared from Thai cultivars. *LWT-Food Science and Technology* **56**, 138–144.

Lespinard, A., Bambicha, R. & Mascheroni, R. (2012) Quality parameters assessment in kiwi jam during pasteurization. Modelling and optimization of the thermal process. *Food and Bioproducts Processing* **90**, 799–808.

Liu, F., Wang, Y., Bi, X., Guo, X., Fu, S. & Liao, X. (2013) Comparison of microbial inactivation and rheological characteristics of mango pulp after high hydrostatic pressure treatment and high temperature short time treatment. *Food and Bioprocess Technology* **6**, 2675–2684.

Liu, F., Wang, Y., Li, R., Bi, X. & Liao, X. (2014) Effects of high hydrostatic pressure and high temperature short time on AOA, antioxidant compounds and color of mango nectars. *Innovative Food Science and Emerging Technologies* **21**, 35–43.

López-Malo, A. & Palou, E. (2005) Ultraviolet light and food preservation. In *Novel Food Processing Technologies*, (eds G.V. Barbosa-Cánovas, M.S. Tapia & M.P. Cano), pp. 405–422. CRC Press, USA.

Lozano, J. (2006) *Fruit Manufacturing. Scientific Basis, Engineering Properties, and Deteriorative Reactions of Technological Importance.* Springer, USA.

Luximon-Ramma, A., Bahorun, T. & Crozier, A. (2003) Antioxidant actions and phenolic and vitamin contents of common Mauritian exotic fruits. *Journal of the Science of Food and Agriculture* **83**, 496–502.

Martín-Belloso, O., Bendicho, S., Elez-Martínez, P. & Barbosa-Cánovas, G.V. (2005) Does high-intensity pulsed electric fields induce changes in enzymatic activity, protein conformation, and vitamin and flavor stability? In *Novel Food Processing Technologies*, (eds G.V. Barbosa-Cánovas, M.S. Tapia & M.P. Cano), pp. 87–104. CRC Press, USA.

Martín-Belloso, O., Marsellés-Fontanet, A.R. & Elez-Martínez, P. (2014) Enzymatic inactivation by pulsed electric fields. In *Emerging Technologies for Food Processing*, (ed S. Da-Wen) 2nd edition, pp. 155–168. Academic Press.

Masson, P., Tonello, C. & Balny, C. (2001) High-pressure biotechnology in medicine and pharmaceutical science. *Journal of Biomedicine and Biotechnology* **1**, 85–88.

McLellan, M.R. & Padilla-Zakour, O.I. 2005. Juice processing. In *Processing Fruits: Science and Technology* (eds D.M. Barrett, L.P. Somogyi & H. Ramaswamy), pp. 72–95. Boca Raton, CRC Press

Mertens-Talcott, S.U., Rios, J., Jilma-Stohlawetz, P., Pacheco-Palencia, L.A., Talcott, B.M.S.T. & Derendorf, H. (2008) Pharmacokinetics of anthocyanins and antioxidant effects after the consumption of anthocyanin-rich açai juice and pulp (*Euterpe oleracea* Mart.) in human healthy volunteers. *Journal of Agriculture and Food Chemistry* **56**, 7796–7802.

Mohapatra, D., Mishra, S., Singh, C.B. & Jayas, D.S. (2011) Post-harvest processing of banana: opportunities and challenges. *Food and Bioprocess Technology* **4**, 327–339.

Mújica-Paz, H., Valdez-Fragoso, A., Tonello-Samson, C., Welti-Chanes, J. & Torres, J.A. (2011) High-pressure processing technologies for the pasteurization and sterilization of foods. *Food and Bioprocess Technology* **4**, 969–985.

Nakasone, H.Y. & Paull, R.E. (1998) *Tropical Fruits*. Cab International, New York, USA.

Narasimham, P. (1990). Breadfruit and jackfruit. In *Fruits of Tropical and Subtropical Origin: Composition, Properties, Uses*. (eds S. Nagy, P.E. Swaw & W.F. Wardowski). pp. 193–259. Florida Science Source.

Noratto, G.D., Angel-Morales, G., Talcott, S.T. & Mertens-Talcott, S.U. (2011) Polyphenolics from (Ac) Acai (Euterpe oleracea Mart.) and red muscadine grape (vitis rotundifolia) protect human umbilical vascular endothelial cells (HUVEC) from glucose- and lipopolysaccharide (LPS)-induced inflammation and target microRNA-126. *Journal of Agriculture and Food Chemistry* **59**, 7999–8012.

Pagán, R., Condón, S., & Raso, J. (2005) Microbial inactivation by pulsed electric fields. In *Novel Food Processing Technologies*, (eds G.V. Barbosa-Cánovas, M.S. Tapia, M.P. Cano), pp. 45–68. CRC Press, USA.

Palou, E., López-Malo, A., Barbosa-Cánovas, G.V., Welti-Chanes, J. & Swanson, B.G. (1999) Polyphenoloxidase activity and color of blanched and high hydrostatic pressure treated banana puree. *Journal of Food Science* **64**, 42–45.

Qiu, W., Jiang, H., Wang, H. & Gao, Y. (2006) Effect of high hydrostatic pressure on lycopene stability. *Food Chemistry* **97**, 516–523.

Ramaswamy, H.S. (2005) Thermal processing of fruits. In. *Processing Fruits, Science and Technology* (eds D.M. Barret, I.L. Somogy & H. Ramaswamy), 2nd edition, pp. 173–200. CRC Press, Boca Raton, Florida, USA.

Rattanathanalerk, M., Chiewchan, N. & Srichumpoung, W. (2005) Effect of thermal processing on the quality loss of pineapple juice. *Journal of Food Engineering* **66**, 259–265.

Rawson, A., Patras, A., Tiwari, B.K. *et al.* (2011) Effect of thermal and nonthermal processing technologies on the bioactive content of exotic fruits and their products: review of recent advances. *Food Research International* **44**, 1875–1887.

Rodrigo, D., Martínez, A. & Rodrigo, M. (2005) Inactivation kinetics of microorganisms by pulsed electric fields. In *Novel Food Processing Technologies* (eds G.V. Barbosa-Cánovas, M.S. Tapia & M.P. Cano), pp. 69–86. CRC Press, USA.

Salomao, B.C.M., Slongo, A.P. & Aragao, G.M.F. (2007) Heat resistance of *Neosartorya fischeri* in varios juices. *LWT – Food Science and Technology* **40**, 676–680.

Sánchez-Moreno, C., De Pascual-Teresa, S., De Ancos, B. & Cano, M.P. (2006) Nutritional values of fruits. In *Handbook of Fruits and Fruit Processing: Science and Technology*. (eds Y.M. Hui, M.P. Cano & J. Barta). Blackwell Publishing, Iowa, USA.

Sánchez-Moreno, C., Pilar-Cano, M., de Ancos, B., Plaza, L., Olmedilla, B., Granado, F. & Martin, A. (2003) High-pressurized orange juice consumption affects plasma vitamin C, antioxidative status and inflammatory markers in healthy humans. *Journal of Nutrition* **133**, 2204–2209.

Sánchez-Moreno, C., Plaza, L., Elez-Martinez, P., de Ancos, B., Martin-Belloso, O. & Cano, M.P. (2005) Impact of high pressure and pulsed electric fields on bioactive compounds and AOA of orange juice in comparison with traditional thermal processing. *Journal of Agricultural and Food Chemistry* **53**, 4403–4409.

Santhirasegaram, V., Razali, Z. & Somasundram, C. (2013) Effects of thermal treatment and sonication on quality attributes of Chokanan mango (*Mangifera indica* L.) juice. *Ultrasonics Sonochemistry* **20** (5), 1276–1282.

Santhirasegaram, V., Razali, Z. & Somasundram, C. (2015) Effects of sonication and ultraviolet-C treatment as a hurdle concept on quality attributes of Chokanan mango (*Mangifera indica* L.) juice. *Food Science and Technology* **21**, 232–241.

Sew, C.C., Ghazali, H.M., Martín-Belloso, O. & Noranizan, M.A. (2014) Effects of combining ultraviolet and mild heat treatments on enzymatic activities and total phenolic contents in pineapple juice. *Innovative Food Science and Emerging Technologies* **26**, 511–516.

Sims, C.A., Bates, R.P. & Arreola, A.G. (1994) Color, polyphenoloxidase, and sensory changes in banana juice as affected by heat and ultrafiltration. *Journal of Food Quality* **17**, 371–379.

Strawn, L.K., Schneider, K.R. & Danyluk, M.D. (2011) Microbial safety of tropical fruits. *Critical Reviews in Food Science and Nutrition* **51**, 132–145.

Sugai, A.Y. & Tadini, C.C. (2006) Thermal inactivation of mango (*Mangifera indica* L., variety Palmer) puree peroxidase. CIGR Section VI International Symposium on Future of Food Engineering, Warsaw, Poland, 26–28 April 2006.

Swi-Bea, J., Chen, H. & Fang, T. (1993) Mango juice. In *Fruit Juice Processing Technology* (eds S. Nagy, C.S. Chen & P.E. Shaw), pp. 620–655. Agscience, Inc., USA.

Tesoriere, L., Butera, D., Pintaudi, A.M., Allegra, M. & Livrea, M.A. (2004) Supplementation with cactus pear (*Opuntia ficus-indica*) fruit decreases oxidative stress in healthy humans: a comparative study with vitamin C1–3. *American Journal of Clinical Nutrition* **80**, 391–395.

Torres, J.A., Sanz, P.D., Otero, L., Pérez-Lamela, C. & Saldaña, M.D.A. (2009) Engineering principles to improve food quality and safety by high pressure processing. In *Processing Effects on Safety and Quality of Foods* (ed E. Ortega-Rivas), pp. 379–414. CRC Taylor and Francis, Inc., Boca Raton FL.

Torres, J.A. & Velazquez, G. (2005) Commercial opportunities and research challenges in the high pressure processing of foods. *Journal of Food Engineering* **67**, 95–112.

US Department of Health and Human Services (2004) Juice HACCP Hazards and Control Guidance. Guidance for Industry. 1st edition. Food and Drug Administration, Center for Food Safety and Applied Nutrition (CDFSAN).

Vásquez-Caicedo, A.L., Schilling, S., Carle, R. & Neidhart, S. (2007) Effects of thermal processing and fruit matrix on β-carotene stability and enzyme inactivation during transformation of mangoes into puree and nectar. *Food Chemistry* **102**, 1172–1186.

Walkling-Ribeiro, M., Noci, F., Cronin, D.A., Lyng, J.G. & Morgan, D.J. (2008) Inactivation of *Escherichia coli* in a tropical fruit smoothie by a combination of heat and pulsed electric fields. *Journal of Food Science* **73**, M395–M399.

Welti-Chanes, J., López-Malo, A., Palou, E., Bermúdez, D., Guerrero-Beltrán, J.A. & Barbosa-Cánovas, G.V. (2005) Fundamentals and applications of high pressure processing to foods. In *Novel Food Processing Technologies*, (eds G.V. Barbosa-Cánovas, M.S. Tapia & M.P. Cano), pp. 157–182 CRC Press, USA.

Welti-Chanes, J., San Martín-González, F. & Barbosa-Cánovas, G.V. (2006) Water and biological structures at high pressure. In *Water Properties of Food, Pharmaceutical, and Biological Materials* (eds P. Buera, J. Welti-Chanes, P. Llilford, & H. Corti), pp. 205–232. CRC Press, Inc., Boca Raton, FL.

Wolfram, R.M., Kritz, H., Efthimiou, Y., Stomatopoulos, J. & Sinzinger, H. (2002) Effect of prickly pear (*Opuntia robusta*) on glucose-and lipid-metabolism in non-diabetics with hyperlipidemia – a pilot study. *Wiener Klinische Wochenschrift* **114**, 840–846.

Xie, Z., Sintara, M., Chang, T. & Ou, B. (2015) Functional beverage of *Garcinia mangostana* (mangosteen) enhances plasma antioxidant capacity in healthy adults. *Food Science and Nutrition* **3**, 32–38.

Yen, G. & Lin, H. (1999) Changes in volatile flavor components of guava juice with high-pressure treatment and heat processing and during storage. *Journal of Agricultural and Food Chemistry* **47**, 2082–2087.

Zhang, Y. & Mittal, G.S. (2005) Inactivation of spoilage microorganisms in mango juice using low energy pulsed electric field in combination with antimicrobials. *Italian Journal of Food Science* **17**, 167–176.

6

Berry Juices

Sze Ying Leong[1] and Indrawati Oey[1,2]*
[1]Department of Food Science, University of Otago, Dunedin, New Zealand
[2]Riddet Institute, Palmerston North, New Zealand

6.1 Introduction

Berry fruits are considered an excellent source of phytochemicals. Besides direct consumption as whole fruit, berry fruits are usually processed into different products such as canned, dried, frozen, freeze-dried, jams, jellies, and juices. This chapter discusses conventional and novel processing techniques usually used to produce juice from blackberries, blueberries, cranberries, currants, elderberries, grapes, pomegranate, raspberries, strawberries, and tomatoes, with a focus on the impact of different processing techniques toward microbial quality, nutritional properties, and organoleptic attributes (flavor, taste, color, and consistency) of berry juice.

6.2 Conventional Processing Techniques

Similar to other horticultural crops, a berry juice processing line comprises three main steps: extraction, clarification, and preservation. Figure 6.1 illustrates the schematic diagram of the conventional juice processing for berries. Extraction is a blend of size reduction and mechanical separation (or pressing under constant pressure) to give liquid yield as juice and the remaining solid fruit in the form of pulp, mash, or pomace. This is followed by a clarification step to obtain a clear appearance of the juice. The last step in juice production is preservation. Pasteurization is used widely in food and beverage industries. Pasteurization involves heating at temperatures above 90 °C in order to prolong the shelf life of juice by inactivating undesirable enzymes and microorganisms. Each of the processing steps involved in the juice production for different types of berry fruits should be controlled to eliminate negative alteration in the nutritional and organoleptic attributes.

*Corresponding author: Indrawati Oey, indrawati.oey@otago.ac.nz

Innovative Technologies in Beverage Processing, First Edition.
Edited by Ingrid Aguiló-Aguayo and Lucía Plaza.
© 2017 John Wiley & Sons Ltd. Published 2017 by John Wiley & Sons Ltd.

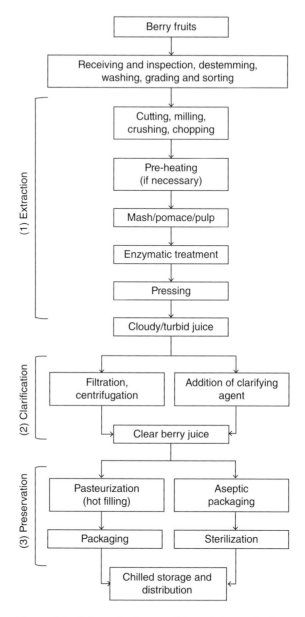

Figure 6.1   Schematic diagram of berry juice production.

6.2.1   Influence on Microbial Quality

Natural microflora such as bacteria, yeasts, and molds attach on the surface of the berries as a result of environmental conditions, soil contamination, wounding, and inappropriate postharvest handling and storage. Ultimately, this poses a potential safety issue (e.g., microbial spoilage and food poisoning) to consumers. Given that berries consist of abundant levels of water, sugars, amino acids, lipids, organic acids, minerals, and vitamins, this would provide an ideal environment to support microbial growth and transmission of hazardous foodborne pathogens and microbial toxin into the juice processing line. The postprocessing survival of some notable spoilage microorganisms such as *Alicyclobacillus acidoterrestris*, *Neosartorya fischeri*, *Escherichia coli* O157:H7, *Salmonella*, and *Listeria monocytogenes* in grape juice (Silva *et al.*, 1999; Rajashekhara *et al.*, 2000; Mazzotta, 2001) and *Saccharomyces cerevisiae*, *E. coli* O157:H7, *Salmonella*, and *Bacillus coagulans* in tomato juice (Shearer *et al.*, 2002; Eribo & Ashenafi, 2003; Yuk & Schneider, 2006; Peng *et al.*, 2012) has been well documented. Several incidences relating to severe foodborne outbreaks caused by microbiologically contaminated fruit juice, attributing to inadequate cleaning and sanitizing of fruit, improperly pasteurized, unhygienic processing facilities, and contamination from infected workers, have been summarized in the works of Beuchat (2002), Harris *et al.* (2003), Lima Tribst *et al.* (2009), and Raybaudi-Massilia *et al.* (2009).

Elimination of spoilage microorganisms, for instance, lactic acid bacteria and acetic bacteria, is an important measure to preserve the inherent flavor of juice, as these microorganisms are capable of producing lactic, acetic and formic acids, ethanol, and carbon dioxide. The initial washing and rinsing steps of berry fruit production could help to remove pesticide residue, soil debris, and other potential contaminants on the fruit surface. These steps are crucial to prevent subsequent carry-over contamination on the juice processing equipment. Later, this step is followed by sanitization, performed either before or postcutting operation, to minimize the microbial load on the surfaces. The use of various sanitizing agents such as hydrogen peroxide (Sapers, 2001), trisodium phosphate (Yu *et al.*, 2001), acetic acid (Yu *et al.*, 2001), cinnamaldehyde (Bevilacqua *et al.*, 2010), and organic acids (São José & Vanetti, 2012), at low or permitted concentrations according to the local authorities' regulations, has been recommended to replace the customary chlorine-based compounds to suppress the growth of microorganisms on various berry fruit, as well as removing pesticide residue.

Because some microorganisms (e.g., *A. acidoterrestris*, *S. cerevisiae*, *B. coagulans*, and *Byssochlamys* spp.) showed greater thermal resistance in the acidic berry juice, conventional thermal treatment such as pasteurization could be insufficient to inactivate these microorganisms in juice (Lima Tribst *et al.*, 2009). Microbial contamination in berry juice after pasteurization is not a rare case. Therefore, apart from washing the whole berry fruit with water that contains chemicals that possess antimicrobial activity prior to the juice extraction step, an antimicrobial agent can be added directly to the final unpasteurized and pasteurized berry juice products to protect against microbial spoilage during distribution and storage, as demonstrated in many studies (refer Table 6.1). On a side note, some berry juice possess antimicrobial properties which imply that the berry juice itself can be utilized as a natural microbial agent to be added into other food systems to inhibit the growth of foodborne pathogens. For instance, grape juice is effective against *Cronobacter sakazakii* (Kim *et al.*, 2010); cranberry juice against *Pseudomonas aeruginosa*, *Salmonella typhimurium*, *L. monocytogenes*, and *Staphylococcus aureus* (Ingham *et al.*, 2006; Côté *et al.*, 2011); and blackberry juice against *L. monocytogenes*, *S. typhimurium*, and *E. coli* O157:H7 (Yang *et al.*, 2014).

6.2.2   Influence on Nutritional Attributes

Majority of the phytochemicals in berry fruit are relatively stable owing to the acidic pH of the fruit in nature. However, variation in the composition and concentration

Table 6.1 Antimicrobial agents used in different types of berry juice for microbial reduction

Berry juice	Antimicrobial agent	Dosage (% w/w)	Juice pH after addition of antimicrobials	Target microorganisms	References
Grape juice	Benzoic acid	0.01	Not reported	Yeast	Pederson et al. (1961)
	Capric and caprylic acid	0.01	Not reported	Yeast	Pederson et al. (1961)
	Sorbic acid	0.01	Not reported	Yeast	Pederson et al. (1961)
Strawberry juice	Citric acid	0.5–2.0	2.5–2.9	Salmonella enteritidis, Escherichia coli O157:H7	Mosqueda-Melgar et al. (2008b)
	Cinnamon bark oil	0.05–0.3	3.15–3.16	Salmonella enteritidis, Escherichia coli O157:H7	Mosqueda-Melgar et al. (2008b)
Tomato juice	Citric acid	0.5–2.0	3.0–3.7	Salmonella enteritidis	Mosqueda-Melgar et al. (2008a)
	Cinnamon bark oil	0.05–0.3	4.17–4.29	Salmonella enteritidis	Mosqueda-Melgar et al. (2008a)
	Clove oil	0.1	4.2	Natural microflora	Nguyen and Mittal (2007)
	Mint extract	1.2	4.2	Natural microflora	Nguyen and Mittal (2007)

of phytochemicals is dependent on the growing conditions, location, cultivar, and maturity. Besides, compositional characteristics of each berry fruit confer a direct impact on the organoleptic properties of juice. For this reason, horticultural researches have been actively conducted for many years to breed fruit varieties or hybrids with improved sensorial attributes and acceptability, environment adaptation, disease resistance, and nutritional quality. For example, (i) introduction of new local cultivars with a high content of anthocyanin, polyphenols, ascorbic acids, soluble solids, and discriminating aroma and flavor qualities to meet the demand of industrial processing into juice for commercial cultivation of tomatoes in Jordan (Ereifej *et al.*, 1997); blackcurrants in Poland (Markowski & Pluta, 2008); grapes in India (Aggarwal & Gill, 2010); pomegranate in Oman (Al-Said *et al.*, 2009), Italy (Cristofori *et al.*, 2011), Turkey (Gundogdu & Yilmaz, 2012), and Morocco (Martínez *et al.*, 2012); and raspberries in Scotland (Harrison *et al.*, 1998); (ii) enhancement in the fruitiness flavor note of tomato juice can be achieved through selection and breeding of cultivars with high levels of glucose, reducing sugars, reducing sugars/glutamic acid ratio, and glutamic acid (Bucheli *et al.*, 1999); (iii) improvement in the color and viscosity of tomato juice through development of transgenic tomatoes with reduced levels of endogenous enzymes involved in degradation of the cell wall polysaccharide, namely pectin methyl esterase (Errington *et al.*, 1998) and polygalacturonase (Kalamaki *et al.*, 2003); (iv) increased viscosity of strawberry juice through development of transgenic strawberries with reduced level of endogenous enzyme responsible for pectin degradation, that is, pectate lyase (Sesmero *et al.*, 2009); (v) enhanced vitamin C content in tomato juice as a result of potassium fertilization during fruit cultivation (Ghebbi Si-smail *et al.*, 2006); and (vi) triggering stress response in plant cells using postharvest ultraviolet-C light treatment on grape berries to induce formation of phytoalexins and hence improve the stilbene resveratrol content in the resultant grape juice (González-Barrio *et al.*, 2009). Therefore, selection of berry cultivars with high polyphenols for juice processing becomes important as different stages of processing might affect the nutritional quality of berry juice (Table 6.2).

Berry fruits are rich in vitamin C and polyphenols, including anthocyanin pigments that give red color to the fruit. In general, mechanical crushing and pressing result in juice and color (i.e., free anthocyanin) extraction from the vacuole of the plant cell in the flesh and skin portions of berries, respectively. Although the resulting berry juice has a satisfactory level of polyphenols, the skin cells or solid press residues are still concentrated with polyphenols, mainly in bound form and conjugated with glucose and other sugars, as shown in the juice-processing wastes from cranberries (Zheng & Shetty, 2000), blackcurrants (Landbo & Meyer, 2001), blueberries (Lee & Wrolstad, 2004), pomegranate (Qu *et al.*, 2010), and grapes (Li *et al.*, 2013). The localization of these bound polyphenols within the compact skin cell walls or those tightly bound to the cell wall material has been reported, and particularly, extensive study has been conducted on grape skin (Pinelo *et al.*, 2006). As plant cell wall composes mainly of polysaccharides (cellulose, xyloglucan, arabinan, mannan, lignin, hemicellulose, and pectin) and proteins, stabilized by strong ionic and covalent linkages, this highly complex framework presents a barrier for the release of polyphenols. Therefore, the addition of pectinolytic enzymes, which usually consist of polygalacturonase, pectin lyase, pectin methyl esterase, and other cell wall degrading enzymes, such as arabinase, hemicellulase, and xylanase, into the berry mash has been widely used in juice processing to degrade cell wall material in order to release the bound polyphenols from the berry skins. Enzyme-assisted release of polyphenols has been proven efficient in maximizing the juice yield and contents of anthocyanin and polyphenols in the juice of blackcurrant

Table 6.2 Changes in the nutritional components of berry juice after conventional processing

Berry juice	Conventional juice processing procedures	Nutritional components	Effect of processing[a]	References
Blackberry	Frozen berries → blanching (95 °C, 3 min) → enzymatic treatment (40 °C, 1 h) → pressing	Total ellagitannin	(−) 69%	Hager et al. (2010)
	Frozen berries → blanching (95 °C, 3 min) → enzymatic treatment (40 °C, 1 h) → pressing → pasteurization (90 °C)	Total ellagitannin	(−) 70%	Hager et al. (2010)
	Frozen berries → blanching (95 °C, 3 min) → enzymatic treatment (40 °C, 1 h) → pressing → clarification	Total ellagitannin	(−) 88%	Hager et al. (2010)
	Frozen berries → blanching (95 °C, 3 min) → enzymatic treatment (40 °C, 1 h) → pressing → clarification → pasteurization (90 °C)	Total ellagitannin	(−) 82%	Hager et al. (2010)
	Frozen berries → blanching (95.5 °C, 5 min) → crushing → pressing → pasteurization (88–93 °C, 40 s)	Total phenolics	(−) 45%	Gancel et al. (2011)
Blackcurrant	Preprocessing conditions and types of raw materials not reported, juice was produced by steam extraction for 60 min	Quercetin	(−) 86%	Hakkinen et al. (2000)
		Myricetin	(−) 70%	
		Vitamin C	(−) 49%	
	Frozen berries → crushing → enzymatic treatment → pressing → thermal (75 °C, 90 s)	Total anthocyanin	(+) 68%	Landbo and Meyer (2004)
	Frozen berries → crushing → enzymatic treatment → pressing	Total phenolics	(+) 52%	Koponen et al. (2008a, b)
		Total anthocyanin	(+) 11%	
	Fresh berries → crushing → enzymatic treatment (50 °C) → pressing	Total flavonol	(+) 8%	Holtung et al. (2011)
		Total phenolics	(−) 24%	
	Fresh berries → crushing → enzymatic treatment (50 °C) → pressing → clarification	Total anthocyanin	(−) 29%	Holtung et al. (2011)
		Total phenolics	(−) 8%	

Process	Compound	Change	Reference
Fresh berries → crushing → enzymatic treatment (50 °C) → pressing → clarification → filtration	Total anthocyanin	(−) 39%	Holtung et al. (2011)
	Total phenolics	(−) 28%	Holtung et al. (2011)
Fresh berries → crushing → heating (50 °C) → enzymatic treatment (50 °C, 3 h) → pressing	Total anthocyanin	(−) 47%	Woodward et al. (2011)
	Total anthocyanin	(=)	Woodward et al. (2011)
Fresh berries → crushing → heating (50 °C) → enzymatic treatment (50 °C, 3 h) → pressing → pasteurization (103 °C, 45 s)	Total anthocyanin	(−) 20%	Woodward et al. (2011)
Fresh berries → crushing → heating (50 °C) → enzymatic treatment (50 °C, 3 h) → pressing → pasteurization (103 °C, 45 s) → filtration	Total anthocyanin	(−) 20%	Woodward et al. (2011)
Blueberry Frozen berries → crushing → enzymatic treatment (43 °C) → pressing	Total anthocyanin	(−) 69%	Skrede et al. (2000)
	Delphinidin-3-galactoside	(−) 96%	
	Delphinidin-3-glucoside	(−) 93%	
	Cyanidin-3-galactoside + Delphinidin-3-arabinoside	(−) 97%	
	Petunidin-3-galactoside	(−) 67%	
	Petunidin-3-glucoside	(−) 76%	
	Peonidin-3-galactoside	(+) 340%	
	Petunidin-3-arabinoside	(−) 59%	
	Malvidin-3-galactoside	(+) 65%	
	Malvidin-3-glucoside	(+) 130%	
	Malvidin-3-arabinoside	(+) 61%	
	Chlorogenic acid	(−) 56%	
	Flavonol glucosides	(−) 43%	
	Procyanidin	(−) 84%	

(continued overleaf)

Table 6.2 (*continued*)

Berry juice	Conventional juice processing procedures	Nutritional components	Effect of processing[a]	References
	Frozen berries → crushing → enzymatic treatment (43 °C) → pressing → pasteurization (90 °C, 1 min)	Total anthocyanin	(−) 65%	Skrede *et al.* (2000)
		Delphinidin-3-galactoside	(−) 59%	
		Delphinidin-3-glucoside	(−) 29%	
		Cyanidin-3-galactoside + delphinidin-3-arabinoside	(−) 43%	
		Petunidin-3-galactoside	(−) 35%	
		Petunidin-3-glucoside	(−) 33%	
		Peonidin-3-galactoside	(+) 270%	
		Petunidin-3-arabinoside	(−) 33%	
		Malvidin-3-galactoside	(+) 25%	
		Malvidin-3-glucoside	(+) 84%	
		Malvidin-3-arabinoside	(+) 33%	
		Chlorogenic acid	(−) 42%	
		Flavonol glucosides	(−) 62%	
		Procyanidin	(−) 54%	
		Total phenolics	(+) 77%	
Blueberry	Frozen berries → blanching (95 °C, 2 min) → crushing → enzymatic treatment → pressing	Delphinidin-3-glucoside	(+) 590%	Lee *et al.* (2002)
		Cyanidin-3-glucoside	(+) 50%	
		Petunidin-3-glucoside	(+) 2350%	
		Peonidin-3-glucoside	(+) 20%	
		Malvidin-3-glucoside	(+) 202%	
		Total anthocyanin	(+) 372%	
		Cinnamic acids	(+) 9%	
		Flavonol glucosides	(+) 25%	
		Total phenolics	(+) 15%	

	Process	Compound	Change	Reference
	Frozen berries → blanching (95 °C, 2 min) → crushing → enzymatic treatment → pressing → clarification	Delphinidin-3-glucoside	(+)1667%	Lee et al. (2002)
		Cyanidin-3-glucoside	(+) 40%	
		Petunidin-3-glucoside	(+) 4400%	
		Peonidin-3-glucoside	(+) 10%	
		Malvidin-3-glucoside	(+) 185%	
		Total anthocyanin	(+) 333%	
		Cinnamic acids	(−) 4%	
		Flavonol glucosides	(+) 12%	
		Total phenolics	(+) 2%	
	Frozen berries → blanching (95 °C, 2 min) → crushing → enzymatic treatment → pressing → clarification → pasteurization	Delphinidin-3-glucoside	(+) 354%	Lee et al. (2002)
		Cyanidin-3-glucoside	(+) 25%	
		Petunidin-3-glucoside	(+) 163%	
		Peonidin-3-glucoside	(+) 20%	
		Malvidin-3-glucoside	(+) 41%	
		Total anthocyanin	(+) 80%	
		Cinnamic acids	(=)	
		Flavonol glucosides	(+) 19%	
		Total phenolics	(+) 7%	
Blueberry	Frozen berries → blanching (steam, 85 °C, 3 min) → crushing → enzymatic treatment (25 °C, 1 h) → pressing → pasteurization (90 °C, 1 min)	Total anthocyanin	(+) 104%	Rossi et al. (2003)

(continued overleaf)

Table 6.2 (*continued*)

Berry juice	Conventional juice processing procedures	Nutritional components	Effect of processing[a]	References
		Total cinnamates	(+) 34%	
		Delphinidin-3-galactoside	(+) 1133%	
		Delphinidin-3-glucoside	(+) 2003%	
		Delphinidin-3-arabinoside	(+) 2672%	
		Cyanidin-3-galactoside	(+) 217%	
		Cyanidin-3-glucoside	(+) 71%	
		Cyanidin-3-arabinoside	(+) 286%	
		Petunidin-3-galactoside	(+) 637%	
		Petunidin-3-glucoside	(+) 526%	
		Petunidin-3-arabinoside	(+) 595%	
		Peonidin-3-galactoside	(+) 179%	
		Peonidin-3-glucoside	(+) 194%	
		Malvidin-3-galactoside	(+) 132%	
		Malvidin-3-glucoside	(+) 120%	
		Malvidin-3-arabinoside	(+) 178%	
		Malvidin	(+) 12%	Brambilla *et al.* (2008)
	Frozen berries → blanching (steam, 85 °C, 3 min) → cooling → crushing → enzymatic treatment (25 °C, 1 h) → pressing → pasteurization (90 °C, 1 min)	Delphinidin	(−) 24%	
		Petunidin	(=)	
		Cyanidin	(=)	
		Peonidin	(=)	

	Process	Compound	Change	Reference
	Frozen berries → blanching (95 °C, 3 min) → enzymatic treatment (40 °C, 1 h) → pressing	Total anthocyanin	(−) 20%	Brownmiller et al. (2008)
	Frozen berries → blanching (95 °C, 3 min) → enzymatic treatment (40 °C, 1 h) → pressing → pasteurization (90 °C)	Total anthocyanin	(−) 28%	Brownmiller et al. (2008)
	Frozen berries → blanching (95 °C, 3 min) → enzymatic treatment (40 °C, 1 h) → pressing → clarification	Total anthocyanin	(−) 54%	Brownmiller et al. (2008)
	Frozen berries → blanching (95 °C, 3 min) → enzymatic treatment (40 °C, 1 h) → pressing → clarification → pasteurization (90 °C)	Total anthocyanin	(−) 59%	Brownmiller et al. (2008)
	Fresh berries → blanching (steam, 95 °C, 2 min) → crushing → pressing → pasteurization (92 °C, 2 min)	Total anthocyanin	(+) 783%	Syamaladevi et al. (2012)
Cranberry	Preprocessing conditions and types of raw materials not reported, juice was produced by pasteurization	Total procyanidin	(+) 1212%	Prior et al. (2001)
	Frozen berries → crushing → enzymatic treatment (55 °C) → pressing	Total phenolics	(−) 97%	Côté et al. (2011)
		Total anthocyanin	(−) 100%	
	Frozen berries → crushing → enzymatic treatment (55 °C) → pressing → clarification	Total phenolics	(−) 78%	Côté et al. (2011)
			(−) 66%	
	Frozen berries → blanching (95 °C, 3 min) → enzymatic treatment (45 °C, 1 h) → pressing → clarification → pasteurization (90 °C, 10 min)	Total anthocyanin	(−) 27%	White et al. (2011)
		Total flavonol	(−) 17%	
		Procyanidin	(−) 12%	

(continued overleaf)

Table 6.2   (*continued*)

Berry juice	Conventional juice processing procedures	Nutritional components	Effect of processing[a]	References
Elderberry	Frozen berries → crushing → blanching (70 °C, 10 min)	Total phenolics	(+) 20%	Galić *et al.* (2009)
		Flavonoids	(+) 15%	
		Nonflavonoids	(−) 29%	
		Flavan-3-ol	(+) 10%	
		Total anthocyanin	(+) 59%	
	Frozen berries → crushing → blanching (70 °C, 10 min) → pressing	Total phenolics	(+) 10%	Galić *et al.* (2009)
		Flavonoids	(+) 49%	
		Nonflavonoids	(=)	
		Flavan-3-ol	(+) 37%	
		Total anthocyanin	(+) 90%	
Grape	Fresh berries → crushing → maceration (18 h) → pressing → filtration → pasteurization (85 °C)	(+)-Catechin	(+) 569%	Fuleki and Ricardo-da-Silva (2002)
		(−)-Epicatechin	(+) 482%	
		Procyanidin B1 dimer	(+) 601%	
		Procyanidin B2 dimer	(+) 22%	
		Procyanidin B3 dimer	(+) 160%	
		Procyanidin B4 dimer	(−) 100%	
		Procyanidin B1-3-dimer-gallate	(+) 1500%	
		Procyanidin B2-3-dimer-gallate	(−) 100%	
		Procyanidin B3-3-dimer-gallate	(+) 960%	
		Procyanidin C1 trimer	(+) 1650%	
		Procyanidin T2 trimer	(+) 393%	

Grape	Fresh berries → crushing → pressing → filtration → pasteurization (85 °C)	(+)-Catechin	(+) 105%	Fuleki and Ricardo-da-Silva (2002)
		(−)-Epicatechin	(+) 135%	
		Procyanidin B1 dimer	(−) 27%	
		Procyanidin B2 dimer	(−) 19%	
		Procyanidin B3 dimer	(−) 85%	
		Procyanidin B4 dimer	(−) 50%	
		Procyanidin	(+) 80%	
		B1-3-dimer-gallate		
		Procyanidin	(−) 100%	
		B2-3-dimer-gallate		
		Procyanidin	(+) 7%	
		B3-3-dimer-gallate		
		Procyanidin C1 trimer	(−) 60%	
		Procyanidin T2 trimer	(−) 60%	
	Fresh berries → crushing → heating (60 °C, 1.5 h) → pressing → filtration → pasteurization (85 °C)	(+)-Catechin	(−) 34%	Fuleki and Ricardo-da-Silva (2002)
		(−)-Epicatechin	(−) 66%	
		Procyanidin B1 dimer	(+) 32%	
		Procyanidin B2 dimer	(+) 56%	
		Procyanidin B3 dimer	(−) 72%	
		Procyanidin B4 dimer	(−) 100%	
		Procyanidin	(+) 9%	
		B1-3-dimer-gallate		

(continued overleaf)

Table 6.2 (*continued*)

Berry juice	Conventional juice processing procedures	Nutritional components	Effect of processing[a]	References
	Fresh berries → crushing → heating (70 °C, 60 min) → pressing → pasteurization (95 °C, 15 min)	Procyanidin B2-3-dimer-gallate	(−) 100%	
		Procyanidin B3-3-dimer-gallate	(+) 1070%	
		Procyanidin C1 trimer	(+) 40%	
		Procyanidin T2 trimer	(−) 23%	
		Total phenolics	(+) 7%	Talcott *et al.* (2003)
		Total anthocyanin	(+) 4%	
		Delphinidin-3,5-diglucoside	(+) 7%	
		Cyanidin-3,5-diglucoside	(=)	
		Petunidin-3,5-diglucoside	(=)	
		Pelargonidin-3,5-diglucoside	(+) 26%	
		Peonidin-3,5-diglucoside	(+) 8%	
		Malvidin-3,5-diglucoside	(+) 17%	
Grape	Fresh berries → crushing → pressing	trans-Resveratrol	(=)	
		trans-Piceid	(=)	
		cis-Piceid	(=)	
		Total stilbene	(=)	Leblanc *et al.* (2008)
	Fresh berries → crushing → heating (60 °C) → pressing	trans-Resveratrol	(+) 10%	
		trans-Piceid	(+) 500%	
		cis-Piceid	(+) 220%	
		Total stilbene	(+) 250%	Leblanc *et al.* (2008)
	Frozen berries → crushing → pressing	trans-Resveratrol	(=)	Leblanc *et al.* (2008)

	Processing	Compound	Change	Reference
	Fresh berries → crushing → enzymatic treatment (7 °C, 14 h) → pressing	trans-Piceid	(+) 580%	Leblanc et al. (2008)
		cis-Piceid	(=)	
		Total stilbene	(+) 100%	
		trans-Resveratrol	(=)	
	Fresh berries → heating (50 °C) → crushing → pressing → pasteurization (107 °C)	trans-Piceid	(+) 450%	Capanoglu et al. (2013)
		cis-Piceid	(=)	
		Total phenolics	(−) 80%	
		Total flavonoids	(−) 82%	
		Delphinidin-3-glucoside	(−) 96%	
		Cyanidin-3-glucoside	(−) 48%	
		Petunidin-3-glucoside	(−) 93%	
		Peonidin-3-glucoside	(−) 23%	
		Malvidin-3-glucoside	(−) 44%	
		Malvidin-acetyl-glucoside	(−) 3460%	
		Peonidin-coumaroyl-glucoside	(−) 80%	
		Malvidin-coumaroyl-glucoside	(−) 85%	
		Quercetin-3-rutinoside	(−) 46%	
		Total catechin	(−) 95%	
		Total epicatechin	(−) 98%	
Grape	Fresh berries → heating (50 °C) → crushing → pressing → pasteurization (107 °C) → clarification	Total phenolics	(−) 83%	Capanoglu et al. (2013)

(continued overleaf)

Table 6.2 (*continued*)

Berry juice	Conventional juice processing procedures	Nutritional components	Effect of processing[a]	References
		Total flavonoids	(−) 54%	
		Delphinidin-3-glucoside	(−) 98%	
		Cyanidin-3-glucoside	(−) 85%	
		Petunidin-3-glucoside	(−) 6000%	
		Peonidin-3-glucoside	(−) 84%	
		Malvidin-3-glucoside	(−) 94%	
		Malvidin-acetyl-glucoside	(−) 3460%	
		Peonidin-coumaroyl-glucoside	(−) 1410%	
		Malvidin-coumaroyl-glucoside	(−) 7420%	
		Quercetin-3-rutinoside	(−) 23%	
		Total catechin	(−) 96%	
		Total epicatechin	(−) 98%	
		Total phenolics	(−) 84%	Capanoglu *et al.* (2013)
	Fresh berries → heating (50 °C) → crushing → pressing → pasteurization (107 °C) → clarification → filtration	Total flavonoids	(−) 94%	
		Delphinidin-3-glucoside	(−) 99%	
		Cyanidin-3-glucoside	(−) 94%	
		Petunidin-3-glucoside	(−) 6000%	
		Peonidin-3-glucoside	(−) 95%	
		Malvidin-3-glucoside	(−) 98%	
		Malvidin-acetyl-glucoside	(−) 3460%	
		Peonidin-coumaroyl-glucoside	(−) 1410%	
		Malvidin-coumaroyl-glucoside	(−) 7420%	
		Quercetin-3-rutinoside	(−) 54%	

	Processing	Compound	Change	Reference
	Preprocessing conditions and types of raw materials not reported, juice was produced by clarification, followed by pasteurization (80 °C, 30 min)	Total catechin Total epicatechin Total phenolics	(−) 94% (−) 94% (−) 3%	Martino et al. (2013)
	Preprocessing conditions and types of raw materials not reported, juice was produced by pasteurization (80 °C, 30 min), followed by clarification	Total phenolics	(+) 8%	Martino et al. (2013)
Pomegranate	Fresh berries → crushing → pressing → heating (50 °C) → cooling (20 °C) → pasteurization (20 min)	Total phenolics	(−) 7%	Alper et al. (2005)
	Fresh berries → crushing → pressing → heating (50 °C) → cooling (20 °C) → clarification (gelatin + bentonite) → pasteurization (20 min)	Total phenolics	(=)	Alper et al. (2005)
	Fresh berries → crushing → pressing → heating (50 °C) → cooling (20 °C) → clarification (gelatin + bentonite + PVPP) → pasteurization (20 min)	Total phenolics	(=)	Alper et al. (2005)
	Fresh berries → crushing → pressing → heating (50 °C) → cooling (20 °C) → ultrafiltration (10 kDa) → pasteurization (20 min)	Total phenolics	(−) 14%	Alper et al. (2005)
	Fresh berries → crushing → pressing → pasteurization (85 °C, 5 min)	Delphnidin-3,5-diglucoside Cyanidin-3,5-diglucoside Pelargonidin-3,5-diglucoside Delphnidin-3-glucoside Cyanidin-3-glucoside	(−) 13% (−) 12% (+) 4% (−) 21% (−) 35%	Alighourchi et al. (2008)

(continued overleaf)

Table 6.2 (*continued*)

Berry juice	Conventional juice processing procedures	Nutritional components	Effect of processing[a]	References
	Fresh berries → crushing → pressing → filtration → clarification	Pelargonidin-3-glucoside	(−) 30%	Turfan *et al.* (2011)
		Total anthocyanin	(−) 15%	
		Total anthocyanin	(−) 5%	
		Delphnidin-3,5-diglucoside	(=)	
		Cyanidin-3,5-diglucoside	(−) 4%	
		Delphnidin-3-glucoside	(−) 38%	
		Pelargonidin-3,5-diglucoside	(−) 6%	
		Cyanidin-3-glucoside	(−) 14%	
		Pelargonidin-3-glucoside	(=)	
	Fresh berries → crushing → pressing → filtration → pasteurization (95 °C, 10 min)	Total anthocyanin	(−) 6%	Turfan *et al.* (2011)
		Delphnidin-3,5-diglucoside	(+) 68%	
		Cyanidin-3,5-diglucoside	(−) 16%	
		Delphnidin-3-glucoside	(+) 60%	
		Pelargonidin-3,5-diglucoside	(−) 26%	
		Cyanidin-3-glucoside	(−) 12%	
		Pelargonidin-3-glucoside	(−) 11%	
Pomegranate	Fresh berries → crushing → pressing → filtration → clarification → pasteurization (95 °C, 10 min)	Total anthocyanin	(−) 16%	Turfan *et al.* (2011)
		Delphnidin-3,5-diglucoside	(+) 55%	
		Cyanidin-3,5-diglucoside	(−) 19.%	
		Delphnidin-3-glucoside	(+) 22%	
		Pelargonidin-3,5-diglucoside	(−) 26%	

Process	Compound	Change	Reference
Fresh berries → steaming (8 min) → pressing → enzymatic treatment (25 °C, 2 h) → double-clarification → pasteurization (90 °C)	Cyanidin-3-glucoside	(−) 26%	Fischer et al. (2011)
	Pelargonidin-3-glucoside	(−) 26%	
	Delphinidin-3,5-diglucoside	(−) 51%	
	Cyanidin-3,5-diglucoside	(−) 55%	
	Pelargonidin-3,5-diglucoside	(−) 32%	
	Delphinidin-3-glucoside	(−) 54%	
	Cyanidin-3-penta-hexoside	(−) 43%	
	Cyanidin-3-glucoside	(−) 54%	
	Pelargonidin-3-glucoside	(−) 53%	
	Cyanidin-3-pentoside	(−) 43%	
	Total anthocyanin	(−) 53%	
	Digalloyl-hexoside	(+) 429%	
	Total gallotannins	(+) 429%	
	Ellagic acid	(−) 33%	
	Ellagic acid derivative	(−) 17%	
	Ellagitannin	(+) 33%	
	Total ellagitannin	(+) 65%	
	Total hydrolysable tannins	(+) 65%	
	Total dihydroflavonols	(−) 82%	
	Total phenolics	(+) 63%	
Pomegranate Fresh berries → pressing → enzymatic treatment (25 °C, 2 h) → double-clarification → pasteurization (90 °C)	Delphinidin-3,5-diglucoside	(−) 19%	Fischer et al. (2011)

(continued overleaf)

Table 6.2 *(continued)*

Berry juice	Conventional juice processing procedures	Nutritional components	Effect of processing[a]	References
		Cyanidin-3,5-diglucoside	(+) 19%	
		Pelargonidin-3,5-diglucoside	(+) 19%	
		Delphinidin-3-glucoside	(−) 15%	
		Cyanidin-3-penat-hexoside	(+) 33%	
		Cyanidin-3-glucoside	(+) 18%	
		Pelargonidin-3-glucoside	(+) 40%	
		Cyanidin-3-pentoside	(=)	
		Total anthocyanin	(=)	
		Total gallotannins	(=)	
		Ellagic acid	(−) 20%	
		Ellagitannin	(−) 48%	
		Total ellagitannin	(−) 53%	
		Total hydrolysable tannins	(−) 53%	
		Total dihydroflavonols	(−) 22%	
		Total phenolics	(−) 51%	
	Fresh berries → pressing → clarification → thermal (65 °C, 30 s)	Vitamin C	(−) 48%	Mena *et al.* (2013)
		Anthocyanin	(−) 5%	
		Punicalagins	(+) 580%	
		Punicalagin-like	(−) 28%	
		Punicalin	(−) 10%	
		Ellagic acid	(−) 25%	
	Fresh berries → pressing → clarification → thermal (90 °C, 5 s)	Vitamin C	(−) 60%	Mena *et al.* (2013)
		Anthocyanin	(−) 7%	
		Punicalagins	(+) 80%	
		Punicalagin-like	(−) 18%	

Raspberry	Fresh berries → crushing → heating (82 °C, 2 min) → enzymatic treatment (6 h) → pasteurization (85 °C, 2 min) → clarification	Punicalin Ellagic acid Anthocyanin	(+) 5% (=) (−) 15%	Versari et al. (1997)
	Frozen berries → blanching (95 °C, 3 min) → enzymatic treatment (1 h) → pressing	Quercetin Ellagic acid Total anthocyanin	(=) (=) (−) 62%	Hager et al. (2008)
	Frozen berries → blanching (95 °C, 3 min) → enzymatic treatment (1 h) → pressing → pasteurization (90 °C)	Total anthocyanin	(−) 69%	Hager et al. (2008)
	Frozen berries → blanching (95 °C, 3 min) → enzymatic treatment (1 h) → pressing → clarification	Total anthocyanin	(−) 65%	Hager et al. (2008)
	Frozen berries → blanching (95 °C, 3 min) → enzymatic treatment (1 h) → pressing → clarification → pasteurization (90 °C)	Total anthocyanin	(−) 73%	Hager et al. (2008)
	Frozen berries → crushing → enzymatic treatment (50 °C, 4 h) → clarification	Cyanidin-3-sophoroside Cyanidin-3-glucorutinoside Cyanidin-3-glucoside Cyanidin-3-rutinoside Total anthocyanin Ascorbic acid Dehydroascorbic acid Total vitamin C Total phenolics	(=) (=) (=) (=) (=) (−) 11% (+) 8% (−) 7% (+) 20%	Verbeyst et al. (2012)

(continued overleaf)

Table 6.2 (*continued*)

Berry juice	Conventional juice processing procedures	Nutritional components	Effect of processing[a]	References
Strawberry	Fresh berries → crushing → heating (82 °C, 2 min) → enzymatic treatment (6 h) → pasteurization (85 °C, 2 min) → clarification	Anthocyanin	(=)	Versari *et al.* (1997)
		Quercetin	(=)	
		Ellagic acid	(+) 113%	
	Frozen berries → crushing → heating (40 °C) → enzymatic treatment (20 min) → pressing	Vitamin C	(−) 25%	Klopotek *et al.* (2005)
		Total phenolics	(−) 39%	
		Total anthocyanin	(+) 2%	
	Frozen berries → crushing → heating (40 °C) → enzymatic treatment (20 min) → pressing → centrifugation	Vitamin C	(−) 34%	Klopotek *et al.* (2005)
		Total phenolics	(−) 47%	
		Total anthocyanin	(+) 3%	
Strawberry	Frozen berries → crushing → heating (40 °C) → enzymatic treatment (20 min) → pressing → centrifugation → clarification	Vitamin C	(−) 36%	Klopotek *et al.* (2005)
		Total phenolics	(−) 50%	
		Total anthocyanin	(−) 10%	
	Frozen berries → crushing → heating (40 °C) → enzymatic treatment (20 min) → pressing → centrifugation → clarification → filtration	Vitamin C	(−) 45%	Klopotek *et al.* (2005)
		Total phenolics	(−) 51%	
		Total anthocyanin	(−) 8%	

Process	Component	Change	Reference
Frozen berries → crushing → heating (40 °C) → enzymatic treatment (20 min) → pressing → centrifugation → clarification → filtration → filling → pasteurization (85 °C, 5 min)	Vitamin C	(−) 64%	Klopotek et al. (2005)
Frozen berries → crushing → heating (45 °C) → enzymatic treatment (90 min) → pressing → hot-filling (85 °C, 5 s)	Total phenolics Total anthocyanin Ascorbic acid	(−) 64% (−) 33% (+) 56%	Hartmann et al. (2008)
Frozen berries → crushing → heating (25 °C) → pressing → Pasteurization (85 °C, 15 min)	Polyphenols Anthocyanin Ascorbic acid	(+) 80% (+) 87% (+) 46%	Hartmann et al. (2008)
Frozen berries → crushing → heating (25 °C) → pressing → hot-filling (85 °C, 5 s)	Polyphenols Anthocyanin Ascorbic acid	(+) 62% (+) 61% (+) 64%	Hartmann et al. (2008)
Frozen berries → crushing → enzymatic treatment (50 °C, 4 h) → clarification	Polyphenols Anthocyanin Pelargonidin-3-glucoside Pelargonidin-3-arabinoside Total anthocyanin Ascorbic acid Dehydroascorbic acid	(+) 65% (+) 71% (=) (=) (=) (−) 8% (+) 10%	Verbeyst et al. (2012)

(continued overleaf)

Table 6.2 (*continued*)

Berry juice	Conventional juice processing procedures	Nutritional components	Effect of processing[a]	References
Tomato	Processing conditions and types of raw materials not reported	Total vitamin C	(−) 4%	Tavares and Rodriguez-Amaya (1994)
		Total phenolics	(+) 21%	
		cis-Phytofluene	(+) 38%	
		13-*cis*-β-Carotene	(+) 100%	Gahler *et al.* (2003)
		trans-β-Carotene	(−) 61%	
		trans-ζ-Carotene	(+) 225%	
		trans-γ-Carotene	(−) 100%	
		cis-Lycopene	(+) 137%	
		trans-Lycopene	(+) 98%	
		Total carotenoid	(+) 86%	
		Vitamin A	(−) 59%	
		Vitamin C	(−) 67%	
	Fresh berries → crushing → sterilization (121 °C, 2 min) → filling → pasteurization (80 °C, 20 min)	Total phenolics	(+) 14%	Lin and Chen (2005)
		All-*trans*-lutein	(+) 100%	
	Fresh berries → crushing → heating (steam, 82 °C, 2 min) → pressing → filtration → thermal (90 °C, 5 min) → canning	9-*cis*-Lutein	(=)	
		13-*cis*-Lutein	(=)	
		Di-*cis*-β-carotene	(=)	
		15-*cis*-β-Carotene	(=)	
		9-*cis*-β-Carotene	(=)	

Tomato	Fresh berries → crushing → heating (steam, 82 °C, 2 min) → pressing → filtration → thermal 1 (steam, 70 °C, 10 min) → canning → thermal 2 (100 °C, 30 min)	All-*trans*-β-carotene	(=)
		cis-β-Carotene	(=)
		13-*cis*-β-Carotene	(=)
		All-*trans*-lycopene	(+) 75%
		Di-*cis*-lycopene (I)	(=)
		15-*cis*-Lycopene	(=)
		13-*cis*-Lycopene	(=)
		Di-*cis*-lycopene (II)	(+) 20%
		9-*cis*-Lycopene	(=)
		5-*cis*-Lycopene	(+) 29%
		All-*trans*-lutein	(+) 125% Lin and Chen (2005)
		9-*cis*-Lutein	(+) 50%
		13-*cis*-Lutein	(+) 50%
		Di-*cis*-β-carotene	(−) 100%
		15-*cis*-β-Carotene	(=)
		9-*cis*-β-Carotene	(=)
		All-*trans*-β-carotene	(=)
		cis-β-Carotene	(=)
		13-*cis*-β-Carotene	(=)
		All-*trans*-lycopene	(+) 50%
		Di-*cis*-lycopene (I)	(+) 7%
		15-*cis*-Lycopene	(+) 27%

(continued overleaf)

Table 6.2 (*continued*)

Berry juice	Conventional juice processing procedures	Nutritional components	Effect of processing[a]	References
	Fresh berries → crushing → heating (steam, 82°C, 2 min) → pressing → filtration → thermal 1 (95°C) → thermal 2 (121°C, 40 s) → thermal 3 (93°C) → canning	13-*cis*-Lycopene	(=)	Lin and Chen (2005)
		Di-*cis*-lycopene (II)	(=)	
		9-*cis*-Lycopene	(+) 33%	
		5-*cis*-Lycopene	(=)	
		All-*trans*-lutein	(=)	
		9-*cis*-Lutein	(+) 200%	
		13-*cis*-Lutein	(+) 500%	
		Di-*cis*-β-carotene	(=)	
		15-*cis*-β-Carotene	(=)	
		9-*cis*-β-Carotene	(=)	
		All-*trans*-β-carotene	(−) 20%	
		cis-β-Carotene	(=)	
		13-*cis*-β-Carotene	(=)	
		All-*trans*-lycopene	(+) 70%	
		Di-*cis*-lycopene (I)	(−) 80%	
		15-*cis*-Lycopene	(=)	
		13-*cis*-Lycopene	(+) 65%	
		Di-*cis*-lycopene (II)	(=)	
		9-*cis*-Lycopene	(=)	
		5-*cis*-Lycopene	(=)	

Food	Preprocessing conditions	Component	Change	Reference
Tomato	Preprocessing conditions and types of raw materials not reported, juice was produced by pasteurization	Total carotenoids	(+) 63%	Sánchez-Moreno et al. (2006)
		Vitamin A	(=)	
		Lutein	(+) 14%	
		Lycopene epoxide	(+) 44%	
		Lycopene	(+) 108%	
		γ-Carotene	(+) 17%	
		β-Carotene	(+) 51%	
		L-ascorbic acid	(−) 42%	
		Total vitamin C	(−) 44%	
	Preprocessing conditions and types of raw materials not reported, juice was produced by thermal treatment (90 °C, 2 min), followed by cold storage for 1 month	5-Methyltetrahydrofolate	(−) 100%	Iniesta et al. (2009)
	Fresh berries → crushing → heating → pasteurization (95 °C, 3 min)	Total carotenoid	(+) 25%	Mendelová et al. (2013)
	Fresh berries → thermal 1 (70 °C, 20 min) → thermal 2 (70 °C, 5 min)	Lycopene	(+) 26%	Kamiloglu et al. (2014)
		Rutin apioside	(−) 100%	
		Rutin	(−) 100%	
		Naringenin	(−) 100%	
		Naringenin chalcone	(−) 100%	
		Chlorogenic acid	(+) 29%	

[a]Processing resulted in a considerable '+' increase, (−) decrease, and (=) no significant change in specified nutritional components when compared to the unprocessed (freshly prepared, raw, or untreated).

(+68% total anthocyanin) (Landbo & Meyer, 2004; Koponen *et al.*, 2008a, b), grape (+450% stilbene, +464% (+)-catechin, +347% (−)-epicatechin) (Fuleki & Ricardo-da-Silva, 2002; Leblanc *et al.*, 2008), pomegranate (+125% total phenolics) (Rinaldi *et al.*, 2013), raspberry (up to +20% total phenolics) (Versari *et al.*, 1997; Verbeyst *et al.*, 2012), and strawberry (up to +113% ellagic acid, +80% total phenolics, +87% total anthocyanin) (Versari *et al.*, 1997; Hartmann *et al.*, 2008; Verbeyst *et al.*, 2012). When adding pectinolytic enzymes for juice extraction from different types of berry fruits, factors important for enzyme reaction such as enzyme types, enzyme dosage, hydrolysis time, and temperature have to be taken into consideration to optimally extract the bound polyphenols from the berry skins. Sometimes, the presence of trace amount of β-glucosidase from the commercial enzyme preparations would assist in the removal of sugar moiety of anthocyanin; hence, it results in the formation of anthocyanidins that are more readily oxidized by endogenous polyphenol oxidase. Therefore, it is highly recommended to determine the presence of β-glucosidase from the commercial enzyme preparations prior to use to avoid greater degradation of anthocyanin in the resulting juice (Wightman & Wrolstad, 1996).

On the other hand, freezing of berries before juice processing may have profound advantages because freeze–thaw cycle leads to severe cell disruption and thus enhances the release of polyphenols from berry skins, as shown in the grape juice made from frozen berries as compared to those from fresh berries (Leblanc *et al.*, 2008).

As illustrated in Fig. 6.1, changes in the content and stability of polyphenols at different stages of berry juice processing, as a result of processing temperature and time and pretreatment, are inevitable (as summarized in Table 6.2). For instance, a short holding time (5 s vs 15 min at 85 °C) for strawberry juice (Hartmann *et al.*, 2008) or a high-temperature–short time for pomegranate juice (Mena *et al.*, 2013) during the pasteurization step is recommended to preserve thermolabile components such as vitamin C, polyphenols, and anthocyanin while controlling the associated-oxidation processes. However, when more processing steps are applied to the juice processing line, pasteurized blackberry (Hager *et al.*, 2010; Gancel *et al.*, 2011), blueberry (Lee *et al.*, 2002; Brownmiller *et al.*, 2008), cranberry (Prior *et al.*, 2001; White *et al.*, 2011), pomegranate (Alper *et al.*, 2005; Alighourchi *et al.*, 2008), raspberry (Hager *et al.*, 2008), and strawberry (Klopotek *et al.*, 2005) juice suffered a greater loss of total phenolics (between 4% and 97%), total anthocyanin (between 8% and 292%), and vitamin C (up to 39%) as compared to their filtered-, clarified-, or pressed-juice counterparts. There are some cases where polyphenols are relatively stable during berry juice processing. For instance, the content of total phenolics (+20%), flavonoids (+15%), flavan-3-ol (+10%), and total anthocyanin (+59%) is higher in pasteurized juice when compared to unprocessed elderberries (Galić *et al.*, 2009). In blueberry juice, a pronounced increase is detected for specific individual anthocyanin compounds such as delphinidin-3-glucoside, delphinidin-3-galactoside, cyanidin-3-galactoside, delphinidin-3-arabinoside, petunidin-3-glactoside, petunidin-3-glucoside, and petunidin-3-arabinoside (Skrede *et al.*, 2000; Rossi *et al.*, 2003; Brambilla *et al.*, 2008; Syamaladevi *et al.*, 2012). For individual polyphenol compounds in blueberry juice, the study of Wu *et al.* (2012) showed that the levels of isoquercetin, quercetin-3-pentoside, syringetin-3-hexoside, and caffeoylquinic acid were higher than other polyphenol compounds. In pomegranate juice, delphinidin-3, 5-diglucoside and delphinidin-3-glucoside are the predominant anthocyanin compounds that are more stable than others (Fischer *et al.*, 2011; Turfan *et al.*, 2011; Syamaladevi *et al.*, 2012). Maximal extraction of stilbene for up to 250% (Leblanc *et al.*, 2008), as well as procyanidin compounds (+)-catechin and (−)-epicatechin (Fuleki & Ricardo-da-Silva, 2002), can be achieved in grape juice making. Similarly, the level of malvidin-3-glucoside and delphinidin-3-acetylglucoside in grape juice (Talcott *et al.*, 2003; Wu *et al.*, 2012) increases after processing, possibly due to inactivation of oxidative enzymes such as polyphenol oxidase (i.e., protect against enzyme-catalyzed degradation of polyphenols) when blanching the fruit mash before performing juice pressing. This process is also known as hot pressing.

Clarification is important in juice processing as it involves transforming the cloudy (hazy and turbid) juice to clear juice, particularly for berries with high amounts of pectin to produce a stable juice product during storage. A number of methods can be applied to achieve juice clarification, including (i) centrifugation or membrane processes (e.g., micro- and ultrafiltration) to remove colloidally suspended particles (e.g., proteins, high molecular weight polyphenols and pectin) in juice, (ii) degrading pectin through addition of pectinolytic enzymes, and (iii) a fining treatment to remove cloudiness with suitable fining or clarifying agents (e.g., bentonite, gelatin, silica gel, and kieselsol). Both the works of Côté *et al.* (2011) and Martino *et al.* (2013) have demonstrated that adding a clarification step on pressed cranberry juice and grape juice, respectively, resulted in a better retention of phenolic compounds (up to +12%). Similar positive results are also obtained on total phenolic content (+16%) of clarified blackcurrant juice but not in the recovery of total anthocyanin content (−18%) (Holtung *et al.*, 2011). The loss of total ellagitannin content (−19%) in clarified blackberry juice has also been reported (Hager *et al.*, 2010). The stability of anthocyanin during juice clarification (centrifugation of 6000g for 10 min) is markedly affected in blueberry juice (up to −39%) (Brownmiller *et al.*, 2008). The work of Lee *et al.* (2002) showed that clarification degraded anthocyanin compounds such as cyanidin-3-glucoside, peonidin-3-glucoside, and malvidin-3-glucoside, whereas the levels of delphinidin-3-glucoside and petunidin-3-glucoside increased significantly to those found in unclarified blueberry juice. In clarified grape juice, a significant loss of petunidin-3-glucoside, peonidin-coumaroyl-glucoside, and malvidin-coumaroyl-glucoside has been reported (Capanoglu *et al.*, 2013). Moreover, a low recovery percentage of anthocyanin is found for pomegranate juice that has been clarified with gelatin (Turfan *et al.*, 2011).

Tomato is a good source of lycopene. This compound plays an important role in reducing the risk of both carcinoma and cardiovascular disease. It was found that the levels of lycopene and other naturally occurring carotenoids such as β- and γ-carotene and lutein increased for both industrial and home-made juice as compared to fresh tomatoes (Tavares & Rodriguez-Amaya, 1994; Lin & Chen, 2005; Sánchez-Moreno *et al.*, 2006; Mendelová *et al.*, 2013), which implies an enhanced bioaccessibility of carotenoids. Carotenoids are readily isomerized during food processing. The work of Lin and Chen (2005) concluded that heating at 121 °C for 40 s resulted in the highest yield of all-*trans* and *cis*-lycopene and lutein (up to +500%), followed by heating at 90 °C for 5 min, and heating in water at 100 °C for 30 min. On the other hand, the levels of phenolic compounds (Kamiloglu *et al.*, 2014), 5-methyltetrahydrofolate (Iniesta *et al.*, 2009), and heat-sensitive vitamin C (Gahler *et al.*, 2003; Sánchez-Moreno *et al.*, 2006) in tomato juice are reported to suffer significant reduction after thermal treatment.

6.2.3   Influence on Organoleptic Attributes

Important organoleptic attributes of berry juice is mostly judged on the basis of color, mouth feel (i.e., viscosity), appearance (i.e., stability of juice, no haze), taste, flavor, and aroma composition. As berry juice processing includes crushing, heating, enzyme treatment, pressing, clarification, and pasteurization, these trigger numerous complex chemical reactions leading to the formation of "processing-induced contaminants." These compounds are usually the products coming from nutrient degradation, lipid oxidation, Maillard reaction, and decomposition of sulfur-containing amino acids (Reineccius, 2006; Ong & Liu, 2010), taking place at high temperature. Therefore, these contaminants may present negative effects on the sensory qualities of berry juice.

6.2.3.1 Changes in Aroma Profile   Berry fruits consist of aroma-active compounds such as aliphatic esters, aldehydes, ketones, terpenes, terpenoids, and alcohols. During blackcurrant juice processing, the levels for most of these volatiles were not affected by crushing, clarification (gelatin and kieselsol mixture), and double-filtration steps (Mikkelsen & Poll, 2002). During the pressing and heating (98 °C) steps, blackcurrant juice suffered greater loss (between 50% and 100%) of the discriminating volatiles of fresh blackcurrant fruit, namely esters, terpenes, terpenoids, and ketones. Although β-damascenone has been exclusively detected at high levels in the pasteurized blackcurrant juice, this volatile compound is associated with an undesirable porridge-like, stewed fruit flavor perception. In another study that compared the aroma profile between pasteurized and enzyme-treated blackcurrant nectar, Iversen et al. (1998) have demonstrated that enzymatic treatment (polygalacturonase, 50 °C, 2 h) diminished the fruity aroma notes in the blackcurrant nectar, possibly because the enzyme reaction resulted in the hydrolysis of esters and some terpenes. No significant degradation in the levels of majority of the volatiles in pasteurized blackcurrant nectar have suggested the ability of pasteurization (88 °C, 27 s) to protect the degradation of these volatiles by inactivating endogenous oxidative enzymes. Although enzyme-assisted juice processing has been shown promising in extracting greater amount of polyphenols, as discussed earlier, this increase implies a negative sensory impact as higher intensity of mouth-drying astringency and bitterness has been perceived for enzyme-treated (pectinase, 47 °C, 4 h) blackcurrant juice when compared to freshly pressed juice (Laaksonen et al., 2013). It appeared that blackcurrant juice processed using enzymatic treatment suffered considerable losses in sugar level, low pH and sugar/acid ratio, high phenolics content, higher degree of polymerization, and lower procyanidin/prodelphinidin ratio in proanthocyanidins (Laaksonen et al., 2014). As a result, the original flavor and aroma of blackcurrant fruit is unable to be preserved.

According to Jensen et al. (2001) and Kaack et al. (2005), the aroma compounds of the freshly pressed elderberry juice can be grouped into six odor classes: elderberry (dihydroedulan), flowery (1-nonanal), fruity (2,3-methyl-1-butanol), grassy (1-hexanal), agrestic (1-octen-3-ol), and miscellaneous odors (1-penten-3-ol, eugenol). In the study of Kaack (2008), elderberry juice processed using enzymatic treatment (0.12% pectinase, 30 min) resulted in loss of esters, but some alcohols, terpenoids, carbonyls, aldehydes, ketones, trans-rose oxide, nerol oxide, and dihydroedulan were detected in high levels. In particular, an increase in selected volatiles of 1-pentanol and cis-rose oxide could be associated with the enzymatic conversion of fatty acids that is triggered by lipoxygenase. Other volatiles, namely (Z)-rose oxide, linalool, benzaldehyde, α-terpineol, and β-damascenone, have shown increments in concentration, which may be due to enzymatic hydrolysis (associated to activity of β-glucosidase) of glucosides in the pulp or in the juice from elderberries. Although enzymatic treatment of fruit juice from elderberries results in hydrolysis of all esters (associated to a fruity note), the presence of a distinct elderberry aroma compound (i.e., dihydroedulan) in the final juice is still prominent.

Regarding the grape juice, there is little information in the literature on the aroma changes during common grape juice production as most studies address the impact of winemaking practices on the aroma profile. The presence of green leaf-associated volatiles (in short known as GLVs), which primarily include hexanal, hexanol, cis-3-hexenol, trans-2-hexenal, and trans-2-hexenol, associated to herbaceous off-flavor in grape juice is of particular interest. The formation of GLVs is attributed to the enzymatic reaction of polyunsaturated fatty acids in the skin cells. During grapes crushing, GLVs can be transformed enzymatically (via lipoxygenase enzymatic action) to form other less aroma-active GLVs (Conde et al., 2007). Therefore, it is critical to control the enzymatic transformation of GLVs during grape juice processing to avoid the enduring green aroma in the juice. The study of Iyer et al. (2010) assessed the impact between hot press (heat to 60 °C, followed by pectinolytic enzymatic treatment) and hot break (heat to 82 °C and cooled to 60 °C, followed by pectinolytic enzymatic treatment) processing on crushed grapes on the levels of GLVs. The level of trans-2-hexenal, a potent aroma-active GLV, was at least sixfold higher for grape juice pretreated with hot break as compared to hot press. This occurrence is undesirable as the enzymatic transformation of trans-2-hexenal to trans-2-hexenol, a less aroma-active GLV, is hindered as a result of enzyme inactivation by initial high-temperature heat treatment (hot-break process). Therefore, conventional hot press processing (i.e., thermal treatment of no more than 60 °C on crushed grapes) is still a preferable processing procedure to decrease the levels of GLVs in grape juice.

Fresh tomato fruits are characterized by high levels of the following volatile chemical classes: terpenes and lactones (linalool and geraniol), carbonyls and alcohols (acetaldehyde and 3-methyl-2-butanol), sulfur compounds (2-isobutyl-thiazole), and free acids (2-methylbutyric acid) (Marković et al., 2007). In all heat-treated (either hot-break or cold-break processing) tomato juice, a common conclusion can be drawn to distinguish the volatile composition changes to those in freshly pressed tomato juice: thermal treatment primarily affected the levels of saturated and unsaturated C_6 alcohols and aldehydes, esters, ketones, terpenes, and carotenoid derivatives (Sieso & Crouzet, 1977; Servili et al., 2000; Sucan & Russell, 2002; Marković et al., 2007). Thermal treatment has accelerated Maillard and Strecker degradation reactions, as well as thermal and oxidative degradation of carotenoids, fatty acids, and amino acids. Meanwhile, complete loss of the characteristic tomato fresh-like volatiles of C_6- and C_9-aldehydes and C_6- and C_9-alcohols, that is, cis-3-hexenal in heat-treated tomato juice, has been reported (Marković et al., 2007). The reason behind this loss is that thermal treatment at high temperature has prevented the enzymatic formation of desirable volatile aldehydes in the heat-treated tomato juice via endogenous lipoxygenase (enzyme acts on fatty acid substrates, following crushing and homogenization of tomatoes) since thermal treatment is often used after destruction of the berries to inactivate endogenous enzymes in order to protect the valuable nutritional components from degradation. At very high processing temperature, formation of heat-induced volatiles, such as the potent dimethyl sulfide, is unavoidable and thus elicits an off-flavor or a cooked aroma note in pasteurized tomato juice (Scherb *et al.*, 2009). Therefore, in order to preserve volatiles correlated to the fresh tomato aroma and at the same time minimize the formation of cooked aroma, selecting an optimal condition for blanching the crushed tomatoes is crucial. In the study of Servili *et al.* (2000), hot-break treatment (typically >85 °C for <5 min) is preferable to prevent the formation of excessive cooked aroma, while cold-break treatment (typically between 65 and 70 °C for 20–30 min) would be beneficial to allow endogenous enzymes that are responsible for formation of fresh tomato aroma to remain active before the final pasteurization step takes place.

6.2.3.2   Changes in Appearance (Color and Juice Clarity)   Enzymatic and nonenzymatic browning during berry juice processing and storage is of great concern. Insufficient heat treatment leads to the formation of polymerized brown pigments owing to enzymatic action, primarily catalysis of polyphenol oxidase on polyphenol substrates in the presence of oxygen, following crushing of berry fruit. Albeit thermal processing applied during juice processing would eliminate enzymatic browning, nonenzymatic reaction may take place during storage. The accumulation of brown color during storage is predominantly caused by caramelization, ascorbic acid degradation, and Maillard reaction. One common strategy to elevate the commercial value and organoleptic properties related to the visual appeal of berry juice is controlling browning using browning inhibitors. For instance, L-ascorbic acid, sodium sulfite, *N*-acetyl-L-cysteine, and L-cysteine have been shown to effectively minimize the brown color development in grape juice (Molnar-Perl & Friedman, 1990), L-ascorbic acid and benzoate in pomegranate juice (García-Viguera & Zafrilla, 2001), and glutathione in grape juice (Wu, 2014).

To deliver a premium quality of berry juice to the consumer, a good retention of aroma and color, as well as consistency in the rheological behavior of the juice (indicated by juice viscosity) and juice clarity, is important. Berry juice with a clear appearance, and does not contain sediment or haze formation during storage, is desirable. From a technological perspective, clarifying and filtering juice from fruit mash, pulp, or pomace is a tedious job because of the presence of suspended solids, haze-active proteins, and turbidity-causing polysaccharides, which include insoluble pectin, cellulose, hemicelluloses, and lignin. Not only would these substances cause severe blockages at the surface of the filter membrane, but they will also limit the efficacy of the filtration process and increase the processing cost. Significant improvement in the clarity of final juice has been achieved through several manipulations in the clarification and filtration step, which is usually performed right after juice extraction. Combined strategy of using pectinolytic enzyme to achieve fruit mash liquefaction before cross-flow microfiltration (mineral tubular membrane with pore diameter of 0.2 μm) on pressed blackberry juice is able to remove a majority of the suspended particles and to obtain a good quality of clarified juice (Vaillant *et al.*, 2001). Alternatively, performing a centrifugation step (10,000*g* for 15 min) would immediately reduce 95% of the turbidity of the freshly pressed blackcurrant juice, followed by an addition of acidic protease enzyme to prevent the development of phenol–protein interactions during storage (Kozák *et al.*, 2008). Similar findings were obtained for pomegranate juice, whereby precentrifugation at a low temperature (8000 rpm, 15 °C for 15 min) is a fast and easy solution to decrease the turbidity level by 20% as opposed to low-temperature overnight sedimentation (0 °C for 18 h) (Gao *et al.*, 2014). Several studies have demonstrated that effective clarification is attainable, specifically on pomegranate juice, whereby gelatin (1 g/l) (Vardin & Fen-ercioglu, 2003) and a combination mixture of gelatin (300 mg/l), bentonite (300 mg/l), and polyvinylpolypyrrolidone (200 mg/l) (Alper *et al.*, 2005) are used as clarifying agents. Microfiltration (hydrophilic mixed cellulose esters flat membranes with pore sizes of 0.22 μm) is more efficient in reducing turbidity of pomegranate juice than ultrafiltration (same membrane material with pore sizes of 0.025 μm) (Mirsaeedghazi *et al.*, 2012). Lastly, cold clarification and active charcoal filtration are able to preserve the attractive color of the clear pomegranate juice (Oziyci *et al.*, 2013).

6.3 Novel Processing Techniques

It is clear that the conventional processing techniques have successfully transformed the berry fruit, which have limited shelf life once harvested, to shelf-stable juice. Unfortunately, the application of combined techniques during berry juice processing has shown promise to some extent with regards to microbial and nutrient qualities but at the same time compromising the organoleptic attributes of the fresh berries. As discussed earlier, some undesirable changes in conventionally processed berry juice would raise major concerns to both the juice processing industry and consumer, for instance:

- Preheating of crushed berries (pulp, pomace, or mash) prior to juice pressing is able to preserve the stability of vitamins and valuable nutritional components with antioxidant properties because of effective inactivation of oxidative enzymes but unable to preserve the original aroma of the fresh fruit.
- Enzyme-assisted juice processing has shown promise in extracting greater amounts of polyphenols, but this increase implies a negative sensory perception (associated with astringency and bitterness) and important volatile compounds such as esters are more readily hydrolyzed.
- Clarification is desirable to produce a berry juice with clear appearance and neither contains sediments nor haze formation during storage but results in poor recovery of anthocyanin compounds.
- Pasteurized and hot-filled berry juice have extended microbiological shelf life but undergo a greater loss of vitamins and valuable nutritional components with antioxidant properties, fresh appearance, and attractive color while inducing the formation of heat-induced volatiles, such as the potent dimethyl sulfide, that elicit an off-flavor or a cooked aroma note.
- Addition of preservatives to enhance the shelf life, fining agents to clarify juice and additives to inhibit juice browning could raise toxicity and environmental concern.

Recently, there has been a growing interest for adopting nonthermal technology in the processing and preservation of berry juice due to high demands of healthy, fresh-like, and safe foods that contain either low amounts or the absence of preserva-tives and additives. Several novel technologies such as pulsed electric fields (PEF), high-pressure processing (HPP), ultrasonication, ultraviolet-C (UV-C) light, ohmic heating, high-pressure homogenization (HPH), and flash vacuum expansion have been considered. Each of these novel technologies play specific roles in assisting the berry juice production from different aspects such as improving the juice yield and extraction of valuable nutritional and aroma volatile components; and some technologies solely served for juice preservation purposes.

6.3.1   Changes in Conventional Methods

Ideally, it is expected that novel processing techniques would reduce the number of processing steps, processing time required in the conventional berry juice production, and more importantly, reducing the thermal degradation effects of the final juice product. In berry juice production, it is always a challenge to find a method to effectively enhance the rate of mass transfer toward the release of juice containing valuable nutritional components from the fruit (in the form of mash, pomace, and pulp). Traditionally, an increase in juice yield and extraction of polyphenols of berries could be achieved with hot pressing, pectinolytic enzyme treatment, and freezing the fruit. A gentle cell membrane rupturing techniques such as PEF, flash vacuum expansion, and ultrasonication on berry fruit is promising to enhance juice extraction. Using these techniques, additional benefits on the production of higher quality juice, in terms of containing higher amounts of nutritional components and juice with clear appearance could be easily achieved.

Preservation of juice is a popular application for most novel technologies in order to replace conventional thermal pasteurization. In this case, there is an opportunity in applying novel technologies such as PEF, HPP, ultrasonication, UV-C light, ohmic heating, and HPH for microbial destruction and extending the shelf life of berry fruit juice. The first commercial application of PEF approved by the United States Food and Drug Administration (FDA) was in 2006 for processing fruit juice. In this operation, Genesis Juice Cooperative was selling PEF-treated juice such as apple, apple–strawberry, carrot, carrot–celery–beet, herbal tonic, strawberry lemonade, and ginger lemonade and their blends in the Portland market. Recently, in the Netherlands, Hoogesteger, Fruity King, and Juicy Line have already adopted PEF technology for juice production with a refrigerated shelf life of 21 days. As compared to PEF, HPP is gaining more popularity for commercial applications. HPP is advertised as a cold-pasteurized technique to preserve the fruit juice by many companies around the globe, for example, Suja Life (USA, 30 days shelf life), Reboot Your Life (Australia, 60 days shelf life), Coldpress (England, 5 months chilled shelf life), Lumi – Love U Mean It (USA), Invo Coconut Water (USA), Villa de Patos (Mexico), Ugo (Czech Republic), Beskyd Fryčovice (Czech Republic), Fruity Line (Netherlands), Preshafood (Australia), FreshBurst Finger Line by Wild Hibiscus Flower Company (Australia), and I'm Real by Pulmuone Co., Ltd. (South Korea), just to name a few.

6.3.2   Processing Berry Juice by Innovative Technologies

6.3.2.1   Improvement in Juice Extraction   Plant material exposed to PEF treatment of moderate field strengths between 0.4 and 3 kV/cm resulted in noticeably raised levels in juice yield (ranging from 2.5% to 45%) compared to that of untreated fruit and vegetables (e.g., apple, carrot, sugar beet, potato, fennel, pepper, alfalfa, and chicory), as reviewed by Donsì *et al.* (2010). This enhancement is due to permeabilization of the cellular membrane that facilitates cell disintegration of PEF-treated plant material to allow leaching of intracellular liquid, that is, juice from the cells during the subsequent cell rupture by mechanical pressing. PEF assists the pressing process of plant material and gives emphasis on a more efficient mass transfer phenomenon that considerably accelerates the juice extraction kinetics. This technology could be used as an alternative to conventional enzyme-assisted juice extraction. In the literature, application of PEF-assisted cold pressing of berry fruit is merely investigated on grapes rather than on other berry fruits. When PEF-assisted cold pressing (0.75 kV/cm and 100 square-wave bipolar pulses, constant press pressure of 5 bar for 45 min) is performed in two different approaches; that is, either as a pretreatment before pressing or as an intermediate treatment during pressing, the former approach is more efficient in improving juice yield (+30%) in grapes (Praporscic et al., 2007). In another study, PEF treatment (0.4 kV/cm and 10 square-wave bipolar pulses) during pressing at progressive pressure increase (from 0 to 1 bar for 60 min) resulted in 8% increase in juice yield and 15% elevation in the content of phenolics in the resulting grape juice (Grimi et al., 2009). An increase in the levels of total phenolics (+19%), chlorogenic acid (+25%), caffeic acid (+87%), p-coumaric acid (+34%), ferulic acid (+33%), caffeic-O-glucoside acid (+17%), naringenin (+40%), and naringenin-7-O-glucoside (+52%) of juice prepared from PEF-treated tomatoes (1 kV/cm and 16 square-wave monopolar pulses) has been demonstrated (Vallverdú-Queralt et al., 2012). The reason behind these increases is related to the increased permeability of the cellular membrane due to PEF processing. This could potentially make the extraction rate of the nutritional components from the fruit to juice more efficient.

Application of flash vacuum expansion has shown to improve the juice yield and extraction of polyphenols and anthocyanin from grapes (Paranjpe et al., 2012). Exposing the heated plant material (60–90 °C) to low pressure, or vacuum, allows the water in the plant tissue to rapidly expand and rupture the cell, forming microchannels inside the plant tissues to enable release of juice and phytochemicals. Ultrasonication treatment (2 W/cm^2 intensity at 74 °C for 13 min) has also been applied on grape mash (Lieu & Le, 2010). Because of the violent agitation induced during ultrasonication, there is an enhancement in the release of cellular content, hence an increase in yield extraction of juice (by 3.4%). The resulting grape juice has improved color density and contains higher levels of sugars, total acids, and total phenolics than the untreated grapes.

6.3.2.2 *Improvement in Juice Clarification*

The raw juice expressed from PEF-treated sugar beet tissue is shown to be easily purified by the means of membrane processes such as ultrafiltration (Loginov et al., 2011; Mhemdi et al., 2012). The raw juice has better qualitative characteristics such as higher purity and low browning levels than the juice obtained from thermally treated plant tissue. The increase in juice clarity and purity after PEF treatment and pressing might be explained by selective permeability of cell membranes after electroporation since lesser amounts of high-molecular-weight impurity compounds (proteins, pectin, and colloids) are found in the raw juice, which prevents the sedimentation of suspended particles and cloudiness in juice over storage time. From an industrial point of view, the use of PEF-assisted cold pressing to obtain higher yields and purity of berry juice could be of major interest to improve the existing technology utilized in berry juice processing. However, further study is required to validate the feasibility of PEF processing to assist berry juice pressing and clarification.

6.3.3   Preservation of Berry Juice by Innovative Technologies

Numerous researches have been conducted to study the effect of novel technologies on preserving berry juice. Table 6.3 summarizes the efficiency of several novel technologies in achieving microbial safety, as well as changes in the content and stability of nutritional components and organoleptic qualities of various berry juices.

6.3.3.1   *Pulsed Electric Field Processing*   PEF involve the application of short pulses of high voltage across a food product (preferably semisolid or liquid form) placed between two conducting electrodes. The effect of PEF on a biological cell can be characterized in four steps: (i) inducing the transmembrane potential across the cytoplasmic membrane, (ii) pore formation initiation, (iii) changes in size and number of pores, and (iv) formation of reversible or irreversible pores, depending on the external field intensity. When the applied electric field is below the critical value, reversible electroporation occurs in which the membrane will return to the initial and normal state by recovering the membrane integrity. However, when electric field strengths applied exceed the critical value, irreversible electroporation takes place leading to the formation of permanent holes in the cell membrane and leakage of cell compounds. Consequently, microorganisms will lose their cell viability and lead to cell death. Reversible electroporation is usually used for PEF biotechnological application, whereas killing microorganisms and improving mass transfer require irreversible electroporation. Electric field strength and treatment time are the two most important factors involved in PEF processing.

High intensity of PEF processing, typically between 30 and 50 kV/cm, is applied to obtain safe berry juice products. Food pathogens and spoilage microorganisms, yeast, and mold are vulnerable under the influence of high-intensity PEF processing, whereby 3.25- to 5-log reduction for vegetative yeast cells and the ascospores of *Zygosaccharomyces bailii* in cranberry and grape juice (Raso *et al.*, 1998b); 6-log reduction for *E. coli* in cranberry juice (Sen Gupta *et al.*, 2005); 4-log reduction for *S. cerevisiae* in grape juice (Garde-Cerdán *et al.*, 2007); and 2- to 4-log reduction for *Kloeckera apiculata, S. cerevisiae, Lactobacillus plantarum, Lactobacillus hilgardii,* and *Gluconobacter oxydans* in grape juice (Marsellés-Fontanet *et al.*, 2009) have been reported. However, in the study of Raso *et al.* (1998a), mold ascospores of *Byssochlamys nivea* and *N. fischeri* in both cranberry and tomato juice are resistant to PEF applied between 30 and 36.5 kV/cm. PEF has the capability to stabilize berry juice microbiologically in conjunction with higher retention of vitamin C, total phenolics, total anthocyanin, and total antioxidant capacity in juice from blueberries (Barba *et al.*, 2012), grapes (Marsellés-Fontanet *et al.*, 2013), strawberries (Odriozola-Serrano *et al.*, 2008a, d, 2009b), and better recovery percentage of carotenoids and lycopene in tomato juice (Min *et al.*, 2003; Odriozola-Serrano *et al.*, 2007, 2008b, c, 2009a) when compared to their thermally treated counterparts.

There is also strong evidence that underpin the feasibility of employing PEF for the processing of berry juice where no vast change or high retention in the organoleptic qualities is desired: fresh-like cranberry volatile profile (Jin & Zhang, 1999) and improved viscosity, better retention of color and fresh fruit characteristic volatiles, and minimal nonenzymatic browning in PEF-treated tomato and strawberry juice (Aguiló-Aguayo *et al.*, 2009a–d, 2010).

Therefore, on the basis of (i) effective inactivation of microorganisms, (ii) similar or higher retention of nutritional quality, and (iii) lowest impact on the organoleptic

Table 6.3 Impact on the microbial, nutritional, and organoleptic qualities of berry juice after subjected to novel preservation technologies

Berry juice	Processing parameters	Remarks	References
(I) Pulsed electric field processing			
Blueberry	36 kV/cm, 100 μs, <60 °C	• 56 days refrigerated (4 °C) shelf life • No changes in ascorbic acid, total phenolics, and total anthocyanin during storage • No color degradation	Barba *et al.* (2012)
Cranberry	36.5 kV/cm, 6.6 μs, 2 pulses	• Ascospores of *Byssochlamys nivea* and *Neosartorya fischeri* were resistant • 6 logs reduction for conidospores of *Byssochlamys fulva*	Raso *et al.* (1998a)
	36.5 kV/cm, 6.6 μs	• 4.6 logs reduction for vegetative cells of *Zygosaccharomyces bailii* • 4.2 logs reduction for ascospores of *Zygosaccharomyces bailii*	Raso *et al.* (1998b)
	40 kV/cm 150 μs, <25 °C, square-wave pulses, 2 μs pulse width, 1000 Hz	• Reduce more viable microbial cells • Not affecting volatile profile, color • No growth of mold, yeast and aerobic bacteria during storage at 4 °C	Jin and Zhang (1999)
	32 kV/cm, 100 l/h, 550 Hz, 1.4 μs pulse width, 47 μs treatment time, <29 °C, bipolar pseudo-square-wave pulses	• Extend shelf life from 14 to 197 days at 22, 37 °C • No color degradation • Slow down microbial growth	Evrendilek *et al.* (2001)

(continued overleaf)

Table 6.3 (*continued*)

Berry juice	Processing parameters	Remarks	References
	PEF + heat 60 °C (hold for 32 s)	• Extend shelf life from 14 to 197 days at 22, 37 °C • No color degradation • Slow down microbial growth	Evrendilek *et al.* (2001)
	32 kV/cm, 100 l/h, 550 Hz, 1.4 µs pulse width, 47 µs treatment time, <29 °C, bipolar pseudo-square-wave pulses		
	40 kV/cm, 60 Pulses, 20 °C, 80 s treatment time	• 6 logs reduction for *Escherichia coli*	Sen Gupta *et al.* (2005)
Grape	35 kV/cm, 4.6 µs	• 5 logs reduction for vegetative cells of *Zygosaccharomyces bailii* • 3.5 logs reduction for ascospores of *Zygosaccharomyces bailii*	Raso *et al.* (1998b)
	35 kV/cm, 1000 µs Treatment time, 4 µs pulse width, 1000 Hz, bipolar square-wave pulses	• 4 logs reduction for *Saccharomyces cerevisiae* • No changes in total content of fatty acids and free amino acids • Decrease in level of lauric acid • Some changes in the free amino acid profile	Garde-Cerdán *et al.* (2007)
	35 kV/cm, 1000 µs Treatment time, 5 µs pulse width, 303 Hz, bipolar square-wave pulses, <30 °C	• 2–4 logs reductions for *Kloeckera apiculata, Saccharomyces cerevisiae, Lactobacillus plantarum, Lactobacillus hilgardii,* and *Gluconobacter oxydans*	Marsellés-Fontanet *et al.* (2009)

	35 kV/cm, 2500 μs Treatment time, 4 μs pulse width, 400 Hz, bipolar square-wave pulses	• Compared with thermal-treated (90 °C) • No changes in soluble solids, pH, acidity, and the electrical conductivity, vitamin C, total polyphenol, cinnamic acid, free catechin and nonflavonoid content	Marsellés-Fontanet et al. (2013)
	9–27 kV/cm, 34–275 μs Treatment time, 60 ml/min, 3 μs pulse width, 120 Hz, monopolar square-wave pulses, <50 °C	• PEF resistance in increasing order: *Saccharomyces cerevisiae* < *Staphylococcus aureus* < *Escherichia coli* DH5α	Huang et al. (2014)
Pomegranate	35–38 kV/cm, 281 μs treatment time, 100 l/h, 1 μs pulse width, 2000 Hz, bipolar square-wave pulses, <55 °C	• Compared with thermal-treated (105 °C, 5 s) • Twelve weeks refrigerated (4 °C) shelf life • Inhibit growth of total aerobic bacteria, yeast and mold • No changes in levels of total phenolics and anthocyanin • Better color retention during storage • Significant higher consumer satisfaction scores	Guo et al. (2014)

(continued overleaf)

Table 6.3 (*continued*)

Berry juice	Processing parameters	Remarks	References
Strawberry	20–35 kV/cm, <2000 μs Treatment time, 60 ml/min, 1 μs pulse width, 232 Hz, bipolar square-wave pulses, <40 °C	• High retention of anthocyanin and vitamin C content • Higher antioxidant capacity	Odriozola-Serrano *et al.* (2008a)
	35 kV/cm, 1000 μs Treatment time, 60 ml/min, 4 μs pulse width, 100 Hz, bipolar square-wave pulses, <40 °C	• Increase in ellagic acid and *p*-coumaric acid • Similar antioxidant capacity	Odriozola-Serrano *et al.* (2008d)
	35 kV/cm, 1700 μs Treatment time, 60 ml/min, 4 μs pulse width, 100 Hz, bipolar square-wave pulses, <40 °C	• Compared with thermal-treated (90 °C, 60 s) • Better retention of color, greater luminosity and redness • Low concentration of 5-hydroxy-methyl furfural • Minimize nonenzymatic brown-ing • Improved viscosity	Aguiló-Aguayo *et al.* (2009a)
	35 kV/cm, 1000 μs Treatment time, 60 ml/min, 1 μs pulse width, 232 Hz, bipolar square-wave pulses, <40 °C	• High retention of anthocyanin content • Higher antioxidant capacity	Odriozola-Serrano *et al.* (2009b)

35 kV/cm, 1700 µs Treatment time, 60 ml/min, 4 µs pulse width, 100 Hz, bipolar square-wave pulses, <40 °C	• Compared with thermal-treated (90 °C, 60 s) • Better retention of 2,5-dimethyl-4-hydroxy-2H-furan-one, ethyl butanoate and 1-butanol	Aguiló-Aguayo et al. (2009b)
35 kV/cm, 1000 µs Treatment time, 60 ml/min, 4.5 µs pulse width, 100 Hz, monopolar square-wave pulses, <40 °C	• Compared with thermal-treated (90 °C, 60 s) • Better retention of color • Low concentration of 5-hydroxy-methyl furfural • Minimize nonenzymatic browning	Aguiló-Aguayo et al. (2009c)
35 kV/cm, 1000 µs Treatment time, 1 µs pulse width, 100 Hz, bipolar square-wave pulses	• Increase in juice viscosity	Aguiló-Aguayo et al. (2009d)
35 kV/cm, 1700 µs Treatment time, 60 ml/min, 4 µs pulse width, 100 Hz, bipolar square-wave pulses, <40 °C	• Compared with thermal-treated (90 °C, 60 s) • Increase in phenylalanine (+27%), glutamic acid (+6.8%), valine (+6.3%), serine (+5.5%) and alanine (+4.8%)	Odriozola-Serrano et al. (2013)

(continued overleaf)

Table 6.3 (*continued*)

Berry juice	Processing parameters	Remarks	References
Tomato	30 kV/cm, 4 μs	• Ascospores of *Byssochlamys nivea* and *Neosartorya fischeri* were resistant • 1 log reduction for conidospores of *Byssochlamys fulva*	Raso *et al.* (1998a)
	40 kV/cm, 57 μs Treatment time	• Compared with thermal-treated (92 °C, 90 s) • 112 days refrigerated (4 °C) shelf life • Better retention of tomato flavor compounds of *trans*-2-hexenal, 2-isobutylthiazole, *cis*-3-hexanol • Higher redness • Minimize nonenzymatic browning • More preferable by consumers	Min and Zhang (2003)
	40 kV/cm, 57 μs Treatment time	• Compared with thermal-treated (92 °C, 90 s) • 112 days refrigerated (4 °C) shelf life • Better retention of vitamin C content • No changes in lycopene content, °Brix, pH, or viscosity • More preferable by consumers in terms of flavor and overall acceptability	Min *et al.* (2003)
	35 kV/cm, 1000 μs Treatment time, 1 μs pulse width, 250 Hz, bipolar square-wave pulses	• Maximal relative lycopene content (131.8%), vitamin C content (90.2%), and antioxidant capacity retention (89.4%)	Odriozola-Serrano *et al.* (2007)

Parameters	Effects	Reference
35 kV/cm, 1500 μs Treatment time, 4 μs pulse width, 100 Hz, bipolar square-wave pulses	• Compared with thermal-treated (90 °C, 60 s) • 77 days refrigerated (4 °C) shelf life • Higher lightness • Better juice viscosity	Aguiló-Aguayo et al. (2008)
20–35 kV/cm, <2000 μs Treatment time, 1 μs pulse width, 250 Hz, bipolar square-wave pulses	• Better retention of contents of lycopene and vitamin C	Odriozola-Serrano et al. (2008b)
35 kV/cm, 1000 μs Treatment time, 60 ml/min, 2.5 μs pulse width, 100 Hz, monopolar square-wave pulses, <40 °C	• Compared with thermal-treated (90 °C, 60 s) • Better retention of color • Low concentration of 5-hydroxy-methyl furfural • Minimize nonenzymatic browning	Aguiló-Aguayo et al. (2009c)
35 kV/cm, 1000 μs Treatment time, 7 μs pulse width, 250 Hz, bipolar square-wave pulses	• Increase in juice viscosity	Aguiló-Aguayo et al. (2009d)
35 kV/cm, 1500 μs Treatment time, 4 μs pulse width, 100 Hz, bipolar square-wave pulses	• Compared with thermal-treated (90 °C, 60 s) • Enhancement in the production of tomato aroma contributors such as linalool, 6-methyl-5-hepten-2-one, geranylacetone, and 2--isobutylthiazole • Preservation of fresh-like aroma flavor during storage	Aguiló-Aguayo et al. (2010)

(continued overleaf)

Table 6.3 (*continued*)

Berry juice	Processing parameters	Remarks	References
	50 kV/cm, 300 Pulses	• 89% Reduction in total microbial count	Sathyanathan *et al.* (2012)
Tomato	35 kV/cm, 1700 μs Treatment time, 60 ml/min, 4 μs pulse width, 100 Hz, bipolar square-wave pulses, <40 °C	• Compared with thermal-treated (90 °C, 60 s) • Increase in total free amino acids (+3%) • Decrease in proline, leucine, valine, isoleucine, arginine, lysine, phenylalanine and methionine with storage time	Odriozola-Serrano *et al.* (2013)
(II) High-pressure processing Blueberry	600 MPa + thermal (100 °C)	• Loss of total anthocyanin (−50%)	Buckow *et al.* (2010)
	600 MPa, <42 °C, 5 min treatment time	• 56 days refrigerated (4 °C) shelf life • No changes in ascorbic acid, total phenolics and total anthocyanin during storage • No color degradation	Barba *et al.* (2012)
	200–600 MPa, 25–42 °C, 5–15 min Treatment time	• Better retention of vitamin C (92%) • Increase in levels of total phenolic and total anthocyanin (+16%) • No changes in total antioxidant capacity • No changes in pH, °Brix and color parameters (a^*, b^*, L^*, and ΔE)	Barba *et al.* (2013)

Product	Process conditions	Effects	Reference
Grape	250 MPa + thermal (40 °C) for 10 min	• 60 days refrigerated (4 °C) shelf life • No changes in color parameters • Similar sensory characteristics with fresh juice • No development of fermentation during storage	Daoudi et al. (2002)
	550 MPa, 44 °C, and 2 min treatment time	• High retention of total antioxidant activity, phenolics and flavonoids	Chauhan et al. (2011)
Pomegranate	400 MPa + Thermal (25 °C)	• Increase in intensity of red color • Better retention of total anthocyanin content • No changes in total phenolics content and aroma profile	Ferrari et al. (2010)
	350–550 MPa, 30–150 s Treatment time	• 35 days refrigerated (4 °C) shelf life • 4 logs reduction for total microbial count • Increase in total phenolic content (up to +12%) • Decrease in total color difference and antioxidant capacity • During the first 15 days, no changes in pH, °Brix and titratable acid	Varela-Santos et al. (2012)
Raspberry	200 MPa, 25 °C, 5 min Treatment time	• Complete inactivation of Escherichia coli	Yan and Zhao (2010)

(continued overleaf)

Table 6.3   (*continued*)

Berry juice	Processing parameters	Remarks	References
Raspberry	300 MPa, 25 °C, 5 min Treatment time	• Complete inactivation of *Salmonella*	Yan and Zhao (2010)
	400 MPa, 25 °C, 15 min Treatment time	• Complete inactivation mold and yeast	Yan and Zhao (2010)
	600 MPa, 25 °C, 25 min Treatment time	• Microorganisms in the juice were not completely killed • Total number of colonies could be reduced below 10 cfu/ml	Yan and Zhao (2010)
	400–700 MPa + Thermal (20–110 °C)	• Breakdown of anthocyanin and vitamin C occurred at constant elevated pressure as temperature increased • No changes in total phenolic content	Verbeyst *et al.* (2012)
Strawberry	300–700 MPa + Thermal (65 °C) for 60 min	• Increase in color parameters of L^*a^*/b^* (8.8%)	Rodrigo *et al.* (2007)
	600 MPa, 4 min Treatment time	• 6 months (4 and 25 °C) shelf life • Decrease in levels of vitamin C (−49%), anthocyanin (−30%) and total phenolics (−16%) after storage • Decrease in total antioxidant capacity by 10% • Decrease in color parameters of L^* and a^* • Increase in browning index • Decrease in juice viscosity (up to −80%)	Cao *et al.* (2012)

400–700 MPa + Thermal (20–110 °C)	• Breakdown of anthocyanin and vitamin C occurred at constant elevated pressure as temperature increased • No changes in total phenolic content	Verbeyst et al. (2012)
Tomato 500–900 MPa for 3–9 min	• Improved juice viscosity and color parameters • Increase in levels of n-hexanal and cis-3-hexenal	Porretta et al. (1995)
800 MPa, <35 °C	• Loss in antimutagenic activities against carcinogen 2-amino-3-methyl-imidazo[4,5-f]quinoline	Butz et al. (1997)
350 MPa + Thermal (50 °C) for 20 min	• Compared with thermal-treated (50 °C, 20 min) • 4.7 logs reduction for Alicyclobacillus acidoterrestris vegetative cells • D values for HPP inactivation of Alicyclobacillus acidoterrestris vegetative cells is 4.4 min versus 18.9 min for thermal	Alpas et al. (2003)
250 MPa + thermal (35 °C) for 15 min	• Compared with thermal-treated (80 °C, 60 s) • 30 days (25 °C) shelf life • Better retention of vitamin C • Loss of total antioxidant capacity by 10% • Minor color change	Dede et al. (2007)

(continued overleaf)

Table 6.3 (*continued*)

Berry juice	Processing parameters	Remarks	References
Tomato	300–700 MPa + Thermal (65 °C) for 60 min	• No changes in color parameters of $L*a*/b*$	Rodrigo *et al.* (2007)
	500 MPa + Thermal (4, 25 °C) for 10 min	• Compared with hot-break (92 °C for 2 min) and cold-break (60 °C for 2 min) • High retention of color, total carotenoids and lycopene, and radical-scavenging capacity	Hsu (2008)
	500 MPa + Thermal (25 °C) for 10 min	• Compared with hot-break (92 °C for 2 min) and cold-break (60 °C for 2 min) 28 days refrigerated (4 °C) shelf life • High retention of the levels of vitamin C, extractable carotenoids and lycopene • No changes in color parameters and viscosity • Significant reduction of total microbial count	Hsu *et al.* (2008)
	600–700 MPa + Thermal (45, 100 °C) for 10 min	• Compared with thermal-treated (100 °C, 35 min) • Increase in lycopene extractability • All-*trans* lycopene are fairly stable to isomerization during processing • No changes in levels of *cis*-isomer carotenoids	Gupta *et al.* (2010)

500–700 MPa + Thermal (30, 100 °C) for 10 min	• Compared with thermal-treated (100 °C, 10 min) • Increase in lycopene extractability (+12%) • Low retention of all *trans*-β-carotene (60–95%)	Gupta *et al.* (2011)
(III) Ultrasonication Blackberry	100% Amplitude (acoustic intensity of 22.8 W/cm² at a constant frequency of 20 kHz, 10 min and pulse durations of 5 s on and 5 s off	• Higher retention of total anthocyanin content (>94%) • Minor color changes (±5%)	Tiwari *et al.* (2009)
Blueberry	60–120 µm Amplitude (acoustic intensity of 14.2–45.8 W/cm²) at a constant frequency of 20 kHz, 3–9 min	• Maintain non-Newtonian dilatant fluid characteristics	Šimunek *et al.* (2014)
Cranberry	20–60 µm Amplitude (acoustic intensity of 15–48.2 W/cm²) at a constant frequency of 20 kHz, 3–9 min	• Maintain non-Newtonian dilatant fluid characteristics	Šimunek *et al.* (2014)
Grape	24.4–61 µm Amplitude at a constant frequency of 20 kHz, 2–10 min and pulse durations of 5 s on and 5 s off, 32–45 °C	• High retention of cyanidin-3-glucoside (97.5%), malvidin-3-glucoside (48.2%) and delphinidin-3-glucoside (80.9%)	Tiwari *et al.* (2010)

(continued overleaf)

Table 6.3 *(continued)*

Berry juice	Processing parameters	Remarks	References
Pomegranate	24.2–61 μm Amplitude at a constant frequency of 20 kHz, 3–9 min at 25 °C	• Loss of total anthocyanin content (up to −10%) • Increase in total phenolic content (+43%) • No changes on total antioxidant capacity and color parameters	Alighourchi *et al.* (2013)
	100% Amplitude at a constant frequency of 20 kHz, 15 min at 25 °C	• 3.5 logs reduction for *Escherichia coli* • 1.9 logs reduction for *Saccharomyces cerevisiae*	Alighourchi *et al.* (2014)
Strawberry	Ultrasound + thermal (20–60 °C) 60, 90, and 120 μm amplitude at a constant frequency of 20 kHz, 3–9 min	• Compared with thermal-treated (85 °C, 2 min) • Higher retention of total anthocyanin content (>85%)	Dubrovic *et al.* (2011)
Tomato	61 μm Amplitude (acoustic intensity of 0.81 W/ml) at a constant frequency of 20 kHz, 10 min, 40 °C	• 5 logs reduction for yeast *Pichia fermentans*	Adekunte *et al.* (2010b)
	24.4–61 μm Amplitude at a constant frequency of 20 kHz, 2–10 min and pulse durations of 5 s on and 5 s off, 32–45 °C	• No changes in pH, °Brix and titratable acidity • Higher retention of color parameters (*L**, *a** and *b**), ascorbic acid content and yeast inactivation	Adekunte *et al.* (2010a)

(IV) Ultraviolet-C light processing			
Cranberry	1.02 l/min Flow rate, 450 kJ/m² UV-C light dose for 30 min	• 2.5 logs reduction for *Saccharomyces cerevisiae* • *D*-value for UV-C inactivation of *Saccharomyces cerevisiae* is 12.2–40.7 min • Improved total color difference (ΔE*)	Guerrero-Beltrán *et al.* (2009)
Grape	1.02 l/min flow rate, 450 kJ/m² UV-C light dose for 30 min	• 0.5 logs reduction for *Saccharomyces cerevisiae* • *D*-value for UV-C inactivation of *Saccharomyces cerevisiae* is 61.7–113 min • Improved total color difference (ΔE*)	Guerrero-Beltrán *et al.* (2009)
	5.24 kJ/m² UV-C light dose, 0.153 cm penetration depth for 3–10 min	• 5.1 logs reduction for *Escherichia coli* K-12	Hakguder and Unluturk (2009)
	3.67 J/ml UV-C light dose, 4000 l/h flow rate	• Up to 4.8 logs reduction for yeasts, lactic and acetic acid bacteria	Fredericks *et al.* (2011)
	25.2 J/ml UV-C light dose	• Complete inactivation of total aerobic count, yeast and mold counts • Higher retention of total phenolics content and antioxidant capacity • Increase in total anthocyanin content and polymeric color	Pala and Toklucu (2013)

(continued overleaf)

Table 6.3 *(continued)*

Berry juice	Processing parameters	Remarks	References
Pomegranate	12.47–62.35 J/ml UV-C light dose	• Compared with thermal-treated (90 °C, 2 min) • No changes in total monomeric anthocyanin content, antioxidant capacity, total phenolics and polymeric color • Decrease in the levels of selected individual anthocyanin between 8.1% to 16.3% • 1.0 total aerobic plate count • 1.5 logs reduction for yeast and mold count • 6.2 logs reduction for surrogate microorganism of *Escherichia coli* 0157:H7	Pala and Toklucu (2011)
(V) High-pressure homogenization Blueberry	250–300 MPa, <2 s, 70–75 °C,	• Feline calicivirus (FCV-F9) and bacteriophage MS2 of human noroviruses surrogates are most sensitive • 0.7 logs reduction for murine norovirus (MNV-1) of human noroviruses surrogates	Horm *et al.* (2012)
Tomato	150 MPa	• Reduce mean particle diameter and particle size distribution • Improved juice consistency and thixotropy	Augusto *et al.* (2012)

100 MPa	• Improved juice consistency and stability • Reduce particle sedimentation and serum separation • Improved in color due to leakage of lycopene from the disrupted cells	Kubo et al. (2013)
25–150 MPa , <40°C	• Improved elastic and viscous behaviors	Augusto et al. (2013a)
150 MPa	• Improved juice storage (G') and loss (G'') moduli • Improved juice consistency • Improve in juice elastic and viscous behavior	Augusto et al. (2013b)
(VI) Ohmic heating Pomegranate	Ohmic heating + thermal (20–90°C) 10–40 V/cm at 50 Hz for up to 12 min	• No changes in rheological properties, color, and total phenolic content	Yildiz et al. (2009)
Tomato	25 V/cm for 30 s	• 5 logs reduction for in *Escherichia coli* O157:H7, *Salmonella typhimurium* and *Listeria monocytogenes* • Better retention of vitamin C content	Lee et al. (2012)
	10–20 V/cm for 90–480 s	• 5 logs reduction for in *Escherichia coli* O157:H7, *Salmonella typhimurium* and *Listeria monocytogenes*	Sagong et al. (2011)

(continued overleaf)

Table 6.3   (*continued*)

Berry juice	Processing parameters	Remarks	References
(VII) Dense phase carbon dioxide processing			
Grape	48.3 MPa at 170 g/kg CO_2, 35 °C	• 6 logs reduction for yeast • No flavor degradation	Gunes *et al.* (2005)
	34.5 MPa at 8% and 16% CO_2 for 6.25 min, 30 °C	• Compared with thermal-treated (75 °C, 15 s) • 5 weeks refrigerated (4 °C) shelf life • Higher retention of the levels of anthocyanin and polyphenols • Higher retention of total antioxidant capacity • No changes in color, flavor, aroma, and overall likeability • More preferable by sensory panelists • Similar inactivation of total microbial, yeast, and mold with thermal-treated juice	Del Pozo-Insfran *et al.* (2006)
(VIII) Irradiation			
Pomegranate	0.2–3 3 kGy Gamma-irradiation dose	• 6.7 logs reduction for *Escherichia coli* • 5.1 logs reduction for *Saccharomyces cerevisiae*	Alighourchi *et al.* (2014)
(IX) Hydrodynamic cavitation			
Tomato	3600 rpm rotor speed, 104 °C	• 3.1 logs reduction for *Bacillus coagulans*	Milly *et al.* (2007)
(X) Pressure-ohmic-thermal sterilization			
Tomato	Ohmic (50 V/cm) + pressure (600 MPa) + thermal (105 °C) for 30 min	• 4.6 logs reduction for *Bacillus amyloliquefaciens* • 5.6 logs reduction for *Geobacillus stearothermophilus*	Park *et al.* (2013)

properties, PEF juice processing has demonstrated to offer advantages over conventional thermal treatment in better preservation of these three aspects of food quality and consumer acceptance after storage. As such, PEF-treated blueberry juice is recommended to be shelf stable for up to 56 days under 4 °C storage (Barba *et al.*, 2012), 197 days under 22 and 37 °C storage for PEF-treated cranberry juice (Evrendilek *et al.*, 2001), PEF-treated pomegranate juice with shelf life of 12 weeks at 4 °C (Guo *et al.*, 2014), and between 56 and 112 days refrigerated (4 °C) shelf life for PEF-treated tomato and strawberry juice (Min & Zhang, 2003; Min *et al.*, 2003; Odriozola-Serrano *et al.*, 2013). This finding indicates the potential commercialization of PEF as preservation technology for berry juice industries.

6.3.3.2 *High Pressure Processing*
Pressure build-up during HPP affects chemical and biochemical reactions and structural changes when a change in the reaction volume involved occurs. Vegetative bacteria cells are inactivated at about 300 MPa combined with ambient temperature. In raspberry juice, complete inactivation of *E. coli* is obtained at 200 MPa, *Salmonella* at 300 MPa, and yeast and mold at 400 MPa (Yan & Zhao, 2010) when HPP processing was performed at room temperature (~25 °C). Inactivation of bacterial spores by pressure alone is not possible; therefore, HPP should be combined with high temperature. For example, pressure (350 MPa) combined with a temperature of 50 °C for 20 min leads to 4.7-log reduction of *A. acidoterrestris* in tomato juice (Alpas *et al.*, 2003). The decimal reduction time (*D*-value) for *A. acidoterrestris* inactivation is 4.4 min under HPP influence as opposed to *D*-value of 18.9 min after thermal treatment (50 °C for 20 min).

Following exposure to pressure of not more than 550 MPa, almost complete retention of nutritional characteristics of fresh berry juice without sacrificing shelf life has been made possible in the case of blueberry juice (Barba *et al.*, 2012, 2013), grape juice (Chauhan *et al.*, 2011), pomegranate juice (Ferrari *et al.*, 2010; Varela-Santos *et al.*, 2012), and tomato juice (Dede *et al.*, 2007; Hsu, 2008; Hsu *et al.*, 2008; Gupta *et al.*, 2010, 2011). As discussed earlier, combined approach of pressure and thermal would ensure a microbiologically safe juice, but adverse impacts on the nutritional quality of juice after HPP processing at elevated pressure (exceeding 600 MPa) as well as temperature increase (>60 °C) have been reported. Breakdown of anthocyanin and vitamin C occurred in raspberry and strawberry juice (Verbeyst *et al.*, 2012), degradation of total anthocyanin content, up to 50%, in blueberry juice (Buckow *et al.*, 2010), and, unfortunately, HPP-treated tomato juice at 800 MPa resulted in loss of antimutagenic activities against carcinogen 2-amino-3-methyl-imidazo[4,5-*f*]quinolone (Butz *et al.*, 1997).

Similar to PEF, HPP-treated juice has an improved juice quality, in terms of organoleptic properties, which include improved juice viscosity in tomato juice (Porretta *et al.*, 1995) and no color changes in grape juice (Daoudi *et al.*, 2002) and tomato juice (Rodrigo *et al.*, 2007). In the study of Porretta *et al.* (1995), increase in tomato aroma contributors of *n*-hexanal and *cis*-3-hexenal HPP-treated tomato juice has suggested the retention of the fresh-like tomato flavor in the resulting juice as opposed to development of cooked flavor after exposure to conventional thermal pasteurization. The recommended refrigerated (4 °C) shelf life of HPP-treated berries is as follows: 56 days for blueberry juice (Barba *et al.*, 2012), 60 days for grape juice (Daoudi *et al.*, 2002), 35 days for pomegranate juice (Varela-Santos *et al.*, 2012), 6 months for strawberry juice (Cao *et al.*, 2012), and 28 days for tomato juice (Hsu *et al.*, 2008). To date, HPP is regarded as the most well-studied food processing technique and has successfully transferred to commercial applications. Recent advances in the research area, in relevance to HPP, have then progressed to high-pressure–high-temperature (HPHT), which could achieve sterilization intensity since inactivation of bacterial spores by pressure at ambient temperature is still a challenge for HPP at room temperature. HPHT employs pressure of more than 600 MPa combined with elevated processing temperature (90–121 °C). Because the impacts of elevated pressure on enhancing and retarding chemical reactions (according to *Le Chatelier* principle) became more obvious and complex for HPHT technology compared to HPP at room temperature, loss of nutritional values in HPHT-treated food products is unavoidable. The detection of different headspace volatiles profile for HPHT-treated food products and their thermally sterilized counterparts has been investigated (Van der Plancken *et al.*, 2012). There is still a gap in knowledge to correlate the volatile profile with changes in flavor development and consumer acceptance toward HPHT-treated fruit juice.

6.3.3.3 *Other Innovative Technologies for Preservation Purposes* Some innovative food preservation technologies such as ultrasonication, UV-C light, HPH, ohmic heating, dense phase carbon dioxide processing, irradiation, and hydrodynamic cavitation have received little attention. Studies have shown that these techniques can be used as an alternative to thermal pasteurization in inactivating microorganisms and maintaining the nutritional components without compromising sensory attributes of juice after processing and during storage. Besides, a combination of preservative factors (known as hurdle, physically, physiochemically, or microbially) to secure microbial safety and stability in the nutritional and sensorial qualities will be discussed.

Ultrasonication Continuous agitation of food material at ultrasonic frequencies (>20 kHz) using an ultrasonic bath or probe could be sufficient to disrupt cell membranes and release cellular contents of biological cells, that is, microorganisms. In some cases, ultrasonication is accompanied with elevated processing temperatures (up to 60 °C) and this is known as thermosonication. After treatment with ultrasound (100% amplitude at a constant frequency of 20 kHz, 15 min at 25 °C), a 3.5-log and 1.9-log reduction for *E. coli* and *S. cerevisiae* was found in pomegranate juice (Alighourchi *et al.*, 2014). In tomato juice, ultrasound treatment of 61 µm amplitude, at constant frequency of 20 kHz for 10 min, resulted in 5 logs reduction for yeast cells *Pichia fermentans* (Adekunte *et al.*, 2010b). Compared to the thermal-treated juice, high retention of total anthocyanin, total phenolics, vitamin C, and color parameters has been reported after ultrasound treatment (60–120 µm for 3–15 min) for blackberry juice (Tiwari *et al.*, 2009), grape juice (Tiwari *et al.*, 2010), pomegranate juice (Alighourchi *et al.*, 2013), strawberry juice (Dubrovic *et al.*, 2011), and tomato juice (Adekunte *et al.*, 2010a).

Ultraviolet-C (UV-C) Light Processing UV-C light with wavelength range from 200 to 280 nm is lethal to most microorganisms by damaging their DNA and induces functional and structural changes on cell components. As UV-C is a widely used technology in postharvest treatment for horticulture crops to disinfect contaminants on the surface of the fruit, the use of UV-C in food processing is very limited. In the literature, there is a discrepancy in the UV-C light dose required to inactivate microorganisms for different types of berry juices, such as 450 kJ/m^2 UV-C light dose allows 2.5-log reduction for *S. cerevisiae* in cranberry juice and 0.5-log reduction for *S. cerevisiae* in grape juice (Guerrero-Beltrán *et al.*, 2009); 5.24 kJ/m^2 UV-C light dose and subtle light penetration depth at 0.153 cm (cf. 0.5 cm) allows 5.1-log reduction for *E. coli* in grape juice (Hakguder & Unluturk, 2009) and between 3.7 and 62.4 J/ml of UV-C light dose allows up to 4.8-log reduction for total yeast, lactic and acetic acid bacteria in grape juice (Fredericks *et al.*, 2011); and 6.2-log reduction for surrogate microorganisms of *E. coli* O157:H7 in pomegranate juice (Pala & Toklucu, 2011). UV-C-treated grape juice and pomegranate juice, between 12.5 and 62.4 J/ml of UV-C light dose, has a higher retention of total phenolics content, total anthocyanin content, and total antioxidant capacity (Pala & Toklucu, 2011, 2013).

High-Pressure Homogenization (HPH) HPH technology is able to handle fluid food products under high pressure (usually up to 300 MPa) at continuous full-scale operation. This creates disintegration of particles that disperse throughout the food matrix in a form of micrometer-sized droplets. Therefore, HPH results in physicochemical changes of fruit juice such as in tomato juice (Augusto *et al.*, 2012, 2013a, b; Kubo *et al.*, 2013). HPH using pressure between 100 and 150 MPa altered the mean particle size and their distribution, improved juice consistency and thixotropy, improved elastic and viscous behavior, and reduced particle sedimentation resulting in a stable tomato juice product. Horm *et al.* (2012) have reported that homogenization pressure between 250 and 300 MPa results in temperature rise up to 75 °C of blueberry juice. The cultivable feline calicivirus and bacteriophage MS2 of human noroviruses surrogates are found to be more sensitive to this type of treatment. Nevertheless, murin noroviruses are highly resistant to HPH (i.e., only 0.7-log reduction after HPH). This provides evidence that HPH is potential for microbial inactivation and a nonthermal solution to reduce juice-related foodborne outbreaks.

Ohmic Heating Ohmic heating is suitable to heat up liquid foods in which the food itself acts as a conductor of electricity. In fruit juice, the presence of the dissolved ionic species such as sugars and fruit acids plays a role as electrical conductor that allows electric current passing through and dissipates heat internally. Reduction, up to 5-log, of common foodborne pathogens such as *E. coli* O157:H7, *S. typhimurium*, and *L. monocytogenes* in tomato juice after ohmic heating (10–25 V/cm for 30–480 s) has been demonstrated in previous studies (Lee *et al.*, 2012; Sagong *et al.*, 2011).

Dense Phase Carbon Dioxide Processing The use of carbon dioxide (CO_2) under pressure, commonly known as dense phase CO_2, may achieve cold pasteurization and/or sterilization of liquid foods. Studies have revealed the ability of using dense phase CO_2 (under pressure 35–48 MPa at 30–35 °C) to inactivate bacteria, yeast, and mold at a similar log reduction as thermal pasteurization (75 °C, 15 s) in grape juice (Gunes *et al.*, 2005; Del Pozo-Insfran *et al.*, 2006). Toward the end of 5-weeks-refrigerated storage, a better quality of grape juice processed by dense phase CO_2 is obtained as compared to heat-pasteurized juice. No flavor degradation, no changes in aroma profile, and color and higher retention of total anthocyanin and phenolic contents were found. Sensory panelists preferred dense phase CO_2 over heat-pasteurized juice.

Irradiation Food irradiation is a controversial processing technology as it is not widely accepted by consumers because they think radioactive source carries long-term health and safety risk. During food irradiation, the product is passed through an enclosed chamber where it is exposed to a certain dose of ionizing radiation (gamma ray emitted by the radioactive material), penetrating the food and inactivating various microorganisms. In pomegranate juice, gamma-irradiation dose of 0.2–3 kGy has led to 6.7-log reduction for *E. coli* and 5.1-log reduction for *S. cerevisiae* (Alighourchi *et al.*, 2014).

Hydrodynamic Cavitation   Formation of gas bubbles in a fluid product due to pressure fluctuations induced by mechanical actions makes hydrodynamic cavitation a unique and innovative preservation technology. When tomato juice inoculated with *B. coagulans* spores was pumped under pressure into a hydrodynamic cavitation reactor and subjected to 3600 rpm rotor speed, the juice went through a transient temperature gradient from 37.8 to 104 °C and a 3.1-log reduction of *B. coagulans* spores was obtained (Milly *et al.*, 2007).

Pressure-Ohmic-Thermal Sterilization   To achieve commercial sterility of tomato juice, a combination of ohmic (50 V/cm), pressure (600 MPa), and thermal (105 °C) can effectively inactivate pressure-thermal-resistant bacteria spores, between 4.6 and 5.6 logs reduction (Park *et al.*, 2013).

Other Hurdle Technology – Addition of Antimicrobial Agents   Several examples of a synergistic enhancement effect on microorganism inactivation using combined approach of innovative technologies and natural antimicrobials or compounds with antimicrobial activities are presented as follows:

- HPP (500 MPa for 5 min) combined with addition of 0.7% sodium chloride resulted in 5-log reduction for *E. coli* in tomato juice (Jordan *et al.*, 2001),
- HPP (400 MPa for 20 min) combined with addition of 0.1% oregano oil resulted in tomato juice with at least 3 weeks shelf life at 15 °C (Mohácsi-Farkas *et al.*, 2002),
- HPP (200 MPa for 10 min) combined with addition of 0.1% thyme oil resulted in tomato juice with at least 3 weeks shelf life at 15 °C (Mohácsi-Farkas *et al.*, 2002),
- PEF (20 pulses of 80 kV/cm at 50 °C) combined with addition of nisin (400 U/ml) resulted in 6.2-log reduction for naturally occurring microorganisms in grape juice (Wu *et al.*, 2005), and
- PEF (35 kV/cm, 100 Hz) combined with addition of 2% citric acid or 0.1% cinnamon bark oil resulted in strawberry and tomato juice with at least 91 days shelf life at 5 °C (Mosqueda-Melgar *et al.*, 2012).

6.4   Relevance for Human Health

Potential health benefits (e.g., prevention of cardiovascular disease, chronic cancer, and age-related diseases) associated with berry fruit consumption due to the biologically active compounds that include polyphenols (anthocyanin, phenolic acids, and stilbenes) as well as carotenoids and vitamin C have been comprehensively reviewed (Beattie *et al.*, 2005; Szajdek & Borowska, 2008; Joseph *et al.*, 2009; Basu *et al.*, 2010).

Phytochemicals in berry fruit are susceptible to degradation; hence, changes in the composition and content of these phytochemicals due to juice processing of berries would influence their bioactivity.

The study of Schmidt et al. (2005) has shown that heat-processed blueberry products such as heat-dried, cooked, juice concentrate, pie filling, and jam have higher total phenolics content and antioxidant activity to those found in fresh blueberries. However, low antiproliferative activity of murine (Hepa1c1c7) hepatoma cells detected for these heat-processed products suggested that some bioactive properties of blueberries may have been compromised by heat processing. On the contrary, hamsters that were fed with grape juice prepared after hot pressing (60 °C) and flash pasteurization at 85 °C for 20 s showed higher efficacy in preventing an early development of atherosclerosis as compared to the hamsters fed with fresh grapes (Décordé et al., 2008). This study shows the positive impact of processing on the bioactive properties of grapes. It is possible that processing could increase the bioavailability of phytochemicals or lead to the formation of compounds having higher antioxidant status.

Regarding tomatoes, the inhibitory effect of tomato fruit on platelet aggregation induced by adenosine 5'-diphosphate, collagen, thrombin receptor activator peptide-6, and arachidonic acid was found insignificantly different from tomato juice (Fuentes et al., 2013). Despite the content of some phenolic acids in tomato juice and fruit (+234% chlorogenic acid, −100% of caffeic acid, p-coumaric acid and ferulic acid) increases during processing, both tomato fruit and its juice have similar antiplatelet activity to prevent cardiovascular diseases. In another study, thermal treatment at 80 °C for 5 min has also been shown to enhance the extraction of strawberry polyphenols, namely ellagitannin and ellagic acid from the fruit (Truchado et al., 2012) but does not alter the bioavailability of polyphenols when healthy individuals received either 200 g of fresh strawberries or the equivalent dose of processed-strawberry product for a 4-week duration.

The potential cancer chemopreventive effect of cranberry extracts isolated at different stages of juice processing, frozen cranberries, mash, depectinized mash (preheat at 55 °C prior to pectinase addition), pomace, raw juice, clarified juice (via a cross-flow membrane filtration), and juice concentrate (50 °Brix, 100 °C under vacuum), has been evaluated (Caillet et al., 2011). In this study, the ability of these cranberry extracts to stimulate the activity of the phase II xenobiotic detoxification enzyme quinone reductase, an enzyme involved in the detoxification of potential carcinogens and xenobiotics, has been reduced significantly after the juice processing. Therefore, this finding indicates that cranberry juice and juice concentrate have reduced cancer chemopreventive effects as compared to the frozen cranberries. In other study, in vitro antiproliferative properties against two colon cancer cell lines of HT-29 and LS-513 of cranberry extracts isolated during cranberry juice processing was assessed (Vu et al., 2012). Cranberry extracts from fruit and depectinized purée (preheated at 55 °C prior to pectinase addition) inhibit the growth of cancer cells, by 50% more efficiently than the raw and filtered (via a cross-flow membrane filtration) juice. Other studies have investigated the health benefits of cranberry juice consumption, including prevention and treatment of urinary tract infections (Jepson et al., 2012) and improving both vascular functions and cholesterol profiles in ovariectomized rats which provide insight into reducing the risk for cardiovascular diseases among postmenopausal women (Yung et al., 2013).

6.5 Conclusions and Future Trends

The underlying mechanism to explain the significance of processing in influencing the biological and physiological functions of phytochemicals associated with health benefits is rather challenging. The improved health benefits could be resulted from multiple phytochemicals found in berry fruit rather than from a single compound alone. In this respect, continuous characterization and monitoring the profile of polyphenols and other individual phytochemicals during juice processing is required. This would assist in the determination of which individual compounds (or best known as processing biomarkers) are better in resisting or are susceptible to the juice processing conditions and contribute to the bioactivity status of the berry juice. It is noteworthy that knowledge on the effects of novel processing technologies on the bioactive properties, in relevance to human health, of berry juice is very limited; therefore, future studies in this area are necessary.

References

Adekunte, A.O., Tiwari, B.K., Cullen, P.J., Scannell, A.G.M. & O'Donnell, C.P. (2010a) Effect of sonication on colour, ascorbic acid and yeast inactivation in tomato juice. *Food Chemistry* **122**, 500–507.

Adekunte, A.O., Tiwari, B.K., Scannell, A.G.M., Cullen, P.J. & O'Donnell, C. (2010b) Modelling of yeast inactivation in sonicated tomato juice. *International Journal of Food Microbiology* **137**, 116–120.

Aggarwal, P. & Gill, M. (2010) Suitability of newly evolved antioxidant rich grape cultivars for processing into juice and beverages. *Indian Journal of Horticulture* **67**, 102–107.

Aguiló-Aguayo, I., Oms-Oliu, G., Soliva-Fortuny, R. & Martín-Belloso, O. (2009a) Changes in quality attributes throughout storage of strawberry juice processed by high-intensity pulsed electric fields or heat treatments. *LWT–Food Science and Technology* **42**, 813–818.

Aguiló-Aguayo, I., Oms-Oliu, G., Soliva-Fortuny, R. & Martín-Belloso, O. (2009b) Flavour retention and related enzyme activities during storage of strawberry juices processed by high-intensity pulsed electric fields or heat. *Food Chemistry* **116**, 59–65.

Aguiló-Aguayo, I., Soliva-Fortuny, R. & Martín-Belloso, O. (2008) Comparative study on color, viscosity and related enzymes of tomato juice treated by high-intensity pulsed electric fields or heat. *European Food Research and Technology* **227**, 599–606.

Aguiló-Aguayo, I., Soliva-Fortuny, R. & Martín-Belloso, O. (2009c) Avoiding non-enzymatic browning by high-intensity pulsed electric fields in strawberry, tomato and watermelon juices. *Journal of Food Engineering* **92**, 37–43.

Aguiló-Aguayo, I., Soliva-Fortuny, R. & Martín-Belloso, O. (2009d) Changes in viscosity and pectolytic enzymes of tomato and strawberry juices processed by high-intensity pulsed electric fields. *International Journal of Food Science and Technology* **44**, 2268–2277.

Aguiló-Aguayo, I., Soliva-Fortuny, R. & Martín-Belloso, O. (2010) Volatile compounds and changes in flavour-related enzymes during cold storage of high-intensity pulsed electric field- and heat-processed tomato juices. *Journal of the Science of Food and Agriculture* **90**, 1597–1604.

Al-Said, F.A., Opara, L.U. & Al-Yahyai, R.A. (2009) Physico-chemical and textural quality attributes of pomegranate cultivars (*Punica granatum* L.) grown in the Sultanate of Oman. *Journal of Food Engineering* **90**, 129–134.

Alighourchi, H., Barzegar, M. & Abbasi, S. (2008) Anthocyanins characterization of 15 Iranian pomegranate (*Punica granatum* L.) varieties and their variation after cold storage and pasteurization. *European Food Research and Technology* **227**, 881–887.

Alighourchi, H., Barzegar, M., Sahari, M. & Abbasi, S. (2013) Effect of sonication on anthocyanins, total phenolic content, and antioxidant capacity of pomegranate juices. *International Food Research Journal* **20**, 1703–1709.

Alighourchi, H., Barzegar, M., Sahari, M.A. & Abbasi, S. (2014) The effects of sonication and gamma irradiation on the inactivation of *Escherichia coli* and *Saccharomyces cerevisiae* in pomegranate juice. *Iranian Journal of Microbiology* **6**, 51–58.

Alpas, H., Alma, L. & Bozoglu, F. (2003) Inactivation of *Alicyclobacillus acidoterrestris* vegetative cells in model system, apple, orange and tomato juices by high hydrostatic pressure. *World Journal of Microbiology and Biotechnology* **19**, 619–623.

Alper, N., Bahçeci, K.S. & Acar, J. (2005) Influence of processing and pasteurization on color values and total phenolic compounds of pomegranate juice. *Journal of Food Processing and Preservation* **29**, 357–368.

Augusto, P.E.D., Ibarz, A. & Cristianini, M. (2012) Effect of high pressure homogenization (HPH) on the rheological properties of tomato juice: Time-dependent and steady-state shear. *Journal of Food Engineering* **111**, 570–579.

Augusto, P.E.D., Ibarz, A. & Cristianini, M. (2013a) Effect of high pressure homogenization (HPH) on the rheological properties of tomato juice: Creep and recovery behaviours. *Food Research International* **54**, 169–176.

Augusto, P.E.D., Ibarz, A. & Cristianini, M. (2013b) Effect of high pressure homogenization (HPH) on the rheological properties of tomato juice: Viscoelastic properties and the Cox–Merz rule. *Journal of Food Engineering* **114**, 57–63.

Barba, F.J., Esteve, M.J. & Frigola, A. (2013) Physicochemical and nutritional characteristics of blueberry juice after high pressure processing. *Food Research International* **50**, 545–549.

Barba, F.J., Jäger, H., Meneses, N., Esteve, M.J., Frígola, A. & Knorr, D. (2012) Evaluation of quality changes of blueberry juice during refrigerated storage after high-pressure and pulsed electric fields processing. *Innovative Food Science and Emerging Technologies* **14**, 18–24.

Basu, A., Rhone, M. & Lyons, T.J. (2010) Berries: Emerging impact on cardiovascular health. *Nutrition Reviews* **68**, 168–177.

Beattie, J., Crozier, A. & Duthie, G.G. (2005) Potential health benefits of berries. *Current Nutrition and Food Science* **1**, 71–86.

Beuchat, L.R. (2002) Ecological factors influencing survival and growth of human pathogens on raw fruits and vegetables. *Microbes and Infection* **4**, 413–423.

Bevilacqua, A., Sinigaglia, M. & Corbo, M.R. (2010) Use of the response surface methodology and desirability approach to model *Alicyclobacillus acidoterrestris* spore inactivation. *International Journal of Food Science and Technology* **45**, 1219–1227.

Brambilla, A., Lo Scalzo, R., Bertolo, G. & Torreggiani, D. (2008) Steam-blanched highbush blueberry (*Vaccinium corymbosum* L.) juice: Phenolic profile and antioxidant capacity in relation to cultivar selection. *Journal of Agricultural and Food Chemistry* **56**, 2643–2648.

Brownmiller, C., Howard, L.R. & Prior, R.L. (2008) Processing and storage effects on monomeric anthocyanins, percent polymeric color, and antioxidant capacity of processed blueberry products. *Journal of Food Science* **73**, H72–H79.

Bucheli, P., Voirol, E., De La Torre, R., López, J., Rytz, A., Tanksley, S.D. & Pétiard, V. (1999) Definition of nonvolatile markers for flavor of tomato (*Lycopersicon esculentum* Mill.) as tools in selection and breeding. *Journal of Agricultural and Food Chemistry* **47**, 659–664.

Buckow, R., Kastell, A., Terefe, N.S. & Versteeg, C. (2010) Pressure and temperature effects on degradation kinetics and storage stability of total anthocyanins in blueberry juice. *Journal of Agricultural and Food Chemistry* **58**, 10076–10084.

Butz, P., Edenharder, R., Fister, H. & Tauscher, B. (1997) The influence of high pressure processing on antimutagenic activities of fruit and vegetable juices. *Food Research International* **30**, 287–291.

Caillet, S., Côté, J., Doyon, G., Sylvain, J.F. & Lacroix, M. (2011) Effect of juice processing on the cancer chemopreventive effect of cranberry. *Food Research International*, **44**, 902–910.

Cao, X., Bi, X., Huang, W., Wu, J., Hu, X. & Liao, X. (2012) Changes of quality of high hydrostatic pressure processed cloudy and clear strawberry juices during storage. *Innovative Food Science and Emerging Technologies* **16**, 181–190.

Capanoglu, E., de Vos, R.C.H., Hall, R.D., Boyacioglu, D. & Beekwilder, J. (2013) Changes in polyphenol content during production of grape juice concentrate. *Food Chemistry* **139**, 521–526.

Chauhan, O.P., Raju, P.S., Ravi, N., Roopa, N. & Bawa, A.S. (2011) Studies on retention of antioxidant activity, phenolics and flavonoids in high pressure processed black grape juice and their modelling. *International Journal of Food Science and Technology* **46**, 2562–2568.

Conde, C., Silva, P., Fontes, N., Dias, A.C.P., Tavares, R.M., Sousa, M.J., Agasse, A., Delrot, S. & Gerós, H. (2007) Biochemical changes throughout grape berry development and fruit and wine quality. *Food* **1**, 1–22.

Côté, J., Caillet, S., Dussault, D., Sylvain, J.F. & Lacroix, M. (2011) Effect of juice processing on cranberry antibacterial properties. *Food Research International* **44**(9), 2922–2929.

Cristofori, V., Caruso, D., Latini, G., Dell'Agli, M., Cammilli, C., Rugini, E., Bignami, C. & Muleo, R. (2011) Fruit quality of Italian pomegranate (*Punica granatum* L.) autochthonous varieties. *European Food Research and Technology* **232**, 397–403.

Daoudi, L., Quevedo, J.M., Trujillo, A.J., Capdevila, F., Bartra, E., Mínguez, S. & Guamis, B. (2002) Effects of high-pressure treatment on the sensory quality of white grape juice. *High Pressure Research* **22**, 705–709.

Décordé, K., Teissèdre, P.L., Auger, C., Cristol, J.P. & Rouanet, J.M. (2008) Phenolics from purple grape, apple, purple grape juice and apple juice prevent early atherosclerosis induced by an atherogenic diet in hamsters. *Molecular Nutrition and Food Research* **52**, 400–407.

Dede, S., Alpas, H. & Bayíndírlí, A. (2007) High hydrostatic pressure treatment and storage of carrot and tomato juices: Antioxidant activity and microbial safety. *Journal of the Science of Food and Agriculture* **87**, 773–782.

Del Pozo-Insfran, D., Balaban, M.O. & Talcott, S.T. (2006) Microbial stability, phytochemical retention, and organoleptic attributes of dense phase CO_2 processed muscadine grape juice. *Journal of Agricultural and Food Chemistry* **54**, 5468–5473.

Donsì, F., Ferrari, G. & Pataro, G. (2010) Applications of pulsed electric field treatments for the enhancement of mass transfer from vegetable tissue. *Food Engineering Reviews* **2**, 109–130.

Dubrovic, I., Herceg, Z., Režek Jambrak, A., Badanjak, M. & Dragovic-Uzelac, V. (2011) Effect of high intensity ultrasound and pasteurization on anthocyanin content in strawberry juice. *Food Technology and Biotechnology*, **49**, 196–204.

Ereifej, K.I., Shibli, R.A., Ajlouni, M.M. & Hussain, A. (1997) Physico-chemical characteristics and processing quality of newly introduced seven tomato cultivars into Jordan in comparison with local variety. *Journal of Food Science and Technology* **34**, 171–174.

Eribo, B. & Ashenafi, M. (2003) Behavior of *Escherichia coli* O157:H7 in tomato and processed tomato products. *Food Research International* **36**, 823–830.

Errington, N., Tucker, G.A. & Mitchell, J.R. (1998) Effect of genetic down-regulation of polygalacturonase and pectin esterase activity on rheology and composition of tomato juice. *Journal of the Science of Food and Agriculture* **76**, 515–519.

Evrendilek, G.A., Dantzer, W.R., Streaker, C.B., Ratanatriwong, P. & Zhang, Q.H. (2001) Shelf-life evaluations of liquid foods treated by pilot plant pulsed electric field system. *Journal of Food Processing and Preservation* **25**, 283–297.

Ferrari, G., Maresca, P. & Ciccarone, R. (2010) The application of high hydrostatic pressure for the stabilization of functional foods: Pomegranate juice. *Journal of Food Engineering* **100**, 245–253.

Fischer, U.A., Dettmann, J.S., Carle, R. & Kammerer, D.R. (2011) Impact of processing and storage on the phenolic profiles and contents of pomegranate (*Punica granatum* L.) juices. *European Food Research and Technology* **233**, 797–816.

Fredericks, I.N., du Toit, M. & Krügel, M. (2011) Efficacy of ultraviolet radiation as an alternative technology to inactivate microorganisms in grape juices and wines. *Food Microbiology*, **28**, 510–517.

Fuentes, E., Forero-Doria, O., Carrasco, G., Maricá, A., Santos, L.S., Alarcó, M. & Palomo, I. (2013) Effect of tomato industrial processing on phenolic profile and antiplatelet activity. *Molecules* **18**, 11526–11536.

Fuleki, T. & Ricardo-da-Silva, J.M. (2002) Effects of cultivar and processing method on the contents of catechins and procyanidins in grape juice. *Journal of Agricultural and Food Chemistry* **51**, 640–646.

Gahler, S., Otto, K. & Böhm, V. (2003) Alterations of vitamin C, total phenolics, and antioxidant capacity as affected by processing tomatoes to different products. *Journal of Agricultural and Food Chemistry* **51**, 7962–7968.

Galić, A., Dragović-Uzelac, V., Levaj, B., Bursać Kovačević, D., Pliestić, S. & Arnautović, S. (2009) The polyphenols stability, enzyme activity and physico-chemical parameters during producing wild elderberry concentrated juice. *Agriculturae Conspectus Scientificus* **74**, 181–186.

Gancel, A.-L., Feneuil, A., Acosta, O., Pérez, A.M. & Vaillant, F. (2011) Impact of industrial processing and storage on major polyphenols and the antioxidant capacity of tropical highland blackberry (*Rubus adenotrichus*). *Food Research International* **44**, 2243–2251.

Gao, Z., Yuan, Y., Yue, T., Bin, Z. & Wang, Y. (2014) Influence of centrifugation and sedimentation on turbidity of pomegranate juice during storage. *Journal of Chinese Institute of Food Science and Technology* **14**, 117–123.

García-Viguera, C. & Zafrilla, P. (2001) Changes in anthocyanins during food processing: influence on color. In: *Chemistry and Physiology of Selected Food Colorants*. (eds J.M. Ames, & T. Hofmann). pp. 56–65. American Chemical Society, Washington DC. USA.

Garde-Cerdán, T., Arias-Gil, M., Marsellés-Fontanet, A.R., Ancín-Azpilicueta, C. & Martín-Belloso, O. (2007) Effects of thermal and non-thermal processing treatments on fatty acids and free amino acids of grape juice. *Food Control* **18**, 473–479.

Ghebbi Si-smail, K., Bellal, M. & Halladj, F. (2006) Effect of potassium supply on the behaviour of two processing tomato cultivars and on the changes of fruit technological characteristics. *X International Symposium on the Processing Tomato* **758**, 269–274.

González-Barrio, R., Vidal-Guevara, M.L., Tomás-Barberán, F.A. & Espín, J.C. (2009) Preparation of a resveratrol-enriched grape juice based on ultraviolet C-treated berries. *Innovative Food Science and Emerging Technologies* **10**, 374–382.

Grimi, N., Lebovka, N.I., Vorobiev, E. & Vaxelaire, J. (2009) Effect of a pulsed electric field treatment on expression behavior and juice quality of Chardonnay grape. *Food Biophysics* **4**, 191–198.

Guerrero-Beltrán, J., Welti-Chanes, J. & Barbosa-Cánovas, G.V. (2009) Ultraviolet-C light processing of grape, cranberry and grapefruit juices to inactivate *Saccharomyces cerevisiae*. *Journal of Food Process Engineering*, **32**, 916–932.

Gundogdu, M. & Yilmaz, H. (2012) Organic acid, phenolic profile and antioxidant capacities of pomegranate (*Punica granatum* L.) cultivars and selected genotypes. *Scientia Horticulturae* **143**, 38–42.

Gunes, G., Blum, L.K. & Hotchkiss, J.H. (2005) Inactivation of yeasts in grape juice using a continuous dense phase carbon dioxide processing system. *Journal of the Science of Food and Agriculture* **85**, 2362–2368.

Guo, M., Jin, T.Z., Geveke, D.J., Fan, X., Sites, J.E. & Wang, L. (2014) Evaluation of microbial stability, bioactive compounds, physicochemical properties, and consumer acceptance of pomegranate juice processed in a commercial scale pulsed electric field system. *Food and Bioprocess Technology* **7**, 2112–2120.

Gupta, R., Balasubramaniam, V.M., Schwartz, S.J. & Francis, D.M. (2010) Storage stability of lycopene in tomato juice subjected to combined pressure–heat treatments. *Journal of Agricultural and Food Chemistry* **58**, 8305–8313.

Gupta, R., Kopec, R.E., Schwartz, S.J. & Balasubramaniam, V.M. (2011) Combined pressure–temperature effects on carotenoid retention and bioaccessibility in tomato juice. *Journal of Agricultural and Food Chemistry* **59**, 7808–7817.

Hager, A., Howard, L.R., Prior, R.L. & Brownmiller, C. (2008) Processing and storage effects on monomeric anthocyanins, percent polymeric color, and antioxidant capacity of processed black raspberry products. *Journal of Food Science* **73**, H134–H140.

Hager, T.J., Howard, L.R. & Prior, R.L. (2010) Processing and storage effects on the ellagitannin composition of processed blackberry products. *Journal of Agricultural and Food Chemistry* **58**, 11749–11754.

Hakguder, B. & Unluturk, S. (2009) Effect of penetration depth and UV dose on the performance of UV-C disinfection process of white grape juice inoculated with *Escherichia coli* K-12. *5th International Technical Symposium on Food Processing, Monitoring Technology in Bioprocesses and Food Quality Management*, 87–92.

Hakkinen, S.H., Karenlampi, S.O., Mykkanen, H.M. & Torronen, A.R. (2000) Influence of domestic processing and storage on flavonol contents in berries. *Journal of Agricultural and Food Chemistry* **48**, 2960–2965.

Harris, L.J., Farber, J.N., Beuchat, L.R., Parish, M.E., Suslow, T.V., Garrett, E.H. & Busta, F.F. (2003) Outbreaks associated with fresh produce: Incidence, growth, and survival of pathogens in fresh and fresh-cut produce. *Comprehensive Reviews in Food Science and Food Safety* **2**, 78–141.

Harrison, R.E., Muir, D.D. & Hunter, E.A. (1998) Genotypic effects on sensory profiles of drinks made from juice of red raspberries (*Rubus idaeus* L.). *Food Research International* **31**, 303–309.

Hartmann, A., Patz, C.-D., Andlauer, W., Dietrich, H. & Ludwig, M. (2008) Influence of processing on quality parameters of strawberries. *Journal of Agricultural and Food Chemistry* **56**, 9484–9489.

Holtung, L., Grimmer, S. & Aaby, K. (2011) Effect of processing of black currant press-residue on polyphenol composition and cell proliferation. *Journal of Agricultural and Food Chemistry* **59**, 3632–3640.

Horm, K.M., Davidson, P.M., Harte, F.M. & D'Souza, D.H. (2012) Survival and inactivation of human norovirus surrogates in blueberry juice by high-pressure homogenization. *Foodborne Pathogens and Disease* **9**, 974–979.

Hsu, K.-C. (2008) Evaluation of processing qualities of tomato juice induced by thermal and pressure processing. *LWT – -Food Science and Technology* **41**, 450–459.

Hsu, K.-C., Tan, F.-J. & Chi, H.-Y. (2008) Evaluation of microbial inactivation and physicochemical properties of pressurized tomato juice during refrigerated storage. *LWT – Food Science and Technology* **41**, 367–375.

Huang, K., Yu, L., Wang, W., Gai, L. & Wang, J. (2014) Comparing the pulsed electric field resistance of the microorganisms in grape juice: Application of the Weibull model. *Food Control* **35**, 241–251.

Ingham, S.C., Schoeller, E.L. & Engel, R.A. (2006) Pathogen reduction in unpasteurized apple cider: Adding cranberry juice to enhance the lethality of warm hold and freeze-thaw steps. *Journal of Food Protection* **69**, 293–298.

Iniesta, M.D., Pérez-Conesa, D., García-Alonso, J., Ros, G. & Periago, M.J. (2009) Folate content in tomato (*Lycopersicon esculentum*). Influence of cultivar, ripeness, year of harvest, and pasteurization and storage temperatures. *Journal of Agricultural and Food Chemistry* **57**, 4739–4745.

Iversen, C.K., Jakobsen, H.B. & Olsen, C.-E. (1998) Aroma changes during black currant (*Ribes nigrum* L.) nectar processing. *Journal of Agricultural and Food Chemistry*, **46**, 1132–1136.

Iyer, M.M., Sacks, G.L. & Padilla-Zakour, O.I. (2010) Impact of harvesting and processing conditions on green leaf volatile development and phenolics in Concord grape juice. *Journal of Food Science* **75**, C297–C304.

Jensen, K., Christensen, L.P., Hansen, M., Jørgensen, U. & Kaack, K. (2001) Olfactory and quantitative analysis of volatiles in elderberry (*Sambucus nigra* L) juice processed from seven cultivars. *Journal of the Science of Food and Agriculture* **81**, 237–244.

Jepson, R.G. Williams, G. & Craig, J.C. (2012) Cranberries for preventing urinary tract infections. *Cochrane Database of Systematic Reviews*, 20. Art. No.: CD001321. doi: 10.1002/14651858.CD001321.pub5.

Jin, Z.T. & Zhang, Q.H. (1999) Pulsed electric field inactivation of microorganisms and preservation of quality of cranberry juice. *Journal of Food Processing and Preservation* **23**, 481–497.

Jordan, S.L., Pascual, C., Bracey, E. & Mackey, B.M. (2001) Inactivation and injury of pressure-resistant strains of *Escherichia coli* O157 and *Listeria monocytogenes* in fruit juices. *Journal of Applied Microbiology* **91**, 463–469.

Joseph, J.A., Shukitt-Hale, B. & Willis, L.M. (2009) Grape juice, berries, and walnuts affect brain aging and behavior. *The Journal of Nutrition* **139**, 1813S–1817S.

Kaack, K. (2008) Aroma composition and sensory quality of fruit juices processed from cultivars of elderberry (*Sambucus nigra* L.). *European Food Research and Technology* **227**, 45–56.

Kaack, K., Christensen, L.P., Hughes, M. & Eder, R. (2005) The relationship between sensory quality and volatile compounds in raw juice processed from elderberries (*Sambucus nigra* L.). *European Food Research and Technology* **221**, 244–254.

Kalamaki, M.S., Harpster, M.H., Palys, J.M., Labavitch, J.M., Reid, D.S. & Brummell, D.A. (2003) Simultaneous transgenic suppression of LePG and LeExp1 influences rheological properties of juice and concentrates from a processing tomato variety. *Journal of Agricultural and Food Chemistry* **51**, 7456–7464.

Kamiloglu, S., Demirci, M., Selen, S., Toydemir, G., Boyacioglu, D. & Capanoglu, E. (2014) Home processing of tomatoes (*Solanum lycopersicum*): Effects on *in vitro* bioaccessibility of total lycopene, phenolics, flavonoids, and antioxidant capacity. *Journal of the Science of Food and Agriculture* **94**, 2225–2233.

Kim, T.J., Weng, W.L., Silva, J.L., Jung, Y.S. & Marshall, D. (2010) Identification of natural antimicrobial substances in red muscadine juice against *Cronobacter sakazakii*. *Journal of Food Science*, **75**, M150–M154.

Klopotek, Y., Otto, K. & Böhm, V. (2005) Processing strawberries to different products alters contents of vitamin C, total phenolics, total anthocyanins, and antioxidant capacity. *Journal of Agricultural and Food Chemistry*, **53**, 5640–5646.

Koponen, J., Buchert, J., Poutanen, K. & Törrönen, A.R. (2008a) Effect of pectinolytic juice production on the extractability and fate of bilberry and black currant anthocyanins. *European Food Research and Technology* **227**, 485–494.

Koponen, J.M., Happonen, A.M., Auriola, S., Kontkanen, H., Buchert, J., Poutanen, K.S. & Törrönen, A.R. (2008b) Characterization and fate of black currant and bilberry flavonols in enzyme-aided processing. *Journal of Agricultural and Food Chemistry* **56**, 3136–3144.

Kozák, Á., Bánvölgyi, S., Vincze, I., Kiss, I., Békássy-Molnár, E. & Vatai, G. (2008) Comparison of integrated large scale and laboratory scale membrane processes for the production of black currant juice concentrate. *Chemical Engineering and Processing: Process Intensification* **47**, 1171–1177.

Kubo, M.T.K., Augusto, P.E.D. & Cristianini, M. (2013) Effect of high pressure homogenization (HPH) on the physical stability of tomato juice. *Food Research International* **51**, 170–179.

Laaksonen, O.A., Mäkilä, L., Sandell, M.A., Salminen, J.-P., Liu, P., Kallio, H.P. & Yang, B. (2014) Chemical-sensory characteristics and consumer responses of blackcurrant juices produced by different industrial processes. *Food and Bioprocess Technology*, **7**, 2877–2888.

Laaksonen, O.A., Mäkilä, L., Tahvonen, R., Kallio, H. & Yang, B. (2013) Sensory quality and compositional characteristics of blackcurrant juices produced by different processes. *Food Chemistry* **138**, 2421–2429.

Landbo, A.-K. & Meyer, A.S. (2001) Enzyme-assisted extraction of antioxidative phenols from black currant juice press residues (*Ribes nigrum*). *Journal of Agricultural and Food Chemistry* **49**, 3169–3177.

Landbo, A.-K. & Meyer, A.S. (2004) Effects of different enzymatic maceration treatments on enhancement of anthocyanins and other phenolics in black currant juice. *Innovative Food Science and Emerging Technologies* **5**, 503–513.

Leblanc, M.R., Johnson, C.E. & Wilson, P.W. (2008) Influence of pressing method on juice stilbene content in muscadine and bunch grapes. *Journal of Food Science*, **73**, H58–H62.

Lee, J., Durst, R.W. & Wrolstad, R.E. (2002) Impact of juice processing on blueberry anthocyanins and polyphenolics: Comparison of two pretreatments. *Journal of Food Science* **67**, 1660–1667.

Lee, J. & Wrolstad, R.E. (2004) Extraction of anthocyanins and polyphenolics from blueberry processing waste. *Journal of Food Science* **69**, 564–573.

Lee, S.Y., Sagong, H.G., Ryu, S. & Kang, D.H. (2012) Effect of continuous ohmic heating to inactivate *Escherichia coli* O157:H7, *Salmonella typhimurium* and *Listeria monocytogenes* in orange juice and tomato juice. *Journal of Applied Microbiology* **112**, 723–731.

Li, Y., Ma, R., Xu, Z., Wang, J., Chen, T., Chen, F. & Wang, Z. (2013) Identification and quantification of anthocyanins in Kyoho grape juice-making pomace, Cabernet Sauvignon grape winemaking pomace and their fresh skin. *Journal of the Science of Food and Agriculture* **93**, 1404–1411.

Lieu, L.N. & Le, V.V.M. (2010) Application of ultrasound in grape mash treatment in juice processing. *Ultrasonics Sonochemistry* **17**, 273–279.

Lima Tribst, A.A., de Souza Sant'Ana, A. & de Massaguer, P.R. (2009) Review: Microbiological quality and safety of fruit juices—past, present and future perspectives. *Critical Reviews in Microbiology* **35**, 310–339.

Lin, C.H. & Chen, B.H. (2005) Stability of carotenoids in tomato juice during processing. *European Food Research and Technology* **221**, 274–280.

Loginov, M., Loginova, K., Lebovka, N.I. & Vorobiev, E. (2011) Comparison of dead-end ultrafiltration behaviour and filtrate quality of sugar beet juices obtained by conventional and "cold" PEF-assisted diffusion. *Journal of Membrane Science* **377**, 273–283.

Marković, K., Vahčić, N., Ganić, K.K. & Banović, M. (2007) Aroma volatiles of tomatoes and tomato products evaluated by solid-phase microextraction. *Flavour and Fragrance Journal* **22**, 395–400.

Markowski, J. & Pluta, S. (2008) Fruit quality and suitability of new Polish blackcurrant cultivars for processing. *Acta Horticulturae* **777**, 521–524.

Marsellés-Fontanet, Á.R., Puig-Pujol, A., Olmos, P., Mínguez-Sanz, S. & Martín-Belloso, O. (2013) A comparison of the effects of pulsed electric field and thermal treatments on grape juice. *Food and Bioprocess Technology* **6**, 978–987.

Marsellés-Fontanet, À.R., Puig-Pujol, A., Olmos, P., Mínguez-Sanz, S. & Martín-Belloso, O. (2009) Optimising the inactivation of grape juice spoilage organisms by pulse electric fields. *International Journal of Food Microbiology* **130**, 159–165.

Martínez, J.J., Hernández, F., Abdelmajid, H., Legua, P., Martínez, R., Amine, A.E. & Melgarejo, P. (2012) Physico-chemical characterization of six pomegranate cultivars from Morocco: Processing and fresh market aptitudes. *Scientia Horticulturae* **140**, 100–106.

Martino, K.G., Paul, M.S., Pegg, R.B. & Kerr, W.L. (2013) Effect of time–temperature conditions and clarification on the total phenolics and antioxidant constituents of muscadine grape juice. *LWT – -Food Science and Technology* **53**, 327–330.

Mazzotta, A.S. (2001) Thermal inactivation of stationary-phase and acid-adapted *Escherichia coli* O157:H7, *Salmonella*, and *Listeria monocytogenes* in fruit juices. *Journal of Food Protection* **64**, 315–320.

Mena, P., Martí, N., Saura, D., Valero, M. & García-Viguera, C. (2013) Combinatory effect of thermal treatment and blending on the quality of pomegranate juices. *Food and Bioprocess Technology* **6**, 3186–3199.

Mendelová, A., Fikselová, M. & Mendel, Ľ. (2013) Carotenoids and lycopene content in fresh and dried tomato fruits and tomato juice. *Acta Universitatis Agriculturae et Silviculturae Mendelianae Brunensis* **61**, 1329–1337.

Mhemdi, H., Bals, O., Grimi, N. & Vorobiev, E. (2012) Filtration diffusivity and expression behaviour of thermally and electrically pretreated sugar beet tissue and press-cake. *Separation and Purification Technology* **95**, 118–125.

Mikkelsen, B.B. & Poll, L. (2002) Decomposition and transformation of aroma compounds and anthocyanins during black currant (*Ribes nigrum* L.) juice processing. *Journal of Food Science* **67**, 3447–3455.

Milly, P.J., Toledo, R.T., Harrison, M.A. & Armstead, D. (2007) Inactivation of food spoilage microorganisms by hydrodynamic cavitation to achieve pasteurization and sterilization of fluid foods. *Journal of Food Science* **72**, M414–M422.

Min, S., Jin, Z.T. & Zhang, Q.H. (2003) Commercial scale pulsed electric field processing of tomato juice. *Journal of Agricultural and Food Chemistry* **51**, 3338–3344.

Min, S. & Zhang, Q.H. (2003) Effects of commercial-scale pulsed electric field processing on flavor and color of tomato juice. *Journal of Food Science* **68**, 1600–1606.

Mirsaeedghazi, H., Mousavi, S.M., Emam-Djomeh, Z., Rezaei, K., Aroujalian, A. & Navidbakhsh, M. (2012) Comparison between ultrafiltration and microfiltration in the clarification of pomegranate juice. *Journal of Food Process Engineering* **35**, 424–436.

Mohácsi-Farkas, C., Kiskó, G., Mészáros, L. & Farkas, J. (2002) Pasteurisation of tomato juice by high hydrostatic pressure treatment or by its combination with essential oils. *Acta Alimentaria* **31**, 243–252.

Molnar-Perl, I. & Friedman, M. (1990) Inhibition of browning by sulfur amino acids. 2. Fruit juices and protein-containing foods. *Journal of Agricultural and Food Chemistry* **38**, 1648–1651.

Mosqueda-Melgar, J., Raybaudi-Massilia, R.M. & Martín-Belloso, O. (2008a) Inactivation of *Salmonella enterica* ser. enteritidis in tomato juice by combining of high-intensity pulsed electric fields with natural antimicrobials. *Journal of Food Science* **73**, M47–M53.

Mosqueda-Melgar, J., Raybaudi-Massilia, R.M. & Martín-Belloso, O. (2008b) Non-thermal pasteurization of fruit juices by combining high-intensity pulsed electric fields with natural antimicrobials. *Innovative Food Science and Emerging Technologies* **9**, 328–340.

Mosqueda-Melgar, J., Raybaudi-Massilia, R.M. & Martín-Belloso, O. (2012) Microbiological shelf life and sensory evaluation of fruit juices treated by high-intensity pulsed electric fields and antimicrobials. *Food and Bioproducts Processing* **90**, 205–214.

Nguyen, P. & Mittal, G.S. (2007) Inactivation of naturally occurring microorganisms in tomato juice using pulsed electric field (PEF) with and without antimicrobials. *Chemical Engineering and Processing: Process Intensification* **46**, 360–365.

Odriozola-Serrano, I., Aguiló-Aguayo, I., Soliva-Fortuny, R., Gimeno-Añó, V. & Martín-Belloso, O. (2007) Lycopene, vitamin C, and antioxidant capacity of tomato juice as affected by high-intensity pulsed electric fields critical parameters. *Journal of Agricultural and Food Chemistry* **55**, 9036–9042.

Odriozola-Serrano, I., Garde-Cerdán, T., Soliva-Fortuny, R. & Martín-Belloso, O. (2013) Differences in free amino acid profile of non-thermally treated tomato and strawberry juices. *Journal of Food Composition and Analysis* **32**, 51–58.

Odriozola-Serrano, I., Soliva-Fortuny, R., Gimeno-Añó, V. & Martín-Belloso, O. (2008a) Kinetic study of anthocyanins, vitamin C, and antioxidant capacity in strawberry juices treated by high-intensity pulsed electric fields. *Journal of Agricultural and Food Chemistry* **56**, 8387–8393.

Odriozola-Serrano, I., Soliva-Fortuny, R., Gimeno-Añó, V. & Martín-Belloso, O. (2008b) Modeling changes in health-related compounds of tomato juice treated by high-intensity pulsed electric fields. *Journal of Food Engineering* **89**, 210–216.

Odriozola-Serrano, I., Soliva-Fortuny, R., Hernández-Jover, T. & Martín-Belloso, O. (2009a) Carotenoid and phenolic profile of tomato juices processed by high intensity pulsed electric fields compared with conventional thermal treatments. *Food Chemistry* **112**, 258–266.

Odriozola-Serrano, I., Soliva-Fortuny, R. & Martín-Belloso, O. (2008c) Changes of health-related compounds throughout cold storage of tomato juice stabilized by thermal or high intensity pulsed electric field treatments. *Innovative Food Science and Emerging Technologies* **9**, 272–279.

Odriozola-Serrano, I., Soliva-Fortuny, R. & Martín-Belloso, O. (2008d) Phenolic acids, flavonoids, vitamin C and antioxidant capacity of strawberry juices processed by high-intensity pulsed electric fields or heat treatments. *European Food Research and Technology* **228**, 239–248.

Odriozola-Serrano, I., Soliva-Fortuny, R. & Martín-Belloso, O. (2009b) Impact of high-intensity pulsed electric fields variables on vitamin C, anthocyanins and antioxidant capacity of strawberry juice. *LWT – Food Science and Technology* **42**, 93–100.

Ong, P.K.C. & Liu, S.Q. (2010) Flavour and sensory characteristics of vegetables. In: *Handbook of Vegetables and Vegetable Processing.* (eds N.K. Sinha, Y.H. Hui, E.Ö. Evranuz, M. Siddiq & J. Ahmed). John Wiley and Sons, Iowa, USA.

Oziyci, H.R., Karhan, M., Tetik, N. & Turhan, I. (2013) Effects of processing method and storage temperature on clear pomegranate juice turbidity and color. *Journal of Food Processing and Preservation*, **37**, 899–906.

Pala, Ç.U. & Toklucu, A.K. (2011) Effect of UV-C light on anthocyanin content and other quality parameters of pomegranate juice. *Journal of Food Composition and Analysis*, **24**, 790–795.

Pala, Ç.U. & Toklucu, A.K. (2013) Effects of UV-C light processing on some quality characteristics of grape juices. *Food and Bioprocess Technology* **6**, 719–725.

Paranjpe, S.S., Ferruzzi, M. & Morgan, M.T. (2012) Effect of a flash vacuum expansion process on grape juice yield and quality. *LWT – Food Science and Technology* **48**, 147–155.

Park, S.H., Balasubramaniam, V.M., Sastry, S.K. & Lee, J. (2013) Pressure–ohmic–thermal sterilization: A feasible approach for the inactivation of *Bacillus amyloliquefaciens* and *Geobacillus stearothermophilus* spores. *Innovative Food Science and Emerging Technologies* **19**, 115–123.

Pederson, C.S., Albury, M.N. & Christensen, M.D. (1961) The growth of yeasts in grape juice stored at low temperature. IV. Fungistatic effects of organic acids. *Applied Microbiology* **9**, 162–167.

Peng, J., Mah, J.-H., Somavat, R., Mohamed, H., Sastry, S. & Tang, J. (2012) Thermal inactivation kinetics of *Bacillus coagulans* spores in tomato juice. *Journal of Food Protection* **75**, 1236–1242.

Pinelo, M., Arnous, A. & Meyer, A.S. (2006) Upgrading of grape skins: Significance of plant cell-wall structural components and extraction techniques for phenol release. *Trends in Food Science and Technology* **17**, 579–590.

Porretta, S., Birzi, A., Ghizzoni, C. & Vicini, E. (1995) Effects of ultra-high hydrostatic pressure treatments on the quality of tomato juice. *Food Chemistry* **52**, 35–41.

Praporscic, I., Lebovka, N.I., Vorobiev, E. & Mietton-Peuchot, M. (2007) Pulsed electric field enhanced expression and juice quality of white grapes. *Separation and Purification Technology* **52**, 520–526.

Prior, R.L., Lazarus, S.A., Cao, G., Muccitelli, H. & Hammerstone, J.F. (2001) Identification of procyanidins and anthocyanins in blueberries and cranberries (*Vaccinium* spp.) using high-performance liquid chromatography/mass spectrometry. *Journal of Agricultural and Food Chemistry* **49**, 1270–1276.

Qu, W., Pan, Z. & Ma, H. (2010) Extraction modeling and activities of antioxidants from pomegranate marc. *Journal of Food Engineering*, **99**, 16–23.

Rajashekhara, E., Suresh, E.R. & Ethiraj, S. (2000) Modulation of thermal resistance of ascospores of *Neosartorya fischeri* by acidulants and preservatives in mango and grape juice. *Food Microbiology* **17**, 269–275.

Raso, J., Calderón, M.L., Góngora, M., Barbosa-Cánovas, G. & Swanson, B.G. (1998a) Inactivation of mold ascospores and conidiospores suspended in fruit juices by pulsed electric fields. *LWT – Food Science and Technology* **31**, 668–672.

Raso, J., Calderón, M.L., Góngora, M., Barbosa-Cánovas, G.V. & Swanson, B.G. (1998b) Inactivation of *Zygosaccharomyces bailii* in fruit juices by heat, high hydrostatic pressure and pulsed electric fields. *Journal of Food Science* **63**, 1042–1044.

Raybaudi-Massilia, R.M., Mosqueda-Melgar, J., Soliva-Fortuny, R. & Martín-Belloso, O. (2009) Control of pathogenic and spoilage microorganisms in fresh-cut fruits and fruit juices by traditional and alternative natural antimicrobials. *Comprehensive Reviews in Food Science and Food Safety* **8**, 157–180.

Reineccius, G. (2006) *Flavor Chemistry and Technology*, 2nd edition, CRC Press, Boca Raton, Florida, USA.

Rinaldi, M., Caligiani, A., Borgese, R., Palla, G., Barbanti, D. & Massini, R. (2013) The effect of fruit processing and enzymatic treatments on pomegranate juice composition, antioxidant activity and polyphenols content. *LWT – Food Science and Technology* **53**, 355–359.

Rodrigo, D., van Loey, A. & Hendrickx, M. (2007) Combined thermal and high pressure colour degradation of tomato puree and strawberry juice. *Journal of Food Engineering* **79**, 553–560.

Rossi, M., Giussani, E., Morelli, R., Lo Scalzo, R., Nani, R.C. & Torreggiani, D. (2003) Effect of fruit blanching on phenolics and radical scavenging activity of highbush blueberry juice. *Food Research International* **36**, 999–1005.

Sagong, H.-G., Park, S.-H., Choi, Y.-J., Ryu, S. & Kang, D.-H. (2011) Inactivation of *Escherichia coli* O157: H7, *Salmonella typhimurium*, and *Listeria monocytogenes* in orange and tomato juice using ohmic heating. *Journal of Food Protection*, **74**, 899–904.

Sánchez-Moreno, C., Plaza, L., de Ancos, B. & Cano, M.P. (2006) Nutritional characterisation of commercial traditional pasteurised tomato juices: carotenoids, vitamin C and radical-scavenging capacity. *Food Chemistry* **98**, 749–756.

São José, J.F.B. & Vanetti, M.C.D. (2012) Effect of ultrasound and commercial sanitizers in removing natural contaminants and *Salmonella enterica typhimurium* on cherry tomatoes. *Food Control* **24**, 95–99.

Sapers, G.M. (2001) Efficacy of washing and sanitizing methods for disinfection of fresh fruit and vegetable products. *Food Technology and Biotechnology* **39**, 305–312.

Sathyanathan, T., Kayalvizhi, V., Sree, V.G. & Sundararajan, R. (2012) Microbial growth reduction in tomato juice by pulsed electric field with co-axial treatment chamber. *Proceedings of the IEEE International Conference on Properties and Applications of Dielectric Materials*, 1–4.

Scherb, J., Kreissl, J., Haupt, S. & Schieberle, P. (2009) Quantitation of S-methylmethionine in raw vegetables and green malt by a stable isotope dilution assay using LC-MS/MS: Comparison with dimethyl sulfide formation after heat treatment. *Journal of Agricultural and Food Chemistry* **57**, 9091–9096.

Schmidt, B.M., Erdman Jr,, J.W. & Lila, M.A. (2005) Effects of food processing on blueberry antiproliferation and antioxidant activity. *Journal of Food Science* **70**, S389–S394.

Sen Gupta, B., Masterson, F. & Magee, T.R.A. (2005) Inactivation of *E. coli* in cranberry juice by a high voltage pulsed electric field. *Engineering in Life Sciences*, **5**, 148–151.

Servili, M., Selvaggini, R., Taticchi, A., Begliomini, A.L. & Montedoro, G. (2000) Relationships between the volatile compounds evaluated by solid phase microextraction and the thermal treatment of tomato juice: Optimization of the blanching parameters. *Food Chemistry* **71**, 407–415.

Sesmero, R., Mitchell, J.R., Mercado, J.A. & Quesada, M.A. (2009) Rheological characterisation of juices obtained from transgenic pectate lyase-silenced strawberry fruits. *Food Chemistry* **116**, 426–432.

Shearer, A.E.H., Mazzotta, A.S., Chuyate, R. & Gombas, D.E. (2002) Heat resistance of juice spoilage microorganisms. *Journal of Food Protection* **65**, 1271–1275.

Sieso, V. & Crouzet, J. (1977) Tomato volatile components: Effect of processing. *Food Chemistry* **2**, 241–252.

Silva, F.M., Gibbs, P., Vieira, M.C. & Silva, C.L.M. (1999) Thermal inactivation of *Alicyclobacillus acidoterrestris* spores under different temperature, soluble solids and pH conditions for the design of fruit processes. *International Journal of Food Microbiology* **51**, 95–103.

Šimunek, M., Jambrak, A.R., Dobrović, S., Herceg, Z. & Vukušić, T. (2014) Rheological properties of ultrasound treated apple, cranberry and blueberry juice and nectar. *Journal of Food Science and Technology* **51**, 3577–3593.

Skrede, G., Wrolstad, R.E. & Durst, R.W. (2000) Changes in anthocyanins and polyphenolics during juice processing of highbush blueberries (*Vaccinium corymbosum* L.). *Journal of Food Science* **65**, 357–364.

Sucan, M.K. & Russell, G.F. (2002) Effects of processing on tomato bioactive volatile compounds. In: *Bioactive Compounds in Foods.* (eds T.C. Lee & C.T. Ho). pp. 155–172, American Chemical Society, Washington, DC, USA.

Syamaladevi, R.M., Andrews, P.K., Davies, N.M., Walters, T. & Sablani, S.S. (2012) Storage effects on anthocyanins, phenolics and antioxidant activity of thermally processed conventional and organic blueberries. *Journal of the Science of Food and Agriculture* **92**, 916–924.

Szajdek, A. & Borowska, E.J. (2008) Bioactive compounds and health-Promoting properties of berry fruits: A review. *Plant Foods for Human Nutrition* **63**, 147–156.

Talcott, S.T., Brenes, C.H., Pires, D.M. & Del Pozo-Insfran, D. (2003) Phytochemical stability and color retention of copigmented and processed muscadine grape juice. *Journal of Agricultural and Food Chemistry* **51**, 957–963.

Tavares, C.A. & Rodriguez-Amaya, D.B. (1994) Carotenoid composition of Brazilian tomatoes and tomato products. *LWT–Food Science and Technology* **27**, 219–224.

Tiwari, B.K., O'Donnell, C.P. & Cullen, P.J. (2009) Effect of sonication on retention of anthocyanins in blackberry juice. *Journal of Food Engineering* **93**, 166–171.

Tiwari, B.K., Patras, A., Brunton, N., Cullen, P.J. & O'Donnell, C.P. (2010) Effect of ultrasound processing on anthocyanins and color of red grape juice. *Ultrasonics Sonochemistry* **17**, 598–604.

Truchado, P., Larrosa, M., García-Conesa, M.T., Cerdá, B., Vidal-Guevara, M.L., Tomás-Barberán, F.A. & Espín, J.C. (2012) Strawberry processing does not affect the production and urinary excretion of urolithins, ellagic acid metabolites, in humans. *Journal of Agricultural and Food Chemistry* **60**, 5749–5754.

Turfan, Ö., Türkyílmaz, M., Yemiş, O. & Özkan, M. (2011) Anthocyanin and colour changes during processing of pomegranate (*Punica granatum* L., cv. Hicaznar) juice from sacs and whole fruit. *Food Chemistry* **129**, 1644–1651.

Vaillant, F., Millan, A., Dornier, M., Decloux, M. & Reynes, M. (2001) Strategy for economical optimisation of the clarification of pulpy fruit juices using crossflow microfiltration. *Journal of Food Engineering* **48**, 83–90.

Vallverdú-Queralt, A., Odriozola-Serrano, I., Oms-Oliu, G., Lamuela-Raventós, R.M., Elez-Martinez, P. & Martin-Belloso, O. (2012) Changes in polyphenol profile of tomato juices processed by pulsed electric fields. *Journal of Agricultural and Food Chemistry* **60**, 9667–9672.

Van der Plancken, I., Verbeyst, L., De Vleeschouwer, K., Grauwet, T., Heiniö, R.-L., Husband, F.A., Lille, M., Mackie, A.R., Van Loey, A., Viljanen, K. & Hendrickx, M. (2012) (Bio)chemical reactions during high pressure/high temperature processing affect safety and quality of plant-based foods. *Trends in Food Science and Technology* **23**, 28–38.

Vardin, H. & Fenercioglu, H. (2003) Study on the development of pomegranate juice processing technology: Clarification of pomegranate juice. *Nahrung* **47**, 300–303.

Varela-Santos, E., Ochoa-Martinez, A., Tabilo-Munizaga, G., Reyes, J.E., Pérez-Won, M., Briones-Labarca, V. & Morales-Castro, J. (2012) Effect of high hydrostatic pressure (HHP) processing on physicochemical properties, bioactive compounds and shelf-life of pomegranate juice. *Innovative Food Science and Emerging Technologies* **13**, 13–22.

Verbeyst, L., Hendrickx, M. & Van Loey, A. (2012) Characterisation and screening of the process stability of bioactive compounds in red fruit paste and red fruit juice. *European Food Research and Technology* **234**, 593–605.

Versari, A., Biesenbruch, S., Barbanti, D., Farnell, P.J. & Galassi, S. (1997) Effects of pectolytic enzymes on selected phenolic compounds in strawberry and raspberry juices. *Food Research International* **30**, 811–817.

Vu, K.D., Carlettini, H., Bouvet, J., Côté, J., Doyon, G., Sylvain, J.-F. & Lacroix, M. (2012) Effect of different cranberry extracts and juices during cranberry juice processing on the antiproliferative activity against two colon cancer cell lines. *Food Chemistry* **132**, 959–967.

White, B.L., Howard, L.R. & Prior, R.L. (2011) Impact of different stages of juice processing on the anthocyanin, flavonol, and procyanidin contents of cranberries. *Journal of Agricultural and Food Chemistry* **59**, 4692–4698.

Wightman, J.D. & Wrolstad, R.E. (1996) β-Glucosidase activity in juice-processing enzymes based on anthocyanin analysis. *Journal of Food Science* **61**, 544–548.

Woodward, G.M., McCarthy, D., Pham-Thanh, D. & Kay, C.D. (2011) Anthocyanins remain stable during commercial blackcurrant juice processing. *Journal of Food Science* **76**, S408–S414.

Wu, S.-B., Dastmalchi, K., Long, C. & Kennelly, E.J. (2012) Metabolite profiling of jabot-icaba (*Myrciaria cauliflora*) and other dark-colored fruit juices. *Journal of Agricultural and Food Chemistry* **60**, 7513–7525.

Wu, S. (2014) Glutathione suppresses the enzymatic and non-enzymatic browning in grape juice. *Food Chemistry* **160**, 8–10.

Wu, Y., Mittal, G.S. & Griffiths, M.W. (2005) Effect of pulsed electric field on the inacti-vation of microorganisms in grape juices with and without antimicrobials. *Biosystems Engineering* **90**, 1–7.

Yan, X.F. & Zhao, Y.B. (2010) Study on effect of ultra-high pressure processing on steriliza-tion of raspberry juice. *Transactions of the Chinese Society for Agricultural Machinery* **41**, 212–215.

Yang, H., Hewes, D., Salaheen, S., Federman, C. & Biswas, D. (2014) Effects of blackberry juice on growth inhibition of foodborne pathogens and growth promotion of *Lactobacil-lus*. *Food Control* **37**, 15–20.

Yildiz, H., Bozkurt, H. & Icier, F. (2009) Ohmic and conventional heating of pomegranate juice: Effects on rheology, color, and total phenolics. *Food Science and Technology Inter-national* **15**, 503–512.

Yu, K., Newman, M.C., Archbold, D.D. & Hamilton-Kemp, T.R. (2001) Survival of *Escherichia coli* O157:H7 on strawberry fruit and reduction of the pathogen population by chemical agents. *Journal of Food Protection* **64**, 1334–1340.

Yuk, H.G. & Schneider, K.R. (2006) Adaptation of *Salmonella* spp. in juice stored under refrigerated and room temperature enhances acid resistance to simulated gastric fluid. *Food Microbiology* **23**, 694–700.

Yung, L.M., Tian, X.Y., Wong, W.T., Leung, F.P., Yung, L.H., Chen, Z.Y., Lau, C.W., Van-houtte, P.M., Yao, X. & Huang, Y. (2013) Chronic cranberry juice consumption restores cholesterol profiles and improves endothelial function in ovariectomized rats. *European Journal of Nutrition*, **52**, 1145–1155.

Zheng, Z. & Shetty, K. (2000) Solid-state bioconversion of phenolics from cranberry pomace and role of *Lentinus edodes* β-glucosidase. *Journal of Agricultural and Food Chemistry* **48**, 895–900.

7
Juice Blends

Francisco J. Barba[1,2]*, Elena Roselló-Soto[2], Francisco Quilez[3], and Nabil Grimi[4]

[1]University of Copenhagen, Faculty of Science, Department of Food Science, Rolighedsvej, Frederiksberg, Copenhagen, Denmark
[2]Universitat de Valencia, Faculty of Pharmacy, Department of Preventive Medicine and Public Health, Toxicology and Forensic Medicine Department, Avda. Vicent Andrés Estellés, Burjassot, Spain
[3]Valencian School for Health Studies (EVES), Professional Training Unit, Juan de Garay, Valencia, Spain
[4]Sorbonne Université, Laboratoire Transformations Intégrées de la Matière Renouvelable (TIMR EA 4297), Centre de Recherche de Royallieu, Université de Technologie de Compiègne, Compiègne Cedex, France

7.1 Introduction

Consumers are increasingly aware of the physicochemical and nutritional quality of the foods they eat. Generally, fresh food provides the nutrients essential to achieve and maintain good health. In this line, juice blends have attracted consumer's attention as they could be a potential tool to obtain healthier products and improve the sensorial characteristics of them (Barba *et al.*, 2012a; Zulueta *et al.*, 2013 ; Bhardwaj & Pandey, 2011). However, fresh food is not always available and preservation becomes necessary to avoid microbial contamination. Thermal treatments have been traditionally used in the conservation of food products, but heating can adversely affect the sensory and nutritional quality of foods, which has led to an increase in the use of nonthermal treatments in the conservation and processing of foodstuff (Barba *et al.*, 2012b,c).

Nonthermal treatments can be defined as those in which temperature is not the main factor of inactivation of microorganisms and enzymes, although a slight increase in temperature may occur. As a result, quality degradative reactions that can be triggered after heat processing are reduced (Barba *et al.*, 2014; Carbonell-Capella *et al.*, 2013). Some of the most promising emerging nonthermal technologies are pulsed electric fields (PEFs) and high-pressure processing (HPP). Some of the most relevant findings are described in the following section.

*Corresponding authors: *Francisco J. Barba, Nabil Grimi*, francisco.barba@food.ku.dk; nabil.grimi@utc.fr

Innovative Technologies in Beverage Processing, First Edition.
Edited by Ingrid Aguiló-Aguayo and Lucía Plaza.
© 2017 John Wiley & Sons Ltd. Published 2017 by John Wiley & Sons Ltd.

7.2   Pulsed Electric Fields

PEF treatment consists in the application of pulses with a high voltage (normally 20–80 kV/cm) and a capacitance ranging from 80 nF to 9.6 mF for short periods of time (<1 s) to foodstuff placed between two electrodes (Senorans *et al.*, 2003). It constitutes an alternative to traditional thermal processing in the inactivation of contaminating and pathogenic microorganisms and enzymes, with the advantage of retaining or minimally modifying sensorial, nutritional, and health-promoting attributes of fruit and vegetable liquid products, in response to consumers demand of fresh-like products with high nutritional quality (Cserhalmi *et al.*, 2006; Saldana *et al.*, 2014; Barba *et al.*, 2012d). It has been shown that the application of PEF increases the temperature of the product. However, the heat generated during the application of PEF is not considered significant because the temperature reached does not usually exceed 40 °C, and it is also possible to incorporate a cooling system.

The effect that PEF exerts on microorganisms is based on the alteration of the cell wall when a difference in potential between the two sides of the membrane (transmembrane potential) is produced by the electric field strength applied. When this difference in potential attains a particular critical value, which varies according to the type of microorganism, an irreversible formation of pores in the cell membrane takes place (electroporation) with the consequent loss of integrity, increase in permeability, and finally, destruction of the cell affected (Toepfl *et al.*, 2006; Vorobiev & Lebovka, 2010).

These treatments are mainly used for food preservation, although other applications may be carried out. For instance, it has been shown that they can increase extraction of various compounds from foods. This is due to the phenomenon of electroporation, which causes loss of membrane integrity, inactivation of proteins, and release of cell components (Vorobiev & Lebovka, 2010; Toepfl *et al.*, 2006).

The effectiveness of PEF processing depends on critical factors including certain parameters related with the treatment and others that depend on the product to be treated. The processing factors include field strength, treatment time, pulse width, pulse frequency, pulse shape, polarity, energy, and temperature applied.

Field strength is defined as the difference in potential between two electrodes divided by the distance between them. However, a field strength that is too high may cause dielectric rupture of the liquid that is processed. The treatment time is obtained by multiplying the pulse width (pulse duration) by the number of pulses applied. These two factors have been identified as the most important factors for defining inactivation of microbes and enzymes by the application of an electric pulse treatment (Elez-Martinez *et al.*, 2005). When square waves are used, their width is equal to their duration, whereas in the case of exponential decay waves, the width corresponds to the time during which the voltage is greater than 37% of the maximum discharge value (Raso & Barbosa-Canovas, 2003). The pulse frequency (number of pulses applied per unit of time) determines the hold time of the food in the treatment chamber once the values of pulse width and treatment time have been fixed.

Among the inherent factors of the food, food matrix has a decisive influence on the effectiveness of PEF treatment because the presence of components such as fats or proteins makes inactivation of microbes by PEF more difficult in comparison with simple suspensions of microorganisms (Toepfl *et al.*, 2006).

The pH and the presence of natural antimicrobials are not factors directly related with the effectiveness of the treatment, but they do contribute to processing, increasing the effectiveness of PEF for preserving the product (Barbosa-Canovas & Sepulveda, 2005). The electric conductivity of a medium is an important variable because foods with high conductivity produce small electric fields and are not suitable for PEF treatment (Barbosa-Canovas *et al.*, 1998). The presence of particles in suspension may cause an increase or a reduction in treatment intensity (Wouters *et al.*, 2001).

Various research groups have studied how electric pulses may affect nutritional parameters in liquid foods such as fruit and vegetable juices, milk, and other beverages because it has been shown that their behavior varies depending on the food matrix (Zulueta *et al.*, 2013; Soliva-Fortuny *et al.*, 2009; Zulueta *et al.*, 2007), and therefore, it is necessary to evaluate each product individually before it can be marketed. The evaluation of the sensory and nutritional quality of beverages of this kind is very important for the consumer because it determines product acceptance (Riener *et al.*, 2009; Soliva-Fortuny *et al.*, 2009).

The most important factor to evaluate the potential of PEF as a preservation technology for juice blends is its ability to achieve 5-log pathogen reduction, required by the FDA regulation (FDA, 2001).

In a previous work, Saldana *et al.* (2014) reviewed the effectiveness of PEF to inactive microbial pathogen and contaminants from liquid foods. However, it is necessary to study the impact of PEF on each food matrix by separate. Some examples about the effects of PEF on juice blends food safety are described in the following section.

7.2.1 Food Safety

The impact of PEF treatments (28.6–35.8 kV/cm/10.2–46.3 μs) on *Lactobacillus plantarum* inactivation in an orange–carrot beverage was studied (Rodrigo *et al.*, 2001). These authors obtained up to 2.5 decimal reductions of *L. plantarum* under these conditions. More recently, it was evaluated the effects of PEF (25 kV/cm/340 μs) on molds and yeast and total flora in this beverage (Rodrigo *et al.*, 2003). They obtained 3.7 decimal reductions in molds and yeast and 2.4 decimal reductions in total flora when they combined PEF with a moderate temperature (63 °C).

On the other hand, Sampedro *et al.* (2007, 2009a) evaluated the impact of PEF treatment (30 kV/cm/2.5 μs bipolar square-wave pulses/50 μs) on microbial inactivation of an orange juice–milk-based beverage. These authors observed similar reductions in bacterial and mold counts of an orange juice–milk-based beverage after applying PEF treatment (4.5 and 5 log CFU/ml, for bacterial and mold inactivation, respectively) and thermal treatments (85 °C, 66 s) (4.5 and 4.1 log CFU/ml, for bacterial and mold inactivation, respectively) (Sampedro *et al.*, 2009a). It was also studied *L. plantarum* CECT 220 inactivation after PEF treatment, obtaining a maximum inactivation of 2.5 log cycles when PEF treatment (40 kV/cm, 130 μs, 2.5 μs pulse, and 1358 kJ/l) was used (Sampedro *et al.*, 2007). In addition, they estimated a shelf life of 2 and 2.5 weeks at 8–10 °C, respectively, for thermal and PEF-treated samples, based on the initial bacterial counts of the control (Sampedro *et al.*, 2009a).

Moreover, Salvia-Trujillo *et al.* (2011a) reported that PEF treatment (35 kV/cm, bipolar 4 μs square-wave pulses at 200 Hz/1800 μs) was needed to achieve 5-log reductions of *Listeria innocua* in a fruit juice mixture (orange (30%), kiwi (25%), mango (10%), and pineapple (10%)) blended with whole (FJ–WM) or skim milk (FJ–SM) (17.5%) beverages. In the same study, the authors also reported that PEF and thermal treatment (90 °C, 60 s) reduced the microbial counts (psychrotrophic bacteria and molds and yeasts) at <1 log CFU/ml immediately after processing.

In another study, it was evaluated the potential of PEF (35 kV/cm/4 μs bipolar pulses at 200 Hz/800 or 1400 μs) as a cold pasteurization technique for the inactivation of *Lactobacillus brevis* (*Lb. brevis*) or *L. innocua* in a formulated fruit juice–soymilk beverage (Morales-de la Pena *et al.*, 2010a,b). A maximal microbial reduction of 5.44

and 5.09 log units was found after applying PEF treatment for 800 or 1400 μs to the beverage inoculated with *Lb. brevis* or *L. innocua*, respectively. The data indicated an increase in the level of inactivation with increasing treatment time. Similarly, total inactivation (<1 log CFU/ml) of both microorganisms was found after applied an equivalent thermal pasteurization treatment (90 °C, 60 s).

7.2.2   Nutritionally Valuable Compounds

Vitamin C is the vitamin that has been studied most in fruit juices because it is one of the most abundant vitamins and at the same time one of the most sensitive to processing conditions. For instance, Torregrosa *et al.* (2006) evaluated the effects of PEF on ascorbic acid stability of an orange–carrot-blended beverage at different electric field strengths (25–40 kV/cm) and treatment times (30–340 μs). They obtained that ascorbic acid value of the beverage decreased with longer treatment times and higher electric field strength. Similarly, Zulueta *et al.* (2010b) found that ascorbic acid degradation after applying PEF (15–40 kV/cm, 40–700 μs) in an orange juice–milk beverage was adjusted to an exponential model. They also observed a lower ascorbic acid concentration when higher electric fields and longer treatment times were applied although these values were always higher than for thermal pasteurization (90 °C, 20 s). Moreover, Zulueta *et al.*, (2013) studied the effects of PEF (25 kV/cm at 57 °C for 280 μs) and subsequent refrigerated storage (42 days/4 °C) on ascorbic acid content of an orange juice–milk beverage. They did not find significant changes immediately after treatment, although as can be expected a significant depletion with storage time was found.

Morales-de la Pena *et al.* (2010a,b) obtained a vitamin C retention of 87–90% immediately after PEF processing (35 kV/cm, 4 μs bipolar pulses at 200 Hz for 800 or 1400 μs) of a fruit juice–soymilk beverage. Similarly to the results found in the orange juice–milk beverage, they found vitamin C depletion during 4 weeks storage at 4 °C, although the vitamin C content in PEF-treated samples was significantly higher compared to thermally treated beverages. The overall conclusion of these studies was that juice blends treated by PEF retained a greater amount of vitamin C than thermally processed foods.

On the other hand, Rivas *et al.* (2007) compared the water-soluble vitamin content (biotin, folic acid, riboflavin, and pantothenic acid) of a PEF-treated blended orange and carrot juice (15–40 kV/cm, 40–700 μs) with that of a heat-treated juice (84 °C and 95 °C, 45 s). The concentration of water-soluble vitamins remaining in the PEF treated juice was similar to that found in the heat-treated juice at 84 °C, although the heat-treated juice at 95 °C showed higher losses. In another study, there were not found significant changes in the concentration of group B vitamins after PEF treatment (35 kV/cm, 4 μs bipolar pulses at 200 Hz for 800 or 1400 μs) and subsequent refrigerated storage of fruit juices mixed with whole or skim milk. However, a significant decrease for riboflavin (vitamin B_2) was observed at equivalent thermal treatment (90 °C, 60 s) (Salvia-Trujillo *et al.*, 2011b).

Moreover, some studies have evaluated the influence of PEF on phenolic compounds or in compounds derived from isoprene in juice blends. For instance, Zulueta (2009) obtained a significant increase (9%) in total phenolics after applying PEF (40 kV/cm for 130 μs) in an orange juice mixed with milk beverage. In addition, Zulueta *et al.* (2013) observed a significant increase in TPC after 42 days storage at 4 °C of a PEF-treated orange juice–milk beverage. Similarly, Morales-de la Pena *et al.*

(2011a,b) also found an increase in phenolic content, especially in hesperidin, after PEF treatment (35 kV/cm for 800 μs and 1400 μs) of a fruit juice–soymilk beverage. Moreover, these authors did not find significant modifications in total phenolic acids or in total flavonoids during 56 days storage at 4 °C of PEF-treated samples compared to untreated beverages, although the behavior of the individual phenolic compounds was not clear. They found that the content of some phenolic compounds increased with time, while others tended to decrease or remained with no significant changes compared to their initial values. These authors reported some reasons in order to explain the changes observed on the phenolic content after PEF treatment of liquid foods, which are as follows: biochemical reactions during the PEF processing, which led to the formation of new phenolic compounds; significant effects on cell membranes or in phenolic complexes with other compounds, releasing some free phenolic compounds after PEF processing and a possible inactivation of PPO after PEF treatment, preventing further loss of phenolic compounds.

Moreover, it was also studied the impact of PEF (35 kV/cm for 800 μs and 1400 μs) and subsequent refrigerated storage on isoflavones of a fruit juice–soymilk beverage (Morales-de la Pena et al., 2010a,b). These authors obtained a different profile in isoflavone-derived compounds after PEF and thermal treatment. Glucosides, especially glucoside, were the predominant compounds in PEF-treated samples while a higher amount of glucosidic forms was found after thermal treatment (90 °C, 60 s). In addition, they did not found significant modifications of total isoflavone content PEF-treated samples. During 56-days storage at 4 °C, these authors found a significant increase in total isoflavone content mainly due to genistein, daidzein, and daidzin, while they observed a slight decrease in genistin.

With regard to fat-soluble compounds, few studies evaluate the effect of PEFs on this kind of compounds. Research has focused, basically, on the effect that PEF may exert on the extractability of carotenoids, some of which show provitamin A activity. Although they can easily be degraded, carotenoids can be retained during industrial processing if good technological practices are observed. Retention of provitamin A is favored during processing at low temperatures, protected from the light, with the exclusion of oxygen and the presence of antioxidants (Torregrosa et al., 2005), studied the effect of pulse treatment on carotenoids in an orange–carrot mixture (80:20, v/v), using different electric field intensities (25, 30, 35, and 40 kV/cm) and treatment times (30–340 μs). In parallel, a convectional heat treatment (98 °C, 21 s) was applied to the juice, and results were compared. They concluded that PEF processing generally caused a significant increase in concentrations of the various carotenoids identified (9-cis-violaxanthin + neoxanthin mixture, antheraxanthin, cis-β-cryptoxanthin, and 9-cis-α-carotene) in the orange–carrot mixture as treatment time increased, whereas when conventional pasteurization was used to process the juice, the concentrations of most of the carotenoids decreased or else showed a nonsignificant increase. With PEF treatment of the orange–carrot mixture at 25 and 30 kV/cm, it was possible to obtain a vitamin A concentration higher than that found in the pasteurized juice.

In addition, the impact of PEF (15–40 kV/cm, 40–700 μs) on the carotenoid content and vitamin A profile of an orange juice–milk beverage was evaluated and compared with thermal pasteurization (90 °C, 20 s) (Zulueta et al., 2010a). Electric field influenced the amount of extracted carotenoids with a slight increase at 15 kV/cm and a slight decrease at 40 kV/cm. However, the results were significantly higher compared to those obtained after thermal treatment. Moreover, these authors also observed similar trend in carotenoid degradation during refrigerated storage (4 and 10 °C) of thermally and PEF-treated beverages.

More recently, Morales-de la Pena *et al.* (2011a,b) evaluated the stability of different carotenoids (*cis*-violaxanthin + anteraxanthin, *cis*-anteraxanthin, anteraxanthin, lutein, and zeaxanthin) from a fruit juice–soymilk beverage after applying PEF treatments (35 kV/cm for 800 µs and 1400 µs) and during subsequent 56 days storage at 4 °C. They found a significant decrease in lutein, zeaxanthin, and β-cryptoxanthin after PEF treatments, although the degradation was lower compared to thermal treatment (90 °C, 60 s). In addition, they found a significant degradation of total carotenoid content after storage, irrespectively of the treatment applied, although, in general, the modifications were less than those obtained after thermal processing.

On the other hand, the effects of PEF on fat content and fatty acid profile of an orange juice–milk beverage were studied (Zulueta *et al.*, 2007). These authors observed a small reduction in fat content after PEF treatment (35–40 kV/cm, 40–180 µs), which can be important to enhance the extractability of fat-soluble compounds. Moreover, in this study, they did not find significant modifications in fatty acid profile in PEF-treated samples. Similar to these findings, Morales-de la Pena *et al.* (2011a,b) did not find significant modifications in most of fatty acids and minerals of a fruit juice–soymilk after PEF treatment (35 kV/cm with 4 µs bipolar pulses at 200 Hz for 800 and 1400 µs), observing only a decrease in elaidic and linoleic acid contents under these conditions.

Finally, the effects of PEF (35 kV/cm with 4 µs bipolar pulses at 200 Hz for 800 and 1400 µs) or thermal pasteurization (90 °C for 60 s) on the free amino acid (AA) profile of a fruit juice–soymilk (FJ–SM) beverage stored at 4 °C were evaluated (Morales-de la Pena *et al.*, 2012). There were no significant changes in total free AA after PEF (800 µs) compared to untreated samples although it was found a significant increase in Val. However, in the same study, PEF-treated samples (1400 µs) had lower values of total free AA, especially in Glu, Gly, Tyr, Val, Leu, Phe, Lys, and Ile contents, while the concentration of Arg, Ala, and Met slightly increased after applying these conditions. Overall, the authors observed higher values of total free AA, during storage of PEF-treated values compared to thermally treated samples (90 °C, 60 s), mainly due to an increase in most individual free AAs as storage time increased when PEF treatment was applied, while His, Tyr, Met, and Leu contents of thermally treated beverages were decreased.

7.3   High-Pressure Processing

In the past two decades, HPP has emerged as a potential tool to preserve liquid foods. Nowadays, more than 200 industrial HPP equipment for food processing are currently installed around the world, the main markets being North America, the European Union, Japan, Korea, Australia, and New Zealand. HPP equipment can also be found in China, Peru, or Chile.

Traditionally, HPP has been used in order to satisfy consumer trends demanding more natural products with fewer preservatives with a longer shelf life (regulatory requirements demanding absence of pathogens such as *Listeria* or *Salmonella* spp.) and for these reasons producing high-quality products.

HPP is characterized by three parameters: temperature T, pressure p, and exposure time t. The third HPP parameter allows great variability in the design of the process

(Heinz & Buckow, 2010). In addition, because of the great versatility of HPP, it is possible to combine this process with some other kind of treatment or other kinds of microbial inactivation agent in order to obtain more effective preservation, as in the cases of HPP combined with low pH (Raso & Barbosa-Canovas, 2003) or with PEFs (Ross *et al.*, 2003) or CO_2 (Spilimbergo *et al.*, 2003). When HPP is combined with mild heat treatment (10–40 °C), it is very suitable for the pasteurization of liquid foods (Deliza *et al.*, 2005; Barbosa-Canovas & Juliano, 2008). In this line, some authors have demonstrated the feasibility of HPP to achieve 5-log pathogen reduction, required by the FDA regulation (FDA, 2001). Some examples are listed in the following sections.

7.3.1 Food Safety

The impact of HPP on microbial inactivation in fruit juice–milk blends has been studied by different authors. In this line, it was reported a 5-log reduction of *L. plantarum* CECT 220 in an orange juice–milk beverage after HPP treatment at 200 MPa for 5 min (Barba *et al.*, 2012a; Sampedro *et al.*, 2009b).

Moreover, in another study, it was evaluated the ability of HPP as a cold pasteurization technique for the inactivation of pathogenic bacteria in a formulated strawberry–blueberry-based beverage (20 g of freeze-dried powder from each and 25 g of sugar in 8 oz, pH 3.70) (Tadapaneni *et al.*, 2014). These authors demonstrated the effectiveness of HPP (600 MPa/4 °C/10 min) to achieve 5-log reduction of the most resistant pathogens of public health concern (*E. coli* O175:H7 (ENT C9490), *Salmonella Typhimurium* (ATCC 14028), and *Listeria monocytogenes* (FRR W2542)).

More recently, Barba *et al.* (2014) evaluated the combined effects of a natural antimicrobial (*Stevia rebaudiana* Bertoni) and HPP on *L. monocytogenes* inactivation in a mixture of juices (papaya, mango, and orange). These authors observed that pressures higher than 300 MPa or times higher than 5 min were enough to inactivate *L. monocytogenes* at least 5 log cycles. The overall conclusion of these authors was that increasing pressure or pressure holding time led to a higher increase in the level of inactivation of the microbial pathogen. Moreover, in the same study, a synergistic effect of HPP with *Stevia* was found. The authors observed an increase in the number of inactivated log cycles when *Stevia* was used.

7.3.2 Nutritionally Valuable Compounds

Several authors have reported that ascorbic acid of fruit juices was minimally affected by HPP at mild temperatures (Barba *et al.*, 2012b,c). For instance, some authors have evaluated the effects of HPP treatments in juice blends obtaining similar results. In this line, Fernandez Garcia *et al.* (2001) evaluated the impact of HPP (500–800 MPa/RT/5 min) and subsequent storage 21 days at 4 °C on vitamin C in an orange–carrot–lemon (OLC) juice blend. These authors did not find significant modifications in vitamin C compared to untreated samples. Similar results were obtained by Tadapaneni *et al.* (2012) when they studied the effects of HPP (200–800 MPa/18–22 °C/1–15 min) on vitamin C stability of a freeze-dried strawberry powder blended with a dairy beverage as well as by Barba *et al.* (2012a) and Zulueta *et al.* (2013) when they evaluated the effects of HPP (100–400 MPa/RT/2–9 min) in an orange juice–milk blend. Moreover, Carbonell-Capella *et al.* (2013) also found that ascorbic acid retention immediately after applying HPP (300–500 MPa/RT/5–15 min) in a fruit juice mixture (mango, papaya, and orange) sweetened with *S. rebaudiana* was higher than 92% for all the HP treatments compared to fresh samples.

Regarding phenolic compounds, there is not a clear trend in the behavior of these components after HPP and most of the published articles reported that total phenolic compounds (TPCs) appeared to be relatively resistant to HPP. Even, some authors obtained an increase in TPC after applying HPP (100 MPa/7 min) in an orange juice–milk beverage (Barba *et al.*, 2012a).

In another study, it was found a decrease in flavonoid of an strawberry–blueberry-based beverage after applying HPP (200 800 MPa/18–22 °C/1–15 min) in comparison to untreated samples (Tadapaneni *et al.*, 2012), although the overall flavonoid content of HPP samples was significantly higher than those obtained after HTST. In the same work, the authors observed a high dependence between the molecular structure and the stability of flavonoids after HPP. Moreover, a significant increase in TPC (22%) was found when HPP (300 MPa/10 min) was applied in a fruit juice mixture sweetened with *Stevia* (1.25%, w/v). In addition, an increase of 18% was found after applying HPP 300 MPa for 5 min *Stevia* at 2.5%, w/v was used in the formulation of the juice blend (Carbonell-Capella *et al.*, 2013; Barba *et al.*, 2014). These authors attributed the higher amount of TPC after HPP to an increased extractability of some of the antioxidant components following HPP.

On the other hand, the effects of HPP on different fat-soluble compounds were evaluated by different authors. For instance, Fernandez Garcia *et al.* (2001) did not find significant modifications in carotene content of an OLC juice blend after HPP (500–800 MPa/RT/5 min) and subsequent storage 21 days at 4 °C. In another study, the modifications in carotenoid content of an orange juice–milk beverage after HPP (100–400 MPa/RT/2–9 min) were evaluated (Barba *et al.*, 2012a). It was found that pressure and time had a significant influence in carotenoid content with a significant increase in these compounds for all the HPP samples (100–400 MPa) at 7 and 9 min in comparison with the untreated blends. In addition, a significant decrease in carotenoid content of an orange juice–milk beverage was found after applying HPP (400 MPa/42 °C/5 min) and subsequent 42 days refrigerated storage, although the carotenoid contents of HPP samples at the end of the storage period were significantly higher to those obtained for the thermally pasteurized (90 °C, 15 s) beverages (Zulueta *et al.*, 2013). In the line of the results previously reported by Barba *et al.* (2012a), Carbonell-Capella *et al.* (2013) also found that pressure had significant influence in carotenoid content of a fruit juice mixture (mango, papaya, and orange) sweetened with *S. rebaudiana*, observing an overall increase in total carotenoid content when HPP at 500 MPa was conducted.

Finally, the impact of HPP (100–400/RT/9 min) on fat content, fatty acid profile, and fat-soluble vitamins (D and E) of an orange juice-milk-based beverage was studied (Barba *et al.*, 2012b,c). It was obtained a decrease in saturated fatty acid levels when HPP was applied (300–400 MPa) as well as a significant increase in oleic acid. In addition, the authors did not find significant modifications in vitamin D content after applying the different treatments. However, it was observed a significant increase in vitamin E activity when pressures >100 MPa were applied, which was mainly attributed to the ability of HPP to induce the disruption of the chloroplasts where α-tocopherol is confined.

7.4 Conclusion

The application of PEF and HPP for processing of juice blends looks promising. These treatments may be an alternative to thermal pasteurization, with the advantage of assuring none or minimal modification of physicochemical properties and degradation of nutritionally valuable compounds. Total antioxidant capacity may also be preserved in juice blends with these technologies. As a result, implementation of PEF and HPP processing at industrial scale may constitute an alternative for improving physicochemical and nutritional quality of final products.

Acknowledgments

F.J. Barba was supported from the Union by a postdoctoral Marie Curie Intra-European Fellowship (Marie Curie IEF) within the 7th European Community Framework Programme (http://cordis.europa.eu/fp7/mariecurieactions/ief_en.html) (project number 626524 – HPBIOACTIVE – Mechanistic modeling of the formation of bioactive compounds in high pressure processed seedlings of Brussels sprouts for effective solution to preserve healthy compounds in vegetables).

References

Barba, F.J., Cortés, C., Esteve, M.J. & Frígola, A. (2012a) Study of antioxidant capacity and quality parameters in an orange juice–milk beverage after high-pressure processing treatment. *Food and Bioprocess Technology* 5, 2222–2232.

Barba, F.J., Criado, M.N., Belda-Galbis, C.M., Esteve, M.J. & Rodrigo, D. (2014) Stevia rebaudiana Bertoni as a natural antioxidant/antimicrobial for high pressure processed fruit extract: processing parameter optimization. *Food Chemistry* **148**, 261–267.

Barba, F.J., Esteve, M.J. & Frigola, A. (2012b) Impact of high-pressure processing on vitamin e (alpha-, delta, and gamma-tocopherol), vitamin D (cholecalciferol and ergocalciferol), and fatty acid profiles in liquid foods. *Journal of Agricultural and Food Chemistry* **60**, 3763–3768.

Barba, F.J., Esteve, M.J. & Frígola, A. (2012c) High pressure treatment effect on physicochemical and nutritional properties of fluid foods during storage: a review. *Comprehensive Reviews in Food Science and Food Safety* **11**, 307–322.

Barba, F.J., Jäger, H., Meneses, N., Esteve, M.J., Frígola, A. & Knorr, D. (2012d) Evaluation of quality changes of blueberry juice during refrigerated storage after high-pressure and pulsed electric fields processing. *Innovative Food Science and Emerging Technologies* **14**, 18–24.

Barbosa-Canovas, G.V. & Juliano, P. (2008) Food sterilization by combining high pressure and thermal energy. In Food Engineering: Integrated Approaches (eds G.F. Gutierrez-Lopez, G.V. Barbosa-Canovas, J. Welti-Chanes & E. Parada-Arias), pp. 9–46). Springer, New York.

Barbosa-Canovas, G.V. & Sepulveda, D. (2005) Present status and the future of PEF technology. In *Novel Food Processing Technologies* (eds G.V. Barbosa-Canovas, M.S. Tapia & M.C. Cano), pp. 1–44. CRC Press, Boca Raton, Florida.

Barbosa-Canovas, G.V. and Pothakamury, U.R. and Palou, E. & Swanson, B.G. (1998) *Nonthermal Preservation of Foods* Marcel Dekker, Inc.

Bhardwaj, R.L. & Pandey, S. (2011) Juice blends – a way of utilization of under-utilized fruits, vegetables, and spices: a review. *Critical Reviews in Food Science and Nutrition* **51**, 563–570.

Carbonell-Capella, J.M., Barba, F.J., Esteve, M.J. & Frígola, A. (2013) High pressure processing of fruit juice mixture sweetened with *Stevia rebaudiana* Bertoni: optimal retention of physical and nutritional quality. *Innovative Food Science and Emerging Technologies* **18**, 48–56.

Cserhalmi, Z., Sass-Kiss, A., Toth-Markus, M. & Lechner, N. (2006) Study of pulsed electric field treated citrus juices. *Innovative Food Science and Emerging Technologies* **7**, 49–54.

Deliza, R., Rosenthal, A., Abadio, F.B.D., Silva, C.H.O. & Castillo, C. (2005) Application of high pressure technology in the fruit juice processing: benefits perceived by consumers. *Journal of Food Engineering* **67**, 241–246.

Elez-Martinez, P., Escola-Hernandez, J., Soliva-Fortuny, R.C. & Martin-Belloso, O. (2005) Inactivation of *Lactobacillus brevis* in orange juice by high-intensity pulsed electric fields. *Food Microbiology* **22**, 311–319.

FDA. (2001) *Guidance for Industry: The Juice HACCP Regulation – Questions & Answers.* http://www.fda.gov/Food/GuidanceRegulation/GuidanceDocumentsRegulatory Information/Juice/ucm072981.htm.

Fernandez Garcia, A., Butz, P., Bognar, A. & Tauscher, B. (2001) Antioxidative capacity, nutrient content and sensory quality of orange juice and an orange-lemon-carrot juice product after high pressure treatment and storage in different packaging. *European Food Research and Technology* **213**, 290–296.

Heinz, V. & Buckow, R. (2010) Food preservation by high pressure. *Journal Fur Verbraucherschutz Und Lebensmittelsicherheit* **5**, 73–81.

Morales-de la Pena, M., Salvia-Trujillo, L., Garde-Cerdan, T., Rojas-Grau, M.A. & Martin-Belloso, O. (2012) High intensity pulsed electric fields or thermal treatments effects on the amino acid profile of a fruit juice–soymilk beverage during refrigeration storage. *Innovative Food Science and Emerging Technologies* **16**, 47–53.

Morales-de la Pena, M., Salvia-Trujillo, L., Rojas-Grau, M.A. & Martin-Belloso, O. (2010a) Impact of high intensity pulsed electric field on antioxidant properties and quality parameters of a fruit juice-soymilk beverage in chilled storage. *LWT – Food Science and Technology* **43**, 872–881.

Morales-de la Pena, M., Salvia-Trujillo, L., Rojas-Grau, M.A. & Martin-Belloso, O. (2010b) Isoflavone profile of a high intensity pulsed electric field or thermally treated fruit juice–soymilk beverage stored under refrigeration. *Innovative Food Science and Emerging Technologies* **11**, 604–610.

Morales-de la Pena, M., Salvia-Trujillo, L., Rojas-Grau, M.A. & Martin-Belloso, O. (2011a) Changes on phenolic and carotenoid composition of high intensity pulsed electric field and thermally treated fruit juice–soymilk beverages during refrigerated storage. *Food Chemistry* **129**(3), 982–990.

Morales-de la Pena, M., Salvia-Trujillo, L., Rojas-Grau, M.A. & Martin-Belloso, O. (2011b) Impact of high intensity pulsed electric fields or heat treatments on the fatty acid and mineral profiles of a fruit juice–soymilk beverage during storage. *Food Control* **22**, 1975–1983.

Raso, J. & Barbosa-Canovas, G.V. (2003) Nonthermal preservation of foods using combined processing techniques. *Critical Reviews in Food Science and Nutrition* **43**, 265–285.

Riener, J., Noci, F., Cronin, D.A., Morgan, D.J. & Lyng, J.G. (2009) Effect of high intensity pulsed electric fields on enzymes and vitamins in bovine raw milk. *International Journal of Dairy Technology* **62**, 1–6.

Rivas, A., Rodrigo, D., Company, B., Sampedro, F. & Rodrigo, M. (2007) Effects of pulsed electric fields on water-soluble vitamins and ACE inhibitory peptides added to a mixed orange juice and milk beverage. *Food Chemistry* **104**, 1550–1559.

Rodrigo, D., Barbosa-Canovas, G.V., Martinez, A. & Rodrigo, M. (2003) Pectin methyl esterase and natural microflora of fresh mixed orange and carrot juice treated with pulsed electric fields. *Journal of Food Protection* **66**, 2336–2342.

Rodrigo, D., Martinez, A., Harte, F., Barbosa-Canovas, G.V. & Rodrigo, M. (2001) Study of inactivation of *Lactobacillus plantarum* in orange-carrot juice by means of pulsed electric fields: comparison of inactivation kinetics models. *Journal of Food Protection*, **64**, 259–263.

Ross, A.I.V., Griffiths, M.W., Mittal, G.S. & Deeth, H.C. (2003) Combining nonthermal technologies to control foodborne microorganisms. *International Journal of Food Microbiology* **89**, 125–138.

Saldana, G., Alvarez, I., Condon, S. & Raso, J. (2014) Microbiological aspects related to the feasibility of PEF technology for food pasteurization. *Critical Reviews in Food Science and Nutrition* **54**, 1415–1426.

Salvia-Trujillo, L., Morales-de la Pena, M., Rojas-Grau, M.A. & Martin-Belloso, O. (2011a) Microbial and enzymatic stability of fruit juice-milk beverages treated by high intensity pulsed electric fields or heat during refrigerated storage. *Food Control* **22**, 1639–1646.

Salvia-Trujillo, L., Morales-de la Pena, M., Rojas-Grau, M.A. & Martin-Belloso, O. (2011b) Changes in water-soluble vitamins and antioxidant capacity of fruit juice-milk beverages as affected by high-intensity pulsed electric fields (HIPEF) or heat during chilled storage. *Journal of Agricultural and Food Chemistry* **59**, 10034–10043.

Sampedro, F., Geveke, D.J., Fan, X., Rodrigo, D. & Zhang, Q.H. (2009a) Shelf-life study of an orange juice–milk based beverage after pef and thermal processing. *Journal of Food Science*, **74** S107–S112.

Sampedro, F., Geveke, D.J., Fan, X. & Zhang, H.Q. (2009b) Effect of PEF, HHP and thermal treatment on PME inactivation and volatile compounds concentration of an orange juice–milk based beverage. *Innovative Food Science and Emerging Technologies* **10**, 463–469.

Sampedro, F., Rivas, A., Rodrigo, D., Martinez, A. & Rodrigo, M. (2007) Pulsed electric fields inactivation of *Lactobacillus plantarum* in an orange juice-milk based beverage: effect of process parameters. *Journal of Food Engineering* **80**, 931–938.

Senorans, F.J., Ibanez, E. & Cifuentes, A. (2003) New trends in food processing. *Critical Reviews in Food Science and Nutrition* **43**, 507–526.

Soliva-Fortuny, R., Balasa, A., Knorr, D. & Martín-Belloso, O. (2009) Effects of pulsed electric fields on bioactive compounds in foods: a review. *Trends in Food Science and Technology* **20**, 544–556.

Spilimbergo, S., Dehghani, F., Bertucco, A. & Foster, N.R. (2003) Inactivation of bacteria and spores by pulse electric field and high pressure CO_2 at low temperature. *Biotechnology and Bioengineering* **82**(1), 118–125.

Tadapaneni, R.K., Banaszewski, K., Patazca, E., Edirisinghe, I., Cappozzo, J., Jackson, L. & Burton-Freeman, B. (2012) Effect of high-pressure processing and milk on the anthocyanin composition and antioxidant capacity of strawberry-based beverages. *Journal of Agricultural and Food Chemistry* **60**, 5795–5802.

Tadapaneni, R.K., Daryaei, H., Krishnamurthy, K., Edirisinghe, I. & Burton-Freeman, B.M. (2014) High-pressure processing of berry and other fruit products: Implications for bioactive compounds and food safety. *Journal of Agricultural and Food Chemistry* **62**, 3877–3885.

Toepfl, S., Mathys, A., Heinz, V. & Knorr, D. (2006) Review: potential of high hydrostatic pressure and pulsed electric fields for energy efficient and environmentally friendly food processing. *Food Reviews International* **22**, 405–423.

Torregrosa, F., Cortes, C., Esteve, M.J. & Frigola, A. (2005) Effect of high-intensity pulsed electric fields processing and conventional heat treatment on orange–carrot juice carotenoids. *Journal of Agricultural and Food Chemistry* **53**, 9519–9525.

Torregrosa, F., Esteve, M.J., Frigola, A. & Cortes, C. (2006) Ascorbic acid stability during refrigerated storage of orange–carrot juice treated by high pulsed electric field and comparison with pasteurized juice. *Journal of Food Engineering* **73**, 339–345.

Vorobiev, E. & Lebovka, N. (2010) Enhanced extraction from solid foods and biosuspensions by pulsed electrical energy. *Food Engineering Reviews* **2**, 95–108.

Wouters, P.C., Alvarez, I. & Raso, J. (2001) Critical factors determining inactivation kinetics by pulsed electric field food processing. *Trends in Food Science and Technology* **12**, 112–121.

Zulueta, A. (2009) *Changes in bioactive and antioxidant compounds in a juice milk-beverage treated by nonthermal technologies*. PhD thesis, University of Valencia.

Zulueta, A., Barba, F.J., Esteve, M.J. & Frígola, A. (2010a) Effects on the carotenoid pattern and vitamin A of a pulsed electric field-treated orange juice–milk beverage and behavior during storage. *European Food Research and Technology* **231**, 525–534.

Zulueta, A., Barba, F.J., Esteve, M.J. & Frígola, A. (2013) Changes in quality and nutritional parameters during refrigerated storage of an orange juice–milk beverage treated by equivalent thermal and non-thermal processes for mild pasteurization. *Food and Bioprocess Technology* **6**, 2018–2030.

Zulueta, A., Esteve, M.J., Frasquet, I. & Frigola, A. (2007) Fatty acid profile changes during orange juice–milk beverage processing by high-pulsed electric field. *European Journal of Lipid Science and Technology* **109**, 25–31.

Zulueta, A., Esteve, M.J. & Frigola, A. (2010b) Ascorbic acid in orange juice–milk beverage treated by high intensity pulsed electric fields and its stability during storage. *Innovative Food Science and Emerging Technologies* **11**, 84–90.

Part II
Non-Alcoholic Beverages

8

Grain-Based Beverages

Aastha Deswal[1]*, Navneet S. Deora[2], and Hari N. Mishra[3]

[1]*Manager, Product Formulation, Bright LifeCare Private Limited, Gurgaon, India*
[2]*Associate Cereal Specialist, Nestle R&D India Centre Private Limited, Gurgaon, India*
[3]*Department of Agricultural and Food Engineering, Indian Institute of Technology, Kharagpur, India*

8.1   Introduction

Prevalence of lactose intolerance and milk protein allergy among a vast proportion of the world population has led to the research for dairy alternatives. Lactose intolerance is the inability or insufficient ability to digest lactose (present in mammalian milk), when ingested. Lactose intolerance is caused by a deficiency of the enzyme lactase that breaks down lactose into two simpler forms of sugar called glucose and galactose, which are then absorbed into the blood stream. The enzyme lactase is present in high quantities in infants and children, whereas as the child grows, a dramatic reduction in the activity of the enzyme after weaning is observed (Vesa *et al.*, 2000; Campbell *et al.*, 2005).

Lactose intolerance is estimated to affect 33% of the global population. It is also estimated that an average of 75% of human adults have decreased intestinal lactase activity after weaning (NDDI, 2012). In India, it has been observed that around 60–70% of the population is lactose intolerant (Obadina *et al.*, 2013). Approximately 2.5% of world's children younger than 3 years of age and 0.3% of adults are allergic to milk (Sampson, 2004).

The global market for dairy alternatives is forecast to grow at a CAGR growth rate of around 15% between 2013 and 2018 to reach a value of $14 billion by the end of 2018 (Markets & Markets, 2013). Dietary management over the intake of dairy sources is the only treatment available for people suffering from lactose intolerance and milk protein allergy, and hence, they look for alternate sources of protein and calcium (Jarvinen-Seppo *et al.*, 2011; Labuschagne & Lombard, 2012). Soya milk has been the first non-dairy functional drink in the market. It is a very popular beverage among Asians and is gaining popularity globally as well. Soya milk is a rich

*Corresponding author: Dr Aastha Deswal, deswalad@gmail.com

Innovative Technologies in Beverage Processing, First Edition.
Edited by Ingrid Aguiló-Aguayo and Lucía Plaza.
© 2017 John Wiley & Sons Ltd. Published 2017 by John Wiley & Sons Ltd.

source of protein, isoflavones, saponins and fibre (Bricarello *et al.*, 2004). Other dairy alternatives have been developed from rice, peanut, coconut and almond. However, there is need to develop non-dairy milk from other sources such as cereals to fulfil the emerging demands in the sector of dairy alternatives.

8.1.1   Soy-Based Beverages

Cereal grains constitute a major source of dietary nutrients all over the world. Soya milk has been the first non-dairy functional drink in the market. Soya milk is a rich source of protein, isoflavones, saponins and fibre (Keshun, 1997). Soya milk is the aqueous extract of whole soya beans, closely resembling dairy milk in physical appearance and composition. Traditionally, soya milk has been made by soaking the beans in water overnight, which are then ground and filtered. Proximate composition of whole soya milk has been reported as 2.86–3.12% protein, 90–93.8% moisture, 1.5–2% fat, 0.27–0.48% ash and 1.5–3.9% carbohydrate (Rosenthal *et al.*, 2003; Johnson & Snyder, 1978). The protein in soya milk is deficient in methionine and cysteine but comparatively rich in lysine. Carbohydrates in soya milk are oligosaccharides, mainly stachyose and raffinose. Soya milk contains 1.3 g of fibre and zero cholesterol and no lactose (Saidu, 2005).

Apart from providing nutritional benefits, soya milk is found to exhibit various health benefits. Research studies have demonstrated that soya protein and isoflavones result in reduced rate of coronary heart disease (Lichtenstein, 1998), prevention of bone loss in postmenopausal women (Potter *et al.*, 1998; Bawa, 2010) and reduced mortality from cancer and other diseases (Ollberding *et al.*, 2012). Clinical trials have documented that soya protein reduces low-density-lipoprotein (LDL) cholesterol and increases high-density-lipoprotein (HDL) cholesterol (Wofford *et al.*, 2012; Rebholz *et al.*, 2013).

Because of its popularity, apart from its beverage form, soya milk is used to prepare fermented drinks, yoghurt, tofu, ice cream and many other milk-substitute products. A fermented soya milk beverage has been developed by LeBlanc *et al.* (2004) implying that soya milk fermentation by *Lactobacillus fermentum* CRL 722 resulted in elimination of undesirable gastrointestinal disorders normally associated with its consumption (LeBlanc *et al.*, 2004). Fermented products with other cultures such as *Bifidobacterium* have been developed, which are shown to have high degree of acceptability among consumers (Shimakawa *et al.*, 2003; Chou & Hou, 2000). Use of soya milk for replacing dairy for production of products like yoghurt, ice cream and tofu (soya paneer) has also been reported with comparable physical, rheological and sensory properties (Poysa & Woodrow, 2002; Farnworth *et al.*, 2007; Friedeck *et al.*, 2003; Kumar & Mishra, 2004; Abdullah *et al.*, 2003).

Cereal milk comprising an aqueous extract of a blend of soya bean, sesame and maize has been optimised. The cereal milk thus made was found to have 20.2% protein, 6.65% fat and 1.32% ash (Suphamityotin, 2011). In another study, non-dairy milk comprising soy-corn was made with 1:3 of corn to soy. This beverage was found to be highly digestible and widely acceptable by both adults and children. The soy-corn milk was found to be even more acceptable than soya milk (Omueti & Ajomale, 2005).

8.1.2   Rice-Based Beverages

A liquid derived from rice, called as 'rice milk', has been used in many parts of the world (southern Asia, China, Taiwan, etc.) as a substitute for milk (Dias-Morse, 2004).

Traditionally, Japanese used to make fermented drink from rice which was called as *amazake*. This traditional process is considered as a base for making rice milk. However, rice milk beverage thus produced had a problem of sour taste. Hence, another method was developed wherein the starch present was gelatinised and liquefied using α-amylase enzyme. Rice milk, thus prepared, is found to have a milk-like texture and is used for the preparation of beverages and non-dairy-based products. Rice milk can be used for preparation of novel frozen dessert or replacement of milk in standard ice cream (Mitchell *et al.*, 1988).

In order to increase the sensory appeal in terms of viscosity, flavour and nutritional properties of milk, external ingredients, namely, soya bean, peanut and sesame have been added and found to have higher sensory scores (Russell & Delahunty, 2004). An organic rice bran-based beverage has also been developed. The beverage viscosity was of the Newtonian standard behaviour, and its sensory preference tests showed positive perspectives. It was found to be an important source of minerals and unsaturated lipids. All essential amino acids were found in this product. Glutamic and aspartic acids were predominant (Faccin *et al.*, 2009). Very few studies have been conducted for rice milk, and there is need for further investigation of the process, quality parameters, shelf life and sensory properties of rice milk.

8.1.3   Oat-Based Beverages

Oat milk is a beverage made from whole oat groats, which looks like dairy milk. Few processes are mentioned in the literature for making oat milk from oats. One process involves soaking and grinding of whole oats followed by homogenisation (Bernat *et al.*, 2014). The other process is an enzymatic process, wherein whole oat groats are converted into a liquid milk-like product involving a number of steps such as flaking, wet milling, amylase hydrolysis, decanting, formulation, ultra-high-temperature (UHT) treatment and aseptic packaging (Ahlden *et al.*, 1997).

An US patent has disclosed the preparation of water-soluble dietary fibre compositions by treating ground oat products with α-amylases. The α-amylase serves to thin the oat starch, and any α-amylase may thus be used. The produced liquid dietary fibre compositions are used as additives in food products, such as fat substitutes. However, these products not only lack desirable aromatics of natural oats but are also deprived of agreeable natural oat flavourings (Inglett, 1991). Another US patent (No. 5,686,123) revealed a homogeneous and stable cereal suspension having the taste and aroma of natural oats. This oat suspension is a milky product, which can be used as an alternative to milk, especially for lactose-intolerant people. It may also be used as the basis of or an additive in the manufacture of ice cream, gruel, yoghurt, milk shakes, health beverages and snack (Ahlden *et al.*, 1997).

Apart from these patents, very little scientific literature is available on oat milk. The effect of different processing and storage conditions on nutritional properties of oat-based beverages has been reported in the literature. In a study, the decanting process caused a 47% increase in vitamin B_6 and a 45–74% loss of phosphorus, zinc, calcium and iron. The steam-injection UHT treatment caused a 60% loss of vitamin D_3 and a 30% loss of vitamin B_{12} (Zhang *et al.*, 2007). Many studies have been done to determine the effect of fermentation on properties of oat-based beverages (Angelov *et al.*, 2006; Gupta *et al.*, 2010; Mårtensson *et al.*, 2002). The relationship between changes in serum lipids and postprandial glucose and insulin concentrations after consumption of beverages with β-glucans from oats has been established in many studies. For example, it has been reported that consumption of oat milk for 5 weeks lowers serum cholesterol and LDL cholesterol in men with moderate hypercholesterolemia (Önning *et al.*, 1998, 2000). Another study reported that compared to the control, 5 g of β-glucans from oats significantly lowered total cholesterol by 7.4% ($p < 0.01$) and post-prandial concentrations of glucose and insulin (Biörklund *et al.*, 2005).

A group of scientists investigated the effect of oat fibres on perceived satiety of beverages and reported that the beverage containing oat β-glucan increased fullness and showed a trend of having a higher satiety index and decreased the 'desire to eat something' more than the beverage without fibre (Lyly *et al.*, 2009). Another group of researchers also found out similar results and reported that viscous fibres, including β-glucan in oat bran, favourably affect satiety and postprandial carbohydrate and lipid metabolism (Juvonen *et al.*, 2009).

In a recent publication (Deswal *et al.*, 2014a), the optimised process conditions for the development of non-dairy-based oat milk by enzymatic process are reported. The effect of temperature and total soluble solids concentration on rheological parameters of developed oat milk has also been reported (Deswal *et al.*, 2014b).

8.2   Conventional Processing Techniques

8.2.1   Heating Methods

Heating is the most common method of food processing, which is used for almost all kinds of food products. In liquid food products such as milk and beverages, it is used in the form of pasteurisation or heat sterilisation to increase their shelf life. Sometimes, heat treatment is also given to inactivate the anti-nutritional factors present in food systems. Heat, which is applied during cooking, does not have an instant impact or influence on the food being cooked; it takes time for heat to make the desired effect that is required of it on food generally. In general, heat has its adverse effect on food. The following are some of the effects on the nutritional values of food: proteins denature, lipids coagulate and starches that are the source of carbohydrates breakdown into simpler components when cooked for a period of time. To this end, time is indeed a very important factor to be considered when processing food. The cooking/heating time of foods plays a paramount role in determining the nutritional parameters of end food products. As a result, two types of heat treatments are prevalent in food industry: low-temperature–high time (LTHT), high-temperature–short time (HTST) and UHT treatments.

8.2.1.1   LTHT   The original type of heat treatment was a batch process in which milk was heated to 63 °C in open vats and held at that temperature for 30 min. This method is called the holder method or low-temperature–long-time (LTLT) method. Nowadays, milk is almost always heat treated in continuous processes such as thermisation, HTST pasteurisation or UHT treatment.

8.2.1.2   HTST   HTST processes, as the name implies, use higher temperatures and shorter times than conventional thermal processes to achieve pasteurisation and sterilisation of foods and beverages. Because the products are exposed to high temperatures for short times, there is minimal degradation of the products. In general, temperatures and times for HTST range from 161 °F for 15 s used for pasteurisation of milk to higher temperatures for shorter times. Use of temperatures above 280 °F is generally referred to as UHT processing. HTST processing has been used successfully for many years. HTST treatment could minimise those undesirable quality changes made by batch heating due to the much less duration of heat treatment.

8.2.2 Fermentation

Cereal grains are considered to be one of the most important sources of dietary proteins, carbohydrates, vitamins, minerals and fibres for people all over the world. However, the nutritional quality of cereals and the sensorial properties of their products are sometimes inferior or poor in comparison with milk and milk products. The reasons behind this are the lower protein content, the deficiency of certain essential amino acids (lysine), the low starch availability, the presence of determined anti-nutrients (phytic acid, tannins and polyphenols) and the coarse nature of the grains. Fermentation of cereals for a limited time improves amino acid composition and vitamin content, increases protein and starch availabilities and lowers the levels of anti-nutrients. Although cereals are deficient in some basic components (e.g. essential amino acids), fermentation may be the most simple and economical way of improving their nutritional value, sensory properties and functional qualities (Mezemir, 2015).

The other major reason for which fermentation is used in grain beverages is its property of improving organoleptic characteristics of beverages. In addition, the most common grain milk available in market is soya milk. Soya milk has become a very interesting food due to its extraordinary nutritive value and health characteristics. Soya milk has limited consumer acceptability due to its undesirable beany flavour. However, its acceptability can be enhanced by the modification of its processing methods. Some of the modifications of cold-water extraction of soya milk include the application of heat, soaking of soya bean in ethanol or alkali and acid grinding. Several researchers have tried to ferment soya milk using either pure or mixed cultures of the following bacteria: *Lactobacillus cellobiosis*, *Lactobacillus plantarum*, *L. fermentum*, *Lactobacillus delbrueckii*, *Lactobacillus fermenti*, *Lactobacillus pentosaceus* and *Lactobacillus bulgaricus* to improve its acceptability due to reduction in objectionable flavour and oligosaccharides such as starchyose and raffinose that cause flatulence (Mital & Steinkraus, 1975; Wang *et al.*, 2002). Soya milk yoghurt fermented with starter cultures has been greatly researched recently because their nutritional attributes may be changed due to the metabolism of microorganisms.

A number of methods have been employed with the aim of ameliorate the nutritional qualities of cereals. Several processing technologies, which include cooking, sprouting, milling and fermentation, have been put into practise to improve the nutritional properties of cereals, although probably the best one is fermentation (Blandino *et al.*, 2003). In general, natural fermentation of cereals leads to a decrease in the level of carbohydrates as well as some non-digestible poly- and oligosaccharides. Certain amino acids may be synthesised and the availability of B group vitamins may be improved. Fermentation also provides optimum pH conditions for enzymatic degradation of phytate which is present in cereals in the form of complexes with polivalent cations such as iron, zinc, calcium, magnesium and proteins. Such a reduction

Table 8.1   Fermented cereal-based beverages with major culture used to make the drink

Beverages substrate	Major culture	References
Oats	*L. plantarum*	Gupta *et al.* (2010)
Oats	*L. plantarum, L. paracasei casei* and *L. acidophilus*	Gokavi *et al.* (2005)
Blueberries and oats	*Lactobacillus plantarum*	Granfeldt and Bjorck (2011)
Soya beans	*Streptococcus thermophilus*	Champagne *et al.* (2009)
Soya beans	*Bifidobacterium breve*	Shimakawa *et al.* (2003)
Soya milk and apple juice	*Lactobacillus acidophilus*	İçier *et al.* (2015)
Soya milk and peanut	*Pediococcus acidilactici, Lactococcus lactis, L. rhamnosus, L. delbrueckii, L. acidophilus* and *Saccharomyces cerevisiae*	Santos *et al.* (2014)
Rice, soya milk and passion fruit fibre	*L. fermentum, L. plantarum, L. acidophilus, L. casei* and *Bifidobacterium* animalis	Espirito-Santo *et al.* (2014)
Soymilk–tea beverage	*S. thermophilus, L. delbrueckii* and *Bifidobacterium* longum	Zhao and Shah (2014)

in phytate may increase the amount of soluble iron, zinc and calcium several folds (Gillooly *et al.*, 1984; Haard, 1999; Mezemir, 2015; Singh *et al.*, 2015).

The preparation of many traditional or indigenous cereal-based foods is normally carried out by natural fermentation involving mixed cultures of bacteria, yeast and/or fungi. The most common fermenting bacteria are of the genera *Leuconostoc, Lactobacillus, Streptococcus, Pediococcus, Micrococcus* and *Bacillus* (Chavan *et al.*, 1989). The yeasts most frequently found are of the genera *Saccharomyces* and *Candida*, although *Zygosaccharomyces, Geotrichum* and *Torulopsis* have also been identified in some foods (Gotcheva *et al.*, 2000). Table 8.1 shows some of the fermented drinks made with various substrates based on cereal alone or their mixture with fruits.

8.2.3   Influence on Microbial Quality

It is reported that different processing temperatures affect quality characteristics of prepared soya milk. Soya bean was processed at varying temperatures (80, 90 and 110 °C), to produce soya milk samples A, B and D, with product processed using the normal boiling temperature of 100 °C (sample C) as standard. Total viable counts decreased significantly ($p < 0.05$) from 2.3×10^3 CFU/ml in sample A to 1.4×10^3 CFU/ml in D, while yeast and mould counts decreased from 1.2×10^2 CFU/ml in A to 0.3×10^2 CFU/ml in D with increase in processing temperature (Ikya *et al.*, 2013). This could be due to the destructive effect of heat. In another study also, it is reported that reduction in microbial load in soya milk is achieved by consecutive blanching and UHT processing of soya beans (Yuan *et al.*, 2008)

The effect of fermentation on the type and number of microbes growing in developed beverages and products of soya and oats is also reported. In a work, a whole-grain oat substrate was fermented with lactic acid bacteria (LAB) to obtain a drink, combining the health benefits of a probiotic culture with the oat prebiotic β-glucan. The levels of several factors, such as starter culture concentration, oat flour and sucrose content, affecting the fermentation process, were established for completing a controlled fermentation for 8 h. The viable cell counts reached at the end of the process were about 7.5×10^{10} cfu/ml. It was found that the addition of artificial sweeteners had no effect on the dynamics of the fermentation process and on the viability of the starter culture during product storage. The shelf life of the oat drink was estimated to 21 days under refrigerated storage (Angelov *et al.*, 2006).

In another study, *L. plantarum* was used for the development of an oat-based fermented drink with growth of 10.4 log CFU/ml (Gupta *et al.*, 2010). *L. delbrueckii*, *Lactobacillus helveticus*, *Lactobacillus rhamnosus* and *Bifidobacterium longum* have been examined for their ability to grow in combination with *Streptococcus thermophilus* cultures in milk and a laboratory soya beverage. Symbiosis with respect to acidification rate was reported between *S. thermophilus* and *L. helveticus* or *B. longum*. The probiotic populations in the mixed culture were influenced by the *S. thermophilus* strain and the time of fermentation (Champagne *et al.*, 2009). Many other studies also report the use of fermentation technology to influence the type and number of microflora in soy-based beverages (Champagne *et al.*, 2010; Farnworth *et al.*, 2007; Buckley *et al.*, 2011).

8.2.4 Influence on Nutritional Attributes

8.2.4.1 Effect of Heat Treatment on Nutritional Composition of Grain Beverages
It is observed and reported that different kinds of heat treatments affect the nutritional quality of soya milk differently. In a study, the effect of heating soya milk at 100 °C for 15, 30 and 45 min on its nutritional characteristics was evaluated. Percentage crude protein had the following results for cooking soya milk for 15, 30 and 45 min: 3.72%, 4.23% and 4.74%, respectively. These results show an increase in percentage crude protein as it was cooked for the time under consideration. The increase in the percentage crude protein could be as a result of decrease in digestibility of soya protein as it was exposed to excessive heat treatment. Soya protein also denatures when exposed to heat treatment, which could also explain why there was an increase in the crude protein in the soya milk samples prepared over different period of time. There was a remarkable decrease in percentage lipid as soya milk was cooked at its boiling temperature from 12% to 6% between 10 and 30 min of cooking. This could be as a result of the removal of a film which is rich in protein and oil called yuba. Lipids seeped up and were removed as yuba as the soya milk was boiled over the period of time. The percentage of carbohydrates in the soya milk was also affected by the cooking over the period of time. The values of carbohydrates obtained during the research are as follows: 80.11%, 83.46% and 83.09% for 10, 15 and 45 min of cooking, respectively, which may be due to the evaporation of water. The minerals were also observed to have undergone some changes as the soya milk was cooked between 15 and 45 min. Calcium decreased in its value from 109.5 to 90.0 ppm, while sodium had a relatively constant value but with a fluctuation. The remaining minerals under consideration were observed to increase in value in the following order: potassium increased from 41.0 mg/100 ml to 62.0 mg/100 ml and magnesium increased from 32.2 to 35.5 ppm for 15 and 45 min, respectively (Orhevba, 2011).

Changes in the chemical and nutritional composition of naturally fermented soya milk were studied at ambient temperature ($27 \pm 2\,^\circ$C) for 72 h. The moisture, carbohydrate and fat contents were found to decrease from 93.45% to 92.70%, 1.52% to 0.60% and 2.18% to 0.87%, respectively, while total solids, ash and protein contents increased from 6.55% to 7.30%, 0.23% to 0.74% and 2.62% to 5.09%, respectively. Results reveal that the calcium, iron and magnesium contents in fermenting soya milk increased from 52.86 to 71.43, 28.00 to 40.00 and 7.66 to 8.87 mg/l, respectively, within time intervals of 0–54 h and then decreased to 65.00, 28.00 and 7.83 mg/l, respectively, till the end of fermentation period while the zinc content increased from 4.42 to 6.75 mg/l throughout the fermentation period. It was evident that there was increase in protein, calcium, magnesium, zinc and iron contents during natural fermentation of soya milk (Obadina *et al.*, 2013).

In processing soya bean products, destruction of anti-nutritional factors is another important concern. Assessment of the overall nutritional quality of a soya bean product depends not only on its nutrient content but also on the anti-nutritional factors present. It has been shown that heat inactivation of anti-nutritional factors such as trypsin inhibitors parallels the nutritive value improvement of the soya protein (Liener, 1981; Rackis, 1972). In soya beans, trypsin inhibitor activity (TIA) is a combination of inhibitor activity of two different proteins with different heat labilities. These are the Kunitz soya bean trypsin inhibitor (KSTI) and the Bowman–Birk inhibitor (BBI). The KSTI is generally heat labile, whereas the BBI is very heat stable. Previous reports on heat activation of trypsin inhibitor in soya milk were mostly in the temperature range 93–121 °C (Kwok *et al.*, 2002; Hackler *et al.*, 1965). Inactivation of 90% of the native TIA in soya milk could be achieved by heating for 60–70 min at 93 °C or for 5–10 min at 121 °C. At 143 and 154 °C (in the UHT range), the heating times required to inactivate 90% of the total TIA in soya milk at pH 6.5 were determined by Kwok *et al.* to be 62 and 29 s, respectively (Kwok *et al.*, 2002). These heating times, which are far higher than those necessary for spore destruction, may also cause considerable heat damage to the sensory and nutritional qualities of soya milk.

8.2.4.2 Effect of Fermentation on Nutritional Composition of Grain Beverages

In a study, *B. longum B6* and *Bifidobacterium infantis CCRC 14633* were grown in soya milk for a period of 48 h. During various fermentation periods, changes in the contents of crude protein, sugars, B-vitamins and acetic and lactic acids in soya milk were examined. Crude protein and titratable acidity were increased during fermentation. The degree of protein hydrolysis, thiamin and riboflavin contents were increased, while niacin content was decreased in soya milk fermented with either *B. infantis or B. longum*. Acetic and lactic acid contents were increased, while the molar ratio of acetic and lactic acids was decreased during fermentation. Stachyose, raffinose and sucrose contents were decreased, with stachyose showing the largest magnitude of reduction. On the other hand, contents of fructose and glucose plus galactose contents were increased during fermentation (Hou *et al.*, 2000; Wang *et al.*, 2003).

In another study, soya milk fermented using *L. delbrueckii* ssp. *bulgaricus* and *S. thermophilus* was evaluated in terms of rat growth, nitrogen balance assays, anti-nutritional factors and microbiological parameters to study the effect of fermentation on these parameters. For the fermented soya yoghurt, the growth and nitrogen balance values were not different from the control diet, but the nitrogen balance values were higher than for the soya milk, without significant difference in terms of growth assays. Compared to the commercial yoghurt, net protein ratio and nitrogen utilisation values were lower, but the protein efficiency ratio, biological value and net protein utilisation values were equivalent, and for digestibility assays, the best results were obtained with the soya yoghurt. The results indicated that soya yoghurt represents a good protein alternative to milk yoghurt and casein. The protein quality of soya milk evidently increased during the fermentation process using *L. delbrueckii* ssp. *bulgaricus* and *S. thermophilus*, including a reduction in trypsin inhibitor levels of about 30% (Silva Júnior *et al.*, 2012).

8.2.5 Influence on Organoleptic Attributes

The typical green-beany flavour of soya milk is considered the major deterrent to acceptability of soya milk. When raw legume seeds are ground with water, undesirable odour and flavour similar to drying oil develop very rapidly. Wilkens *et al.* (1967) observed that soya milk prepared from whole soya beans develops beany and rancid flavours, whereas soya milk from defatted beans has a bland cereal-like flavour (Wilkens *et al.*, 1967). They showed that lipoxygenase causes off-flavour, and rapid hydration and high-temperature grinding prevent off-flavour development. Several studies have confirmed this concept (Yuan & Chang, 2007; Wolf, 1975; Davies *et al.*, 1987). Daidzein and genistein are responsible for the objectionable flavour of soya milk (Matsuura *et al.*, 1989).

Heating has been the most successful process for the inactivation of undesirable compounds in soya milk including lipoxygenase. Nelson *et al.* (1978) studied various methods of blanching whole soya beans to inactivate lipoxygenase. Blanching fully soaked beans at 99 °C for 10 min, blanching drybeans for 20 min or soaking and blanching in 0.5% sodium bicarbonate for tender beans were suggested depending on the ultimate use of the soya beans. It is reported that when soya milk is subjected to various heat treatments at 95, 121 and 140 °C for various lengths of time, the contents of the aglycones of isoflavone (daidzein, glycitein and genistein) of the soya milk effected. Genistein showed greater stability to heat treatment than daidzein and glycitein. Both the daidzein and glycitein contents decreased rapidly during the early stage of heating, but on continued heating, the rates of decrease were much slower. Heating may cause an increase or decrease in the genistein content of soya milk depending on the temperature and time used. Upon heating at 95 and 121 °C, there was an increase in the genistein content in the early stage of heating, possibly due to the conversion of genistin to genistein. Heating at 140 °C for more than 15 s and prolonged heating at 95 and 121 °C, however, caused a slow decline in the genistein content (Huang *et al.*, 2006).

8.3 Novel Processing Techniques

8.3.1 High and Ultra-High-Pressure Homogenisation

High-pressure homogenisation (HPH) is considered to be one of the most promising non-thermal technologies proposed for preservation of fruit juices and beverages. The primary mechanisms of HPH have been identified as a combination of spatial pressure and velocity gradients, turbulence, impingement, cavitation and viscous shear, which leads to the microbial cell disruption and food constituent modification during the

HPH process. HPH has shown its ability to increase the safety and shelf life of fruit juices including orange juice, apple juice and apricot juice (Pathanibul *et al.*, 2009; Patrignani *et al.*, 2009; Welti-Chanes *et al.*, 2009). The effectiveness of the treatment depends on many parameters including processing factors such as pressure, temperature, number of passes and medium factors such as type of juice and microorganisms. For example, up to 350 MPa processing pressure was required to achieve an equivalent 5-log inactivation of *Listeria innocua*; however, less pressure is required for *Escherichia coli* (>250 MPa) in carrot juices (Pathanibul *et al.*, 2009).

Ultra-high-pressure homogenisation (UHPH) has been studied as an alternative to thermal technologies in order to improve several chemical, physical and sensory parameters of food. Most of the studies have been focussed on a set of parameters such as emulsion stability and microbial and enzymatic activity (Cruz *et al.*, 2007; Poliseli-Scopel *et al.*, 2012). The ability of UHPH for preserving and stabilising of liquid colloidal foods is possible due to the special design of equipments, which may reach pressures up to 350 MPa. UHPH increases the colloidal stability and reduces microbial inactivation since it promotes different physical phenomena such as collapse, shear, cavitation and turbulence of particles passing through the high-pressure valve. From a physical point of view, it causes reduction of fat globules and facilitates the interaction with aqueous phase of macromolecules, leading a better dispersion capacity. From a hygienic point of view, it breaks microbial cells (Ferragut *et al.*, 2011). The progression towards UHPH has also opened the door to new sterilisation opportunities, which might go past the initial theory that it might not be possible to inactivate spores by HPH (Popper & Knorr, 1990).

In published studies, many reports have studied the profile of volatile compounds and changes due to the treatment applied in soya milk samples. However, only one study has been reported on the effect of UHPH on volatile composition of food, in that example, milk. The effect of UHPH at 200 and 300 MPa in combination with different inlet temperatures (55, 65 and 75 °C) on soya milk was studied. UHPH-treated soya milk was compared with the base product (untreated), with pasteurised (95 °C for 30 s) and with ultra-high-temperature (UHT; 142 °C for 6 s)-treated soya milks. Microbiological (total aerobic meshophilic bacteria, aerobic spores and *Bacillus cereus*), physical (dispersion stability and particle size distribution) and chemical (lipoxygenase activity, hydroperoxide index and TIA) parameters of special relevance in soya milk were studied. Microbiological results showed that pressure and inlet temperature combination had a significant impact on the lethal effect of UHPH treatment. While most of the UHPH treatments applied produced high quality of soya milks better than that pasteurised, the combination of 300 MPa and 75 °C produced a commercially sterile soya milk. UHPH treatments caused a significant decrease in particle size resulting in a high physical stability of samples compared with conventional heat treatments. UHPH treatment produced lower values of hydroperoxide index than heat-treated soya milks although TIA was lower in UHT-treated products.

8.3.2   High-Pressure Processing

Thermal processing, that is pasteurisation and sterilisation, is widely used for shelf-life prolongation of food products; on the other hand, this technology is not able to preserve their natural colour, flavour and nutrients (Barba *et al.*, 2013). Thus, non-thermal technologies such as high-pressure processing (HPP) are preferred to maintain the sensory attributes and nutritional values of the products (Cruz *et al.*, 2007; Poliseli-Scopel *et al.*, 2012).

HPP is a method of food processing where food is subjected to elevated pressures (up to 87,000 lb/in^2 or approximately 600 MPa), with or without the addition of heat, to achieve microbial inactivation or to alter the food attributes in order to achieve consumer-desired qualities. The technology is also referred to as high hydrostatic pressure processing (HHP) and ultra-high-pressure processing (UHP) in the literature. Fluid foods such as juices can be processed in batch or semi-continuous mode. The equipment typically consists of the following:

1. Pressure vessel.
2. Top and bottom end closures.
3. Yoke (structure for restraining end closures).
4. High-pressure pump and intensifier for generating target pressures.
5. Process control and instrumentation.
6. A handling system for loading and removing the product.

The effects of HP are explained by two principles. One is that of Le Chatelier, according to which any phenomenon accompanied by a decrease in volume will be enhanced by pressure (Gross & Jaenicke, 1994). The other is the isostatic principle as shown in Fig. 8.1, which states that the pressure is transmitted instantly and uniformly independent of the size and geometry of the food (Smelt, 1998). The key effects of HP include the inactivation of microorganisms, protein denaturation, enzyme activation or inactivation and retention of quality and freshness retention. HP does not affect covalent bonds but does affect other types of chemical bonds, including hydrogen, hydrophobic and ionic bonds. Ion pairs in aqueous solution are strongly destabilised by HP. A decrease in volume favours the dissociation of ionic interactions as each ion arranges water molecules in its vicinity more densely than bulk water. Similarly, the exposure of hydrophobic groups to water disturbs the loosely packed structure of pure water and leads to a hydrophobic solvation layer which is assumed to be more densely packed. HPP exhibits several benefits over thermal processing of foods (Gross & Jaenicke, 1994).

The mechanism(s) responsible for microbial inactivation by HPP are still not completely understood. Many cellular targets have been reported to be involved, including the cell wall, cytoplasmic membrane, nucleic acids, ribosomes and various proteins and enzymes. Pressure affects cellular morphology, for example intracellular gas vacuoles can collapse and some organisms form long filaments. Motile organisms may lose motility due to loss of flagella (Brul *et al.*, 2000; Ritz *et al.*, 2002; Smelt, 1998; Spilimbergo *et al.*, 2002).

The main benefits of HPP are the extension of shelf life while maintaining the fresh qualities of the food product. Unlike thermal processing, pressure treatment is not time or mass dependant; therefore, processing time is reduced. As pressure is transmitted instantly and uniformly the food is preserved evenly, without any particle escaping preservation (Norton & Sun, 2008). Post-processing contamination can be eliminated by pre-packaging food products prior to processing, therefore increasing safety and shelf life. As HPP does not affect covalent bonds (in contrast to heat), many small molecules in foods, such as many flavour compounds and vitamins, are left intact (Tewari *et al.*, 1999). Hence, the taste and quality of many HP-treated foods are superior to those that are thermally processed.

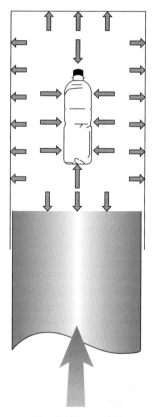

Applied pressure

Figure 8.1   Principle of isostatic pressure.

8.3.3   Pulsed Electric Field

Pulsed electric field (PEF) processing involves the application of short pulses (μs) of high voltage (kV/cm) to foods placed between two electrodes. The application of PEF is restricted to food products that can withstand high electric fields, have low electrical conductivity and do not contain or form bubbles. The particle size of the food may also be a limitation (Cullen & Tiwari, 2011).

In general, PEF treatment systems are composed of PEF treatment chambers, a pulse generator, a fluid-handling system and monitoring systems (Min *et al.*, 2007). The diagrammatic representation of pulsed electric effect is shown in Fig. 8.2. The treatment chamber is used to house electrodes and deliver a high voltage to the food material. It is generally composed of two electrodes held in position by insulating material, thus forming an enclosure containing the food material. Therefore, the proper design of the treatment chamber is an essential component for the efficiency of the PEF technology. The feasibility of processing fluid foods by PEF is related to the potential of this technology for inactivating microorganisms and quality-related enzymes, while maintaining sensory and nutritional properties. The critical factors determining the efficiency of processing foods by PEF can be classified as (i) treatment parameters;

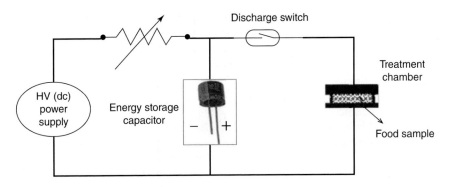

Figure 8.2 Arrangement of pulse electric field system.

(ii) microorganism, enzyme and quality-related compounds characteristics and (iii) product parameters (Mohamed *et al.*, 2012).

Many studies have focussed on the causes and reasons for cell death due to electric field process. Exposure of microorganisms to a PEF directly affects the integrity of the cell by a mechanism known as electroporation. It is generally accepted that cell electroporation is caused by mechanical breakdown of the cell membrane due to the compression exerted by the accumulation of free charges at both sides of the membrane (Zimmermann, 1986). Such instability leads to appreciable morphological changes in the cell envelope and/or cytoplasm disorganisation and, after that, it is followed by electropermeabilisation, consisting of shrinkage and leakage of intracellular content, such as cytoplasmic and nuclear material (Choi *et al.*, 2008) and pore formation in the cell wall and/or cell membrane (García *et al.*, 2007).

The application of PEFs technology has been successfully demonstrated for the pasteurisation of foods such as juices, milk, yoghurt, soups and liquid eggs. The application of PEF processing is restricted to food products with no air bubbles and low electrical conductivity. The maximum particle size in the liquid must be smaller than the gap of the treatment region in the chamber in order to ensure proper treatment. PEF is a continuous processing method, which is not suitable for solid food products that are not pumpable. PEF processing has been successful in a variety of fruit juices with low viscosity and electrical conductivity such as orange, apple and cranberry juice (Qin *et al.*, 1998; Evrendilek *et al.*, 2000). In addition, the colour change in fruit juices (subject to prolonged storage) was reportedly less in juices treated by PEF, as in a recent study of PEF-treated orange juice stored at 4 °C for 112 days; there was less browning than thermally pasteurised juice, which was attributed to conversion of ascorbic acid to furfural (Yeom *et al.*, 2000).

8.3.4 Enzymatic Techniques

Starch is the common storage carbohydrate in plants, which is broken down in the liquefaction processes to produce starch hydrolysates to be used in processes where elimination of starch is necessary. There are three stages in the conversion of starch: gelatinisation, liquefaction and saccharification (Vijayagopal *et al.*, 1988; van der Maarel *et al.*, 2002). Liquefaction is a process of dispersion of insoluble starch granules in aqueous solution followed by partial hydrolysis using either acid or thermostable amylases (Aiyer, 2005).

The acid hydrolysis method for the production of glucose has been replaced recently by enzymatic treatment with three or four different enzymes (Crabb & Mitchinson, 1997; Crabb & Shetty, 1999). In a conventional enzymatic hydrolysis process, slurry containing 15–35% starch is gelatinised, where it is heated to 105 °C to physically disrupt the granule and open the crystalline structure for the enzyme action (Singh & Soni, 2001). This increases the viscosity of the slurry by 20-fold (Robertson *et al.*, 2006), which is liquefied with thermostable α-amylase, and is then saccharified with glucoamylase at a much lower temperature of 50–60 °C.

Amylases are enzymes which hydrolyse internal α-D-(1,4)-glucosidic linkages in starch molecules to give diverse products including dextrins and progressively smaller polymers composed of glucose units (van der Maarel *et al.*, 2002). The microbial α-amylases (EC 3.2.1.1) for industrial purposes are derived mainly from *Bacillus licheniformis*, *Bacillus amyloliquefaciens*, *Bacillus subtilis* and *Aspergillus oryzae* (Nigam & Singh, 1995; Robyt & French, 1963; Pandey *et al.*, 2000). In the past few decades, α-amylases from *B. subtilis* have received considerable attention from researchers all over the world in the area of starch processing and have been found to be suitable for starch liquefaction in food industries (Konsula & Liakopoulou-Kyriakides, 2004; Konsoula & Liakopoulou-Kyriakides, 2006). α-Amylase from *B. subtilis* has been shown to digest potato starch, corn starch, chestnut starch, oat starch (Hayashida *et al.*, 1988; Konsula & Liakopoulou-Kyriakides, 2004) and wheat starch (Sodhi *et al.*, 2005). The α-amylase from *B. subtilis* is found to be quite thermostable. In a study on crude α-amylase characterisation, optimum activity was found at pH 8.0 and 70 °C. The enzyme was quite stable for 1 h at 60 and 70 °C, while at 80 and 90 °C, 12% and 48% of the original activities were lost, respectively. It was also reported that α-amylase is a calcium metalloenzyme and hence was activated by calcium ions (Asgher *et al.*, 2007).

The application of enzymatic liquefaction process has been shown to increase the yield of jicama (*Pachyrhizus erosus*) tuberous root juice (Ramos-de-la-Peña *et al.*, 2012), saccharification of corn starch (Carr *et al.*, 1982), increase in yield of apple juice (Will *et al.*, 2000), recovery of a carotene-rich functional food ingredient from carrot pomace (Stoll *et al.*, 2003) and conversion of corn starch for fat mimicking (Ma *et al.*, 2006). The enzymatic liquefaction has been used for the reduction in viscosity and facilitating the process of filtration (Aiyer, 2005). The process also results in the increase in the amount of total soluble solids in produced oat milk as a result of action of amylases on starch which results in production of maltodextrins (Moore *et al.*, 2005).

8.3.5 Changes in Conventional Methods

The conventional heat treatment techniques employing temperature in the lower range have been modified to UHT treatments. UHT processing technique is nowadays used in place of HTST and other heating methods because it has several advantages over the traditional methods. UHT processing involves heating fluid in a continuous-flow system to a high temperature (135–145 °C) and holding it at that temperature for a short time (1–10 s) followed by rapid cooling. This produces a 'commercially sterile' product, that is, a product in which bacterial growth is highly unlikely to occur under normal storage conditions. The rapid heat transfer rate minimises undesirable changes in the taste and nutritional quality of the resulting product. Since UHT is a continuous process, it produces uniform product quality that does not depend on the size of a container, unlike in-container sterilisation. This attribute is especially important for products containing heat-sensitive ingredients and highly viscous products with poor heat transfer properties (Tewari & Juneja, 2008).

In addition, the homogenisation methods have been modified, and UHPH is the new technique which is used nowadays for processing of fluid foods and UHPH treatment of vegetable milks improves hygienic and colloidal stability.

8.4 Processing Grain-Based Beverages by Innovative Techniques

8.4.1 Enzymatic Techniques

The application of enzymes for preparation of non-dairy milk has been mentioned by few researchers for production of rice milk and other cereal milks (Mitchell *et al.*, 1988; Suphamityotin, 2011; Mårtensson *et al.*, 2000; Önning *et al.*, 2000). Production of cereal milk comprising an aqueous extract of a blend of soya bean, sesame and maize has been mentioned in the literature (Suphamityotin, 2011). Pectinase enzyme was used in this process for extraction. It was found that the enzymatic-assisted method resulted in a higher amount of reducing sugar, protein, fat and antioxidant activity. Thus, disrupting the plant cell wall component network by using cell wall degrading enzymes can increase extractability of protein, fat and antioxidant activity in cereal milk (Suphamityotin, 2011).

In another study, a method has been mentioned wherein the starch present in rice is gelatinised which is then liquefied, preferably with α-amylase enzymes, and then treated with relatively high levels of glucosidase enzyme and/or a β-amylase enzyme in a saccharifying step for production of rice milk. Rice milk thus developed was found to have a milk-like texture and can be used for the preparation of its beverages and frozen desserts. The rice milk thus produced was characterised by its absence of a rice-like flavour (Mitchell *et al.*, 1988).

In a study, the application of enzymes in soya milk production was investigated and it was reported that enzymes improve the yield of soya milk as compared to traditional process. Of the different enzymes tested, the neutral proteinase was found to produce superior results at neutral pH. At 0.5% level, the protein and solid yields obtained at 1 h of reaction were 73% and 66%, respectively, compared to 33% and 42%, respectively, of the control at pH 6.7 (Eriksen, 1983). Hence, the application of enzymes, namely, pectinases, proteinase and amylases, have been shown to yield better quality non-dairy milk. Enzymatic process has also been shown to increase the extraction yield and sensory properties of developed non-dairy milk.

The major portion of oat consists of starch (50–60%), which has a gelatinisation temperature in the range 44.7–73.7 °C (Tester & Karkalas, 1996). Therefore, the viscosity is extremely high following gelatinisation. Thermostable α-amylase is used as a thinning agent, which brings about reduction in viscosity and partial hydrolysis of starch. In the recent years, oat starch has attracted most attention among cereal starches. It offers untypical properties such as small size of granules, well-developed granule surface and high lipid content (Hoover *et al.*, 2003; Mirmoghtadaie *et al.*, 2009). Many patents are available for preparing oat starch cereal suspensions using α-amylase for enzymatic hydrolysis (Inglett, 1991; Ahlden *et al.*, 1997; Triantafyllou, 2001). Enzymatic hydrolysis kinetics of oat flour aqueous dispersions under varying conditions has been studied, and the study reported an extensive reduction in viscosity when the α-amylase was applied for starch hydrolysis (Patsioura *et al.*, 2011). It has also been reported that amylodextrins obtained by enzymatic hydrolysis of oat starch can be used as fat substitute based on their fatty consistency and mouthfeel in products such as fat-free milk, cakes, salad cream and ice cream (Lee *et al.*, 2005; Vatanasuchart & Stonsaovapak, 2000; Pszczola, 1996; Inglett, 1990; Hoffman *et al.*, 1985). An enzymatic liquefaction method has been developed and optimised successfully to prepare oat milk from oat groats. The prepared oat milk maintains its viscosity even at high temperature of processing and does not set into a gel as all the starch has been liquefied (Deswal *et al.*, 2014a,b).

8.4.2   Fermentation

Increasing awareness among consumers about the health benefits of high-fibre diets has emphasised the importance of developing enriched-fibre food products. Cereals contain biologically active ingredients such as dietary and functional fibres, contributing to about 50% of the fibre intake in western countries (Angelov *et al.*, 2006; Luana *et al.*, 2014). Cereals are one of the most suitable substrates for the development of foods containing probiotic microorganisms (in most cases LAB or bifidobacteria) (Ara *et al.*, 2002; Obadina *et al.*, 2013) and may also have prebiotic properties due to the presence of non-digestible components of cereal matrix. Oats (*Avena sativa*) have received considerable interest for their high content of soluble and insoluble fibres and for their high fermentability upon applying probiotic LAB. In a study, *L. plantarum* has been used for the development of a oat-based fermented drink using 5.5% oats, 1.25% sugar and 5% inoculum to prepare a fermented drink to obtain a growth of 10.4 log CFU/ml (Gupta *et al.*, 2010). Many other studies report the use of fermentation technology for production of oat-based beverages (Angelov *et al.*, 2006; Luana *et al.*, 2014).

Cereal (rice, barley, emmer and oat) and soya flours and concentrated red grape have been used for making vegetable yoghurt-like beverages (VYLBs). Two selected strains of *L. plantarum* were used for lactic acid fermentation, according to a process which included the flour gelatinisation. Beverages made with the mixture of rice and barley or emmer flours seemed to possess the best combination of textural, nutritional and sensory properties (Coda *et al.*, 2012).

Flavour is one of the most important characteristics in the sensory profile and acceptability of cereal foods and beverages. Fermentation of cereal-based beverage products using LAB can strongly influence their flavour profiles. The changes in aroma and flavour vary considerably depending on the type of starter culture used, the cereal substrate being fermented and other beverage processing steps. In general, LAB contribute to cereal product flavour by producing organic acids through sugar metabolism, thus lowering the pH resulting in an increase in sourness and decreased sweetness. The decreased pH can also affect endogenous cereal enzymes involved in flavour compound generation or their precursors.

In one example, the selective exclusion of some natural occurring LAB and yeasts, and optimisation of the starter culture to ferment the Turkish beverage, bouza, showed an improvement in the sensory qualities of this highly viscous traditional cereal beverage. Hereby, a starter culture consisting of *Saccharomyces cerevisiae*, *Leuc. mesenteroides* subsp. *mesenteroides* and *Lc. confusus* was found to positively influence the organoleptic and rheological properties of the end product. Fermentation has been shown to improve the sensory properties of soya milk as well (Ara *et al.*, 2002; Obadina *et al.*, 2013).

8.4.3 Ultra-High-Pressure Homogenisation

The effect of UHPH at 200 and 300 MPa in combination with different inlet temperatures (55, 65 and 75 °C) on soya milk was studied. UHPH-treated soya milk was compared with the base product (untreated), pasteurised (95 °C for 30 s) and UHT (142 °C for 6 s)-treated soya milks. Microbiological (total aerobic meshophilic bacteria, aerobic spores and *B. cereus*), physical (dispersion stability and particle size distribution) and chemical (lipoxygenase activity, hydroperoxide index and TIA) parameters of special relevance in soya milk were studied. Microbiological results showed that pressure and inlet temperature combination had a significant impact on the lethal effect of UHPH treatment. While most of UHPH treatments applied produced high quality of soya milks better than that pasteurised, the combination of 300 MPa and 75 °C produced a commercially sterile soya milk. UHPH treatments caused a significant decrease in particle size resulting in a high physical stability of samples compared with conventional heat treatments. UHPH treatment produced lower values of hydroperoxide index than heat-treated soya milks although TIA was lower in UHT-treated products (Poliseli-Scopel *et al.*, 2012).

The effect of UHPH on the volatile profile of soya milk was studied and compared with conventional treatments. Soya milk was treated at 200 MPa combined with two inlet temperatures (55 or 75 °C) and treated at 300 MPa at 80 °C inlet temperature. UHPH-treated soya milks were compared with base product (untreated sample), pasteurised soya milk (90 °C, 30 s) and UHT (142 °C, 6 s)-treated samples. Volatile compounds were extracted by solid-phase microextraction and were identified by gas chromatography coupled with mass spectrometry. Pasteurisation and UHPH treatments at 200 MPa produced few changes in the volatile composition, reaching similar values to untreated soymilk. UHT treatment produced the most important effects on volatile profile compared to UHPH at 300 MPa and 80 °C. Hexanal was the most abundant compound detected in all treatments. The effect of UHPH technology on volatile profile induced modifications depending on the combinations of processing parameters.

8.5 Preservation of Grain-Based Beverages by Innovative Technologies

8.5.1 High-Pressure Processing

In a study, the effects of pressure (400, 500 and 600 MPa), dwell time (1 and 5 min) and temperature (25 and 75 °C) on microbial quality and protein stability of soymilk during 28 days of storage (4 °C) were evaluated under aerobic and anaerobic conditions. After processing and during storage, there were significant differences in total bacterial count (TBC), numbers of psychrotrophs (PSY) and Enterobacteriaceae (ENT) and protein stability between untreated (control) and pressurised samples. After 28 days of refrigerated storage, both aerobic and anaerobic pressurised samples had better or similar stability as the control on day one of storage. Soymilk control samples were spoiled after 7 days, whereas pressurisation increased soymilk shelf life by at least 2 weeks. Pressure (600 MPa) at 75 °C for 1 min not only significantly reduced initial microbial populations and increased the microbial shelf life but also extended the protein stability of soymilk ($P < 0.05$) (Smith *et al.*, 2009).

In another study, the effects of high-pressure treatment on the modifications of soya protein in soymilk were studied using various analytical techniques. Spectrofluorimetry revealed that the soya protein exhibited more hydrophobic regions after high-pressure treatment. High-pressure-induced tofu gels could be formed that had gel strength and a cross-linked network microstructure. This provided a new way to process soya milk for making tofu gels (Zhang *et al.*, 2005).

High hydrostatic pressure was applied to soya beans and soymilk to assess its effect on isoflavone content, profile and water extractability. Combined pressure and mild thermal treatment modified the isoflavone distribution. At 75 °C, the isoflavone profile shifted from malonylglucosides towards β-glucosides, which was correlated to the effect of adiabatic heating. When pressure was applied to the hydrated soya beans, the soymilk isoflavone concentration varied between 4.32 and 6.06 μmol/g. The content of protein decreased and fat increased in soymilks prepared from pressurised soya beans with increasing pressure level (Jung *et al.*, 2008). High-pressure treatments have also shown to inactivate trypsin inhibitor and lipoxygenase (Wang *et al.*, 2008; van der Ven *et al.*, 2005).

Fluid foods such as juices and beverages, with their high acidity and high water activity (a_w), can be pasteurised at ambient or chilled temperatures using pressures in the range 400–600 MPa and process times generally under 10 min (Farkas & Hoover, 2000). High-pressure pasteurisation treatments inactivate pathogenic and spoilage bacteria, yeasts, molds and viruses. However, the treatment has limited efficacy against spores and enzymes. The extent of bacterial inactivation also depends on the type of microorganism, food composition, pH and water activity. Cells in the exponential phase of growth have been found to be less resistant than those in the stationary phase. Gram-positive organisms are more resistant than Gram negatives. Significant variations in pressure resistances can be seen among strains. Pressure treatment alone has little effect on the destruction of spores. It has been found in several studies that combined pressure–temperature treatment has a synergistic effect on spore inactivation (Cheftel, 1995; Ananta *et al.*, 2001). At ambient temperatures, bacterial spores can survive pressures above 1000 MPa (Farkas & Hoover, 2000). During typical pressure-assisted thermal processing (PATP) (also referred to as pressure-assisted thermal sterilisation or PATS), the food is subjected to a combination of elevated pressures and moderate heat for short duration. One of the unique advantages of PATP is its ability to provide a rapid and uniform increase in the temperature of treated food samples. Uniform compression heating and expansion cooling on decompression help to reduce the severity of thermal effects encountered with conventional processing techniques.

8.5.2 Pulsed Electric Field

The inactivation of soya bean lipoxygenase by PEFs has been studied, and it is reported that residual activity of soya bean lipoxygenase decreased with the increase in treatment time, pulse strength, pulse frequency and pulse width. The maximum inactivation of soya bean lipoxygenase by PEF achieved 88% at 42 kV/cm for 1036 µs with 400 Hz of pulse frequency and 2 µs of pulse width at 25 °C (Li *et al.*, 2008).

The effects of high-intensity pulsed electric fields (HIPEFs) treatment on the microbial stability, quality parameters and antioxidant properties of a fruit juice–soymilk (FJ–SM) beverage along the storage time at 4 °C were compared to those obtained by thermal pasteurisation (90 °C, 60 s). HIPEF processing for 800 µs ensured the microbial stability of the beverage during 31 days; however, longer microbial shelf life (56 days) was achieved by increasing the treatment time to 1400 µs or by applying a thermal treatment. Peroxidase and lipoxygenase of HIPEF-treated beverages were inactivated by 17.5–29% and 34–39%, respectively, whereas thermal treatment achieved 100% and 51%. During the storage, vitamin C content and antioxidant capacity depleted with time, and they were higher in FJ–SM beverages processed by HIPEF than in those thermally treated. Instead, total phenolic content of beverages did not present significant changes over the time, and it was higher in the 1400-µs HIPEF-treated beverages. In general, colour, soluble solids, pH and acidity values were not significantly affected by the processing treatments. Beverage viscosity increased over time, regardless of the treatment applied. Hence, the application of HIPEF may be a good alternative treatment to thermal processing in order to ensure microbiological stability, high nutritional values and fresh-like characteristics of FJ–SM beverages (Morales-de la Peña *et al.*, 2011).

8.6 Relevance for Human Nutrition

Consumers are more and more concerned about the nutritional and health-related characteristics of the food they eat. The processing of foods is becoming more sophisticated and diverse in response to the growing demand for quality foods. Consumers today expect food products to provide fresh-like appearance, convenience, variety, appropriate shelf life and caloric content, reasonable cost, environmental soundness, high nutritional and functional quality. Non-thermal processing of foods has been revealed as a useful tool to extend their shelf life and quality as well as to preserve their nutritional and functional characteristics. In the past 10 years, there has been an increasing interest in non-thermal technologies as HPP and PEFs to preserve various kinds of food products without the quality and nutritional damage caused by heat treatments. The use of non-thermal treatments for processing foods is very relevant to the human nutrition as more nutritious food without the need of synthetic preservative will be available to the human beings.

8.7   Conclusion and Future Trends

Traditional method of processing is principally thermal based, which extends the shelf life of food and provides other processing benefits, but they actually deteriorate the nutritional and sensory properties of food under processing. Non-thermal technologies are in focus due to consumer demand for food products that are minimally processed, of high quality, and are convenient and safe. Non-thermal processes offer shelf-life extension without the use of preservatives or additives, while still retaining colour, flavour, texture, nutritive and functional qualities. However, the non-thermal processing techniques of processing food products sometimes become very costly. In future, the combined use of non-thermal processes with the traditional method of food processing should be explored to achieve an optimum process which gives desired product characteristics along with being affordable. In addition, combined treatments are advantageous because many individual treatments alone are not adequate to ensure food safety or stability.

References

Abdullah, M., Rehman, S., Zubair, H., Saeed, H., Kousar, S. & Shahid, M. (2003) Effect of skim milk in soymilk blend on the quality of ice cream. *Pakistan Journal of Nutrition* **2**, 305–311.

Ahlden, I., Lindahl, L., Oste, R. & Sjoholm, I. (1997) Homogeneous and stable cereal suspension and a method of making the same. US Patent No. 5,685,123.

Aiyer, P.V. (2005) Amylases and their applications. *African Journal of Biotechnology* **4**, 1525–1529.

Ananta, E., Heinz, V., Schlüter, O. & Knorr, D. (2001) Kinetic studies on high-pressure inactivation of *Bacillus stearothermophilus* spores suspended in food matrices. *Innovative Food Science & Emerging Technologies* **2**, 261–272.

Angelov, A., Gotcheva, V., Kuncheva, R. & Hristozova, T. (2006) Development of a new oat-based probiotic drink. *International Journal of Food Microbiology* **112**, 75–80.

Ara, K., Yoshimatsu, T., Ojima, M., Kawai, S. & Okubo, K. (2002) Effect of new lactic acid fermentation on sensory taste of fermented soymilk. *Journal-Japanese Society of Food Science and Technology* **49**, 377–387.

Asgher, M., Asad, M.J., Rahman, S.U. & Legge, R.L. (2007) A thermostable α-amylase from a moderately thermophilic *Bacillus subtilis* strain for starch processing. *Journal of Food Engineering* **79**, 950–955.

Barba, F.J., Esteve, M.J. & Frigola, A. (2013) Physicochemical and nutritional characteristics of blueberry juice after high pressure processing. *Food Research International* **50**, 545–549.

Bawa, S.E. (2010) *The significance of soy protein and soy bioactive compounds in the prophylaxis and treatment of osteoporosis. Journal of Osteoporosis* **21**.

Bernat, N., Cháfer, M., González-Martínez, C., Rodríguez-García, J. & Chiralt, A. (2014) Optimisation of oat milk formulation to obtain fermented derivatives by using probiotic *Lactobacillus reuteri* microorganisms. *Food Science and Technology International* **1**, 1–13.

Biörklund, M., van Rees, A., Mensink, R. & Önning, G. (2005) Changes in serum lipids and postprandial glucose and insulin concentrations after consumption of beverages with β-glucans from oats or barley: a randomised dose-controlled trial. *European Journal of Clinical Nutrition* **59**, 1272–1281.

Blandino, A., Al-Aseeri, M., Pandiella, S., Cantero, D. & Webb, C. (2003) Cereal-based fermented foods and beverages. *Food Research International* **36**, 527–543.

Bricarello, L.P., Kasinski, N., Bertolami, M.C., Faludi, A., Pinto, L.A., Relvas, W.G., Izar, M.C., Ihara, S.S., Tufik, S. & Fonseca, F.A. (2004) Comparison between the effects of soy milk and non-fat cow milk on lipid profile and lipid peroxidation in patients with primary hypercholesterolemia. *Nutrition* **20**, 200–204.

Brul, S., Rommens, A. & Verrips, C. (2000) Mechanistic studies on the inactivation of Saccharomyces cerevisiae by high pressure. *Innovative Food Science & Emerging Technologies* **1**, 99–108.

Buckley, N.D., Champagne, C.P., Masotti, A.I., Wagar, L.E., Tompkins, T.A. & Green-Johnson, J.M. (2011) Harnessing functional food strategies for the health challenges of space travel—Fermented soy for astronaut nutrition. *Acta Astronautica* **68**, 731–738.

Campbell, A.K., Waud, J.P. & Matthews, S.B. (2005) The molecular basis of lactose intolerance. *Science Progress* **88**, 157–202.

Carr, M., Black, L. & Bagby, M. (1982) Continuous enzymatic liquefaction of starch for saccharification. *Biotechnology and Bioengineering* **24**, 2441–2449.

Champagne, C. P., Green-Johnson, J., Raymond, Y., Barrette, J. & Buckley, N. (2009) Selection of probiotic bacteria for the fermentation of a soy beverage in combination with *Streptococcus thermophilus*. *Food Research International* **42**, 612–621.

Champagne, C.P., Tompkins, T.A., Buckley, N.D. & Green-Johnson, J.M. (2010) Effect of fermentation by pure and mixed cultures of Streptococcus thermophilus and Lactobacillus helveticus on isoflavone and B-vitamin content of a fermented soy beverage. *Food Microbiology* **27**, 968–972.

Chavan, J., Kadam, S. & Beuchat, L.R. (1989) Nutritional improvement of cereals by fermentation. *Critical Reviews in Food Science & Nutrition* **28**, 349–400.

Cheftel, J.C. (1995) Review: high-pressure, microbial inactivation and food preservation/revision: alta-presion, inactivacion microbiologica y conservacion de alimentos. *Food Science and Technology International* **1**, 75–90.

Choi, J., Wang, D., Namihira, T., Katsuki, S., Akiyama, H., Lin, X., Sato, H., Seta, H., Matsubara, H. & Saeki, T. (2008) Inactivation of spores using pulsed electric field in a pressurized flow system. *Journal of Applied Physics* **104**, 094701.

Chou, C.-C. & Hou, J.-W. (2000) Growth of bifidobacteria in soymilk and their survival in the fermented soymilk drink during storage. *International Journal of Food Microbiology* **56**, 113–121.

Coda, R., Lanera, A., Trani, A., Gobbetti, M. & Di Cagno, R. (2012) Yogurt-like beverages made of a mixture of cereals, soy and grape must: microbiology, texture, nutritional and sensory properties. *International Journal of Food Microbiology* **155**, 120–127.

Crabb, D.W. & Mitchinson, C. (1997) Enzymes involved in the processing of starch to sugars. *Trends in Biotechnology* **15**, 349–352.

Crabb, W.D. & Shetty, J. K. (1999) Commodity scale production of sugars from starches. *Current Opinion in Microbiology* **2**, 252–256.

Cruz, N., Capellas, M., Hernández, M., Trujillo, A., Guamis, B. & Ferragut, V. (2007) Ultra high pressure homogenization of soymilk: microbiological, physicochemical and microstructural characteristics. *Food Research International* **40**, 725–732.

Cullen, P.J. & Tiwari, B.K. (2011). *Novel Thermal and Non-Thermal Technologies for Fluid Foods*. Academic Press.

Davies, C., Nielsen, S. & Nielsen, N. (1987) Flavor improvement of soybean preparations by genetic removal of lipoxygenase-2. *Journal of the American Oil Chemists Society* **64**, 1428–1433.

Deswal, A., Deora, N. & Mishra, H. (2014a) Optimization of enzymatic production process of oat milk using response surface methodology. *Food and Bioprocess Technology* **7**, 610–618.

Deswal, A., Deora, N.S. & Mishra, H.N. (2014b) Effect of concentration and temperature on the rheological properties of oat milk. *Food and Bioprocess Technology* **7**, 2451–2459.

Dias-Morse, P.N. (2004). *Effect of Formulation and Processing Conditions on the Rheological and Sensory Properties of Rice Milk*. University of Arkansas, Fayetteville.

Eriksen, S. (1983) Application of enzymes in soy milk production to improve yield. *Journal of Food Science* **48**, 445–447.

Espirito-Santo, A.P.D., Mouquet-Rivier, C., Humblot, C., Cazevieille, C., Icard-Vernière, C., Soccol, C.R. & Guyot, J.-P. (2014) Influence of cofermentation by amylolytic *Lactobacillus* strains and probiotic bacteria on the fermentation process, viscosity and microstructure of gruels made of rice, soy milk and passion fruit fiber. *Food Research International* **57**, 104–113.

Evrendilek, G.A., Jin, Z., Ruhlman, K., Qiu, X., Zhang, Q. & Richter, E. (2000) Microbial safety and shelf-life of apple juice and cider processed by bench and pilot scale PEF systems. *Innovative Food Science & Emerging Technologies* **1**, 77–86.

Faccin, G.L., Miotto, L.A., Vieira, L.D.N., Barreto, P.L.M. & Amante, E.R. (2009) Chemical, sensorial and rheological properties of a new organic rice bran beverage. *Rice Science* **16**, 226–234.

Farkas, D.F. & Hoover, D.G. (2000) High pressure processing. *Journal of Food Science* **65**, 47–64.

Farnworth, E., Mainville, I., Desjardins, M.-P., Gardner, N., Fliss, I. & Champagne, C. (2007) Growth of probiotic bacteria and bifidobacteria in a soy yogurt formulation. *International Journal of Food Microbiology* **116**, 174–181.

Ferragut, V., Hernández-Herrero, M., Poliseli, F., Valencia, D., Guamis, B., Yanniotis, S., Taoukis, P., Stoforos, N. & Karathanos, V. (2011). Ultra high pressure homogenization (UHPH) treatment of vegetable milks: improving hygienic and colloidal stability. *In:* Proceedings of the 11th International Congress on Engineering and Food (ICEF11)–Food Process Engineering in a Changing World, 1193-4.

Friedeck, K.G., Aragul-Yuceer, Y. & Drake, M. (2003) Soy protein fortification of a low-fat dairy-based ice cream. *Journal of Food Science* **68**, 2651–2657.

García, D., Gómez, N., Mañas, P., Raso, J. & Pagán, R. (2007) Pulsed electric fields cause bacterial envelopes permeabilization depending on the treatment intensity, the treatment medium pH and the microorganism investigated. *International Journal of Food Microbiology* **113**, 219–227.

Gillooly, M., Bothwell, T., Charlton, R., Torrance, J., Bezwoda, W., MacPhail, A., Derman, D., Novelli, L., Morrall, P. & Mayet, F. (1984) Factors affecting the absorption of iron from cereals. *British Journal of Nutrition* **51**, 37–46.

Gokavi, S., Zhang, L., Huang, M. K., Zhao, X. & Guo, M. (2005) Oat-based Symbiotic Beverage Fermented by *Lactobacillus plantarum, Lactobacillus paracasei* ssp. *casei,* and *Lactobacillus acidophilus. Journal of Food Science* **70**, M216–M223.

Gotcheva, V., Pandiella, S.S., Angelov, A., Roshkova, Z.G. & Webb, C. (2000) Microflora identification of the Bulgarian cereal-based fermented beverage boza. *Process Biochemistry* **36**, 127–130.

Granfeldt, Y.E. & Bjorck, I. (2011) A bilberry drink with fermented oatmeal decreases postprandial insulin demand in young healthy adults. *Nutrition Journal* **10**, 57.

Gross, M. & Jaenicke, R. (1994) Proteins under pressure. *European Journal of Biochemistry* **221**, 617–630.

Gupta, S., Cox, S. & Abu-Ghannam, N. (2010) Process optimization for the development of a functional beverage based on lactic acid fermentation of oats. *Biochemical Engineering Journal* **52**, 199–204.

Haard, N.F. (1999) *Fermented Cereals: A Global Perspective*. Food & Agriculture Org.

Hackler, L., Buren, J., Steinkraus, K., Rawi, I. & Hand, D. (1965) Effect of heat treatment on nutritive value of soymilk protein fed to weanling rats. *Journal of Food Science* **30**, 723–728.

Hayashida, S., Teramoto, Y. & Inoue, T. (1988) Production and characteristics of raw-potato-starch-digesting α-amylase from *Bacillus subtilis* 65. *Applied and Environmental Microbiology* **54**, 1516–1522.

Hoffman, S., Lenchin, J.M. & Trubiano, P.C. (1985) *Converted starches for use as a fat-or oil-replacement in foodstuffs*.

Hoover, R., Smith, C., Zhou, Y. & Ratnayake, R. (2003) Physicochemical properties of Canadian oat starches. *Carbohydrate Polymers* **52**, 253–261.

Hou, J.-W., Yu, R.-C. & Chou, C.-C. (2000) Changes in some components of soymilk during fermentation with bifidobacteria. *Food Research International* **33**, 393–397.

Huang, H., Liang, H. & Kwok, K.C. (2006) Effect of thermal processing on genistein, daidzein and glycitein content in soymilk. *Journal of the Science of Food and Agriculture* **86**, 1110–1114.

İçier, F., Gündüz, G.T., Yılmaz, B. & Memeli, Z. (2015) Changes on some quality characteristics of fermented soy milk beverage with added apple juice. *LWT – Food Science and Technology* **63**, 57–64.

Ikya, J.K., Gernah, I., Ojobo, E. & Oni, K. (2013) Effect of cooking temperature on some quality characteristics of soy milk. *Advance Journal of Food Science and Technology* **5.5**, 543–546.

Inglett, G. (1990) Oatrim cuts fat, cholesterol in ice cream. *Science of Food and Agriculture* **2**, 4–5.

Inglett, G.F. (1991) Method for making a soluble dietary fiber composition from oats. US Patent No. 4,996,063.

Jarvinen-Seppo, K. M., Sicherer, S.H. & TePas, E. (2011) Milk allergy: management. *Last Literature Review* **version 19**.

Johnson, K. & Snyder, H. (1978) Soymilk: a comparison of processing methods on yields and composition. *Journal of Food Science* **43**, 349–353.

Jung, S., Murphy, P.A. & Sala, I. (2008) Isoflavone profiles of soymilk as affected by high-pressure treatments of soymilk and soybeans. *Food Chemistry* **111**, 592–598.

Juvonen, K.R., Purhonen, A.-K., Salmenkallio-Marttila, M., Lähteenmäki, L., Laaksonen, D.E., Herzig, K.-H., Uusitupa, M.I., Poutanen, K.S. & Karhunen, L.J. (2009) Viscosity of oat bran-enriched beverages influences gastrointestinal hormonal responses in healthy humans. *The Journal of Nutrition* **139**, 461–466.

Keshun, L. (1997). *Soybeans: Chemistry, Technology, and Utilization*, Chapman & Hall.

Konsoula, Z. & Liakopoulou-Kyriakides, M. (2006) Starch hydrolysis by the action of an entrapped in alginate capsules α-amylase from Bacillus subtilis. *Process Biochemistry* **41**, 343–349.

Konsula, Z. & Liakopoulou-Kyriakides, M. (2004) Hydrolysis of starches by the action of an α-amylase from *Bacillus subtilis*. *Process Biochemistry* **39**, 1745–1749.

Kumar, P. & Mishra, H. (2004). Mango soy fortified set yoghurt: effect of stabilizer addition on physicochemical, sensory and textural properties. *Food Chemistry* **87**, 501–507.

Kwok, K.C., Liang, H.H. & Niranjan, K. (2002) Mathematical modelling of the heat inactivation of trypsin inhibitors in soymilk at 121–154° C. *Journal of the Science of Food and Agriculture* **82**, 243–247.

Labuschagne, I. & Lombard, M.J. (2012) Understanding lactose intolerance and the dietary management thereof. *South African Family Practice* **54**, 496–498.

LeBlanc, J., Garro, M., Giori, G.S. & Valdez, G. (2004) A novel functional soy-based food fermented by lactic acid bacteria: effect of heat treatment. *Journal of Food Science* **69**, M246–M250.

Lee, S., Kim, S. & Inglett, G.E. (2005) Effect of shortening replacement with oatrim on the physical and rheological properties of cakes. *Cereal Chemistry* **82**, 120–124.

Li, Y.-Q., Chen, Q., Liu, X.-H. & Chen, Z.-X. (2008) Inactivation of soybean lipoxygenase in soymilk by pulsed electric fields. *Food Chemistry* **109**, 408–414.

Lichtenstein, A. H. (1998) Soy protein, isoflavones and cardiovascular disease risk. *The Journal of Nutrition* **128**, 1589–1592.

Liener, I. (1981) Factors affecting the nutritional quality of soya products. *Journal of the American Oil Chemists' Society* **58**, 406–415.

Luana, N., Rossana, C., Curiel, J. A., Kaisa, P., Marco, G. & Rizzello, C. G. (2014) Manufacture and characterization of a yogurt-like beverage made with oat flakes fermented by selected lactic acid bacteria. *International Journal of Food Microbiology* **185**, 17–26.

Lyly, M., Liukkonen, K.-H., Salmenkallio-Marttila, M., Karhunen, L., Poutanen, K. & Lähteenmäki, L. (2009) Fibre in beverages can enhance perceived satiety. *European Journal of Nutrition* **48**, 251–258.

Ma, Y., Cai, C., Wang, J. & Sun, D.-W. (2006) Enzymatic hydrolysis of corn starch for producing fat mimetics. *Journal of Food Engineering* **73**, 297–303.

van der Maarel, M.J., van der Veen, B., Uitdehaag, J., Leemhuis, H. & Dijkhuizen, L. (2002) Properties and applications of starch-converting enzymes of the α-amylase family. *Journal of Biotechnology* **94**, 137–155.

Markets & Markets (2013). Dairy Alternative (Beverage) Market: Global Trends & Forecast to 2018. http://www.prnewswire.com/news-releases/global-dairy-alternative-beverage-market---global-trends--forecast-to-2018-239482521.html.

Mårtensson, O., Öste, R. & Holst, O. (2000) Lactic acid bacteria in an oat-based non-dairy milk substitute: fermentation characteristics and exopolysaccharide formation. *LWT-Food Science and Technology* **33**, 525–530.

Mårtensson, O., Öste, R. & Holst, O. (2002) The effect of yoghurt culture on the survival of probiotic bacteria in oat-based, non-dairy products. *Food Research International* **35**, 775–784.

Matsuura, M., Obata, A. & Fukushima, D. (1989) Objectionable flavor of soy milk developed during the soaking of soybeans and its control. *Journal of Food Science* **54**, 602–605.

Mezemir, S. (2015) Probiotic potential and nutritional importance of teff (Eragrostis tef (Zucc) Trotter) enjerra – A review. *African Journal of Food Agriculture Nutrition and Development* **15**, 9964–9981.

Min, S., Evrendilek, G.A. & Zhang, H.Q. (2007) Pulsed electric fields: processing system, microbial and enzyme inhibition, and shelf life extension of foods. *IEEE Transactions on Plasma Science* **35**, 59–73.

Mirmoghtadaie, L., Kadivar, M. & Shahedi, M. (2009) Effects of cross-linking and acetylation on oat starch properties. *Food Chemistry* **116**, 709–713.

Mital, B. & Steinkraus, K. (1975) Utilization of oligosaccharides by lactic acid bacteria during fermentation of soy milk. *Journal of Food Science* **40**, 114–118.

Mitchell, C.R., Mitchell, P.R. & Nissenbaum, R. (1988). Nutritional rice milk production. US Patent No. 4,744,992.

Mohamed, M.E., Ayman, H. & Eissa, A. (2012). *Pulsed Electric Fields for Food Processing Technology*, INTECH Open Access Publisher.

Moore, G.R.P., Canto, L.R.D., Amante, E.R. & Soldi, V. (2005) Cassava and corn starch in maltodextrin production. *Quimica Nova* **28**, 596–600.

Morales-de la Peña, M., Salvia-Trujillo, L., Rojas-Graü, M.A. & Martín-Belloso, O. (2011) Changes on phenolic and carotenoid composition of high intensity pulsed electric field and thermally treated fruit juice–soymilk beverages during refrigerated storage. *Food Chemistry* **129**, 982–990.

NDDI (2012) *Lactose Intolerance Statistics, (National Digestive Diseases Information)* [Online], USA Today, Available: http://www.statisticbrain.com/lactose-intolerance-statistics/ [9 May 2014].

Nelson, A., Wei, L. & Steinbe, M. (1978) Food products from whole soybeans. In *Whole Soybeans for Home and Village Use*. INSOY Series 14, pp. 21–24. National Technical Information Service (NTIS).

Nigam, P. & Singh, D. (1995) Enzyme and microbial systems involved in starch processing. *Enzyme and Microbial Technology* **17**, 770–778.

Norton, T. & Sun, D.-W. (2008) Recent advances in the use of high pressure as an effective processing technique in the food industry. *Food and Bioprocess Technology* **1**, 2–34.

Obadina, A.O., Akinola, O.J., Shittu, T.A. & Bakare, H.A. (2013) Effect of natural fermentation on the chemical and nutritional composition of fermented soymilk nono. *Nigerian Food Journal* **31**, 91–97.

Ollberding, N.J., Lim, U., Wilkens, L.R., Setiawan, V.W., Shvetsov, Y.B., Henderson, B.E., Kolonel, L.N. & Goodman, M.T. (2012) Legume, soy, tofu, and isoflavone intake and endometrial cancer risk in postmenopausal women in the multiethnic cohort study. *Journal of the National Cancer Institute* **104**, 67–76.

Omueti, O. & Ajomale, K. (2005) Chemical and sensory attributes of soy-corn milk types. *African Journal of Biotechnology* **4**, 847–851.

Önning, G., Åkesson, B., Öste, R. & Lundquist, I. (1998) Effects of consumption of oat milk, soya milk, or cow's milk on plasma lipids and antioxidative capacity in healthy subjects. *Annals of Nutrition and Metabolism* **42**, 211–220.

Önning, G., Wallmark, A., Persson, M., Åkesson, B., Elmståhl, S. & Öste, R. (2000) Consumption of oat milk for 5 weeks lowers serum cholesterol and LDL cholesterol in free-living men with moderate hypercholesterolemia. *Annals of Nutrition and Metabolism* **43**, 301–309.

Orhevba, B. (2011) The effects of cooking time on the nutritional parameters of soya milk. *American Journal of Food Technology* **6**, 298–305.

Pandey, A., Nigam, P., Soccol, C., Soccol, V., Singh, D. & Mohan, R. (2000) Advances in microbial amylases. *Biotechnology and Applied Biochemistry* **31**, 135–152.

Pathanibul, P., Taylor, T.M., Davidson, P.M. & Harte, F. (2009) Inactivation of *Escherichia coli* and *Listeria innocua* in apple and carrot juices using high pressure homogenization and nisin. *International Journal of Food Microbiology* **129**, 316–320.

Patrignani, F., Vannini, L., Kamdem, S. L.S., Lanciotti, R. & Guerzoni, M.E. (2009) Effect of high pressure homogenization on Saccharomyces cerevisiae inactivation and physico-chemical features in apricot and carrot juices. *International Journal of Food Microbiology* **136**, 26–31.

Patsioura, A., Gekas, V., Lazaridou, A. & Biliaderis, C. (2011) Kinetics of heterogeneous amylolysis in oat flour and characterization of hydrolyzates. *In:* Proceedings of the 11th International Congress on Engineering and Food: Food Process Engineering in a Changing World, 22–6.

Poliseli-Scopel, F.H., Hernández-Herrero, M., Guamis, B. & Ferragut, V. (2012) Comparison of ultra high pressure homogenization and conventional thermal treatments on the microbiological, physical and chemical quality of soymilk. *LWT – Food Science and Technology* **46**, 42–48.

Popper, L. & Knorr, D. (1990) Applications of high-pressure homogenization for food preservation: high-pressure homogenization can be used alone or combined with lytic enzyme or chitosan to reduce the microbial population and heat treatment damage in foods. *Food Technology* **44**, 84–89.

Potter, S.M., Baum, J.A., Teng, H., Stillman, R.J., Shay, N.F. & Erdman, J. (1998) Soy protein and isoflavones: their effects on blood lipids and bone density in postmenopausal women. *The American Journal of Clinical Nutrition* **68**, 1375S–1379S.

Poysa, V. & Woodrow, L. (2002) Stability of soybean seed composition and its effect on soymilk and tofu yield and quality. *Food Research International* **35**, 337–345.

Pszczola, D.E. (1996) Oatrim finds application in fat-free, cholesterol-free milk. *Food Technology* **50**.

Qin, B.-L., Barbosa-Cánovas, G.V., Swanson, B.G., Pedrow, P.D. & Olsen, R.G. (1998) Inactivating microorganisms using a pulsed electric field continuous treatment system. *IEEE Transactions on Industry Applications* **34**, 43–50.

Rackis, J. (1972) Biologically active components. In *Soybeans: Chemistry and Technology* **1**, pp. 158–202. The AVI Publisher Corporation.

Ramos-de-la-Peña, A.M., Renard, C.M.G.C., Wicker, L., Montañez, J., de la Luz Reyes-Vega, M., Voget, C. & Contreras-Esquivel, J.C. (2012) Enzymatic liquefaction of jicama (*Pachyrhizus erosus*) tuberous roots and characterization of the cell walls after processing. *LWT – Food Science and Technology* **49**, 257–262.

Rebholz, C., Reynolds, K., Wofford, M., Chen, J., Kelly, T., Mei, H., Whelton, P. & He, J. (2013) Effect of soybean protein on novel cardiovascular disease risk factors: a randomized controlled trial. *European Journal of Clinical Nutrition* **67**, 58–63.

Ritz, M., Tholozan, J., Federighi, M. & Pilet, M. (2002) Physiological damages of Listeria monocytogenes treated by high hydrostatic pressure. *International Journal of Food Microbiology* **79**, 47–53.

Robertson, G.H., Wong, D.W., Lee, C.C., Wagschal, K., Smith, M.R. & Orts, W.J. (2006) Native or raw starch digestion: a key step in energy efficient biorefining of grain. *Journal of Agricultural and Food Chemistry* **54**, 353–365.

Robyt, J. & French, D. (1963). Action pattern and specificity of an amylase from Bacillus subtilis. *Archives of Biochemistry and Biophysics* **100**, 451–467.

Rosenthal, A., Deliza, R., Cabral, L., Cabral, L.C., Farias, C.A. & Domingues, A.M. (2003) Effect of enzymatic treatment and filtration on sensory characteristics and physical stability of soymilk. *Food Control* **14**, 187–192.

Russell, K. & Delahunty, C. (2004) The effect of viscosity and volume on pleasantness and satiating power of rice milk. *Food Quality and Preference* **15**, 743–750.

Saidu, J.E.P. (2005) *Development, Evaluation and Characterization of Protein-Isoflavone Enriched Soymilk.* Louisiana State University.

Sampson, H.A. (2004) Update on food allergy. *Journal of Allergy and Clinical Immunology* **113**, 805–819.

Santos, C.C.A.D.A., Libeck, B.d.S. & Schwan, R.F. (2014) Co-culture fermentation of peanut-soy milk for the development of a novel functional beverage. *International Journal of Food Microbiology* **186**, 32–41.

Shimakawa, Y., Matsubara, S., Yuki, N., Ikeda, M. & Ishikawa, F. (2003) Evaluation of *Bifidobacterium breve* strain Yakult-fermented soymilk as a probiotic food. *International Journal of Food Microbiology* **81**, 131–136.

Silva Júnior, S., Tavano, O., Demonte, A., Rossi, E. & Pinto, R. (2012) Nutritional evaluation of soy yoghurt in comparison to soymilk and commercial milk yoghurt. Effect of fermentation on soy protein. *Acta Alimentaria* **41**, 443–450.

Singh, H. & Soni, S. K. (2001) Production of starch-gel digesting amyloglucosidase by *Aspergillus oryzae* HS-3 in solid state fermentation. *Process Biochemistry* **37**, 453–459.

Singh, A.K., Rehal, J., Kaur, A. & Jyot, G. (2015) Enhancement of attributes of cereals by germination and fermentation: a review. *Critical Reviews in Food Science and Nutrition* **55**, 1575–1589.

Smelt, J. (1998) Recent advances in the microbiology of high pressure processing. *Trends in Food Science & Technology* **9**, 152–158.

Smith, K., Mendonca, A. & Jung, S. (2009) Impact of high-pressure processing on microbial shelf-life and protein stability of refrigerated soymilk. *Food Microbiology* **26**, 794–800.

Sodhi, H.K., Sharma, K., Gupta, J.K. & Soni, S.K. (2005) Production of a thermostable α-amylase from Bacillus sp. PS-7 by solid state fermentation and its synergistic use in the hydrolysis of malt starch for alcohol production. *Process Biochemistry* **40**, 525–534.

Spilimbergo, S., Elvassore, N. & Bertucco, A. (2002) Microbial inactivation by high-pressure. *The Journal of Supercritical Fluids* **22**, 55–63.

Stoll, T., Schweiggert, U., Schieber, A. & Carle, R. (2003) Process for the recovery of a carotene-rich functional food ingredient from carrot pomace by enzymatic liquefaction. *Innovative Food Science & Emerging Technologies* **4**, 415–423.

Suphamityotin, P. (2011) Optimizing enzymatic extraction of cereal milk using response surface methodology. *Songklanakarin Journal of Science & Technology* **33**.

Tester, R. F. & Karkalas, J. (1996) Swelling and gelatinization of oat starches. *Cereal Chemistry* **73**, 271–277.

Tewari, G. & Juneja, V. 2008. *Advances in Thermal and Non-Thermal Food Preservation.* John Wiley & Sons.

Tewari, G., Jayas, D. & Holley, R. (1999) High pressure processing of foods: an overview. *Sciences des Aliments* **19**, 619–661.

Triantafyllou, A.O. (2001) *Enzyme preparations for modifying cereal suspensions. U.S. Patent No. 6,190,708.* Washington, DC: U.S. Patent and Trademark Office.

Vatanasuchart, N. & Stonsaovapak, S. (2000) Oatrim-5 as fat substitute in low calorie salad cream: nutritional and microbiological qualities. *Katsetsart Journal: Natural Sciences* **34**, 500–509.

van der Ven, C., Matser, A.M. & van den Berg, R.W. (2005) Inactivation of soybean trypsin inhibitors and lipoxygenase by high-pressure processing. *Journal of Agricultural and Food Chemistry* **53**, 1087–1092.

Vesa, T.H., Marteau, P. & Korpela, R. (2000) Lactose intolerance. *Journal of the American College of Nutrition* **19**, 165S–175S.

Vijayagopal, K., Balagopalan, C. & Moorthy, S. (1988) Gelatinisation and liquefaction of cassava flour: effect of temperature, substrate and enzyme concentrations. *Starch-Starke* **40**, 300–302.

Wang, Y.-C., Yu, R.-C. & Chou, C.-C. (2002) Growth and survival of bifidobacteria and lactic acid bacteria during the fermentation and storage of cultured soymilk drinks. *Food Microbiology* **19**, 501–508.

Wang, Y.-C., Yu, R.-C., Yang, H.-Y. & Chou, C.-C. (2003) Sugar and acid contents in soymilk fermented with lactic acid bacteria alone or simultaneously with bifidobacteria. *Food Microbiology* **20**, 333–338.

Wang, R., Zhou, X. & Chen, Z. (2008) High pressure inactivation of lipoxygenase in soy milk and crude soybean extract. *Food Chemistry* **106**, 603–611.

Welti-Chanes, J., Ochoa-Velasco, C. & Guerrero-Beltrán, J. (2009) High-pressure homogenization of orange juice to inactivate pectinmethylesterase. *Innovative Food Science & Emerging Technologies* **10**, 457–462.

Wilkens, W., Mattick, L. & Hand, D. (1967) Effect of processing method on oxidative off-flavors of soybean milk. *Food Technology* **21**, 86–92.

Will, F., Bauckhage, K. & Dietrich, H. (2000) Apple pomace liquefaction with pectinases and cellulases: analytical data of the corresponding juices. *European Food Research and Technology* **211**, 291–297.

Wofford, M., Rebholz, C., Reynolds, K., Chen, J., Chen, C., Myers, L., Xu, J., Jones, D., Whelton, P. & He, J. (2012) Effect of soy and milk protein supplementation on serum lipid levels: a randomized controlled trial. *European Journal of Clinical Nutrition* **66**, 419–425.

Wolf, W.J. (1975) Lipoxygenase and flavor of soybean protein products. *Journal of Agricultural and Food Chemistry* **23**, 136–141.

Yeom, H.W., Streaker, C.B., Zhang, Q.H. & Min, D.B. (2000) Effects of pulsed electric fields on the quality of orange juice and comparison with heat pasteurization. *Journal of Agricultural and Food Chemistry* **48**, 4597–4605.

Yuan, S. & Chang, S.K.-C. (2007) Selected odor compounds in soymilk as affected by chemical composition and lipoxygenases in five soybean materials. *Journal of Agricultural and Food Chemistry* **55**, 426–431.

Yuan, S., Chang, S.K., Liu, Z. & Xu, B. (2008) Elimination of trypsin inhibitor activity and beany flavor in soy milk by consecutive blanching and ultrahigh-temperature (UHT) processing. *Journal of Agricultural and Food Chemistry* **56**, 7957–7963.

Zhang, H., Li, L., Tatsumi, E. & Isobe, S. (2005) High-pressure treatment effects on proteins in soy milk. *LWT – Food Science and Technology* **38**, 7–14.

Zhang, H., Önning, G., Triantafyllou, A.Ö. & Öste, R. (2007) Nutritional properties of oat-based beverages as affected by processing and storage. *Journal of the Science of Food and Agriculture* **87**, 2294–2301.

Zhao, D. & Shah, N.P. (2014) Antiradical and tea polyphenol-stabilizing ability of functional fermented soymilk–tea beverage. *Food Chemistry* **158**, 262–269.

Zimmermann, U. (1986) Electrical breakdown, electropermeabilization and electrofusion, *Reviews of Physiology, Biochemistry and Pharmacology* **105** 175–256.

9

Soups

Begoña de Ancos* and Concepción Sánchez-Moreno

Institute of Food Science, Technology and Nutrition (ICTAN), Spanish National Research Council (CSIC), Madrid, Spain

9.1 Introduction

9.1.1 Processed Foods

At present, both fresh and processed foods are an important part of the food supply of people living in the twenty-first century. However, food and nutrition scientists, public health professionals, agricultural experts and others dedicated to the study of food and the nutritional needs of people around the world recognise that fresh local food cannot meet all the nutritional needs of consumers worldwide at present nor in the future (Floros *et al.*, 2010). Therefore, processed foods are needed. Several studies have demonstrated that processed food contributes very significantly to providing the quantity of safe, nutritious and high-quality food that worldwide consumers need to lead a healthy and enjoyable life (Weaver *et al.*, 2014). On the other hand, the main objectives of national governments and international institutions in terms of food and health are to achieve a reduction in malnutrition in underdeveloped countries as well as acting against the over-nutritional diseases such as obesity, cardiovascular disease, diabetes, some cancers and chronic respiratory diseases, which are more prevalent in developed countries. These health and nutritional goals can be reached through processed foods which have been produced safely, with high quality and according to nutritional recommendations, economically accessible and socially accepted (Floros *et al.*, 2010).

Food processing means to use a series of mechanical and/or chemical operations to change or preserve it. Processed foods have permitted humanity to have food for prolonged periods of time avoiding seasonality, while providing safe food that reduces the risk of foodborne diseases. The food processing technologies have allowed mankind to advance their technological development and increase their life expectancy.

---

*Corresponding author: Begoña de Ancos, ancos@ictan.csic.es

Innovative Technologies in Beverage Processing, First Edition.
Edited by Ingrid Aguiló-Aguayo and Lucía Plaza.
© 2017 John Wiley & Sons Ltd. Published 2017 by John Wiley & Sons Ltd.

The Institute of Food Technologist scientific review described processing as 'one or more of a range of operations, including washing, grinding, mixing, cooling, storing, heating, freezing, filtering, fermenting, extracting, extruding, centrifuging, frying, drying, concentrating, pressurising, irradiating, microwaving and packaging' (Floros *et al.*, 2010). Processed foods can be classified according to the level of processing that they have been submitted to such as unprocessed, minimally processed and highly processed.

Foods can be processed for economical and logistical reasons in order to improve their commercial shelf life and digestibility, in accordance with the consumer habits of each country or to facilitate the consumption by special groups (children, pregnant women, elderly adults, patients with certain pathologies, etc.). In addition to traditional thermal processing, such as frozen, canned or pasteurised products, there is a growing interest in the development of new processing systems that minimally modify or improve the nutritional and health properties related to the consumption of fresh foods, mainly fruit and vegetables. Among these new food processing technologies, researchers, industrialists and distributors have focused on the development of minimal processing technologies for producing foods with minimally modified sensorial and nutritional characteristics such as 'minimally processed', 'ready-to-eat meals', 'fresh-cut vegetables' and 'ready-to-eat processed vegetables' (Oms-Oliu *et al.*, 2010).

Food processing is becoming increasingly sophisticated and complex in response to the growing demands of the twenty-first century consumer for fresh appearance, highly nutritious, shelf-stable food with guaranteed safety and also easy to prepare with appropriate shelf life, low caloric content and reasonable cost, consistent with environmental and high nutritional and functional quality. Strategies to achieve these demands include modifications to existing food processing techniques and the adoption of new processing technologies (thermal and non-thermal) that will allow promoting fresh but stable foods. Several alternative preservation technologies have been developed in the past 25 years, which include both novel thermal technologies such as microwave (MW), radio frequency (RF) and ohmic heating (OH), and nonthermal technologies that use physical methods for microbial inactivation such as high-pressure processing (HPP), pulsed electric fields (PEFs), ultrasonic waves (U), high-intensity pulsed light (HIPL), irradiation (IR), ultraviolet light (UV) and others (Da-Wen, 2005).

9.1.2 Ready-to-Eat Meals: Soups

Today, ready-to-eat meals such as soups are replacing standard home-cooked meals. People are living increasingly hectic lifestyles, and working hours are replacing free time. Moreover, families are smaller and women are working as much as men. The combination of all of these trends has increased the need for convenience foods.

The term ready-to-eat meal, originally referred to a frozen or chilled meal that comes in an individual package, contains all the elements of a single-serving meal and requires little preparation. Today any cooked or semi-cooked food that needs little preparation to be consumed is called a ready-to-eat meal regardless they are marketed refrigerated, frozen, pasteurized or sterilized. Premium and health qualities of products are the two major drivers of the market given the price premium paid for convenience. Ready-to-eat meals come in a variety of sizes and forms and can be packaged in aluminium trays, metal cans, plastic containers, paperboard containers and many more.

Soups are mixtures of vegetables, meat, poultry, fish or any combination of them or solely of these that are boiled in water until the flavours are extracted, forming a broth. Many types of soups can be found, such as clear soups, thick soups, bouillons, consommés, bisques and veloutés. Commercial soup became popular in the 1900s with the invention of canning. Soups can sometimes contain particles which can complicate the processing of the product. In soup manufacturing, the product is held at a certain temperature for a certain amount of time to kill off pathogens. Then, the manufacturer cools the soup back down. Nutrients may be lost during the high-temperature phase, and colour may be affected negatively. For these reasons, new processing technologies are needed to better retain colour, nutrients and other characteristics in soup (Gelski, 2014).

Soup production is a rapidly developing and innovative category worldwide, with an already wide and growing range of value-added soup products with homemade taste and ready-to-eat convenience. Large markets are converting from homemade consumption and these place a high demand on production and product innovation to develop products that match each market's unique preference in terms of flavours and level of convenience, from soup components to ready-to-eat soups.

Optimal solutions enable the food industry to produce a wide range of products: from smooth to particulate soups; from clear soups such as broths and bouillons to thick soups based on milk, cream or vegetable purees and dehydrated soups. This permits flexibility in packaging and distribution options, from dehydrated to bottled and cartoned soups and from chilled to ambient distribution. Commercial product examples to be produced according to the abovementioned considerations are as follows: clear soups or cream soups based in fish, meat or vegetables alone or mixes of them and others such as *gazpacho* soup or goulash soup. Some traditional vegetable soups are mushroom, tomato, pea or asparagus soup and meat based such as oxtail soup.

Growing trends in soup production are shown in the following list:

- Natural products without preservatives or additives but safe and with fresh appearance.
- High-quality soups that bring culinary tradition to the dinner table, matching local taste preferences and cooking and consumption habits.
- Greater variety of convenient soup components that save time cooking at home.
- Growing sophistication of markets with regards to demand of value-added products such as functional soups with health-promoted characteristics in addition to safe, nutritious, natural and home-made appearance.

A food product can only be considered functional if together with the basic nutritional impact it has beneficial effects on one or more functions of the human organism, thus either improving the general and physical conditions or/and decreasing the risk of the evolution of diseases. The amount of intake and form of the functional food should be as it is normally expected for dietary purposes (Siró *et al.,* 2008).

From a product point of view, the functional property can be included in different ways: (i) fortified foods by the addiction of vitamins/and or minerals (vitamin E, B, C, folic acid, zinc, iron and calcium), micronutrients and phytochemicals (omega-3 fatty acid, fibre, phytosterols, isoflavones, lycopene, phenolic compounds and others); (ii) enriched food by the addition of new nutrients or components not normally found in a particular food; (iii) altered foods in which a deleterious component has been removed, reduced or replaced with another substance with beneficial effects and (iv) enhanced foods in which one of the components has been naturally enhanced through different ways such as special growing conditions, new feed composition, genetic manipulation, new processing technologies and others (Siró *et al.*, 2008).

Functional foods have been developed in virtually all food categories, mainly in dairy products, soft drinks, bakery, baby-food, etc. Soups are a good support for the production of functional foods and also can be easily consumed by populations with special physiological conditions (infants, children, pregnant women, older adults, athletes and allergy suffers) or specific health requirements (diabetes, cardiovascular or cancer patients) that could improve their health by the consumption of functional foods.

Nowadays, it is possible to find different commercial functional soups such as a soup with combination of vegetables and fruits rich in vitamins, proteins and antioxidants provided by NutraHelix (2012); soups produced by Baxters and commercialised in Tesco and Walmart that includes different combinations of meat, vegetables and cheese (Baxter, 2015); Vital Fini-Mini Podravka soups with different functional ingredients added such as natural fibres inulin and fructooligosaccharide, having prebiotic and beneficial effect on intestinal flora, and vitamins B, C and E (Podravka, 2015); Hain Kitchen Prescription Soups with echinacea or St John's wort added (Hain Pure Foods, 2015) and others.

In addition to responding to consumer demands for high-quality food with fresh appearance, easy preparation, appropriate shelf life, low caloric content and high nutritional quality, alternative or novel food processing technologies (thermal and non-thermal) should encourage the preservation of health-promoted characteristics of functional food mainly when these products have to be processed and storage. Some of these new processing technologies that are the technological response to the undesirable changes induced by the traditional thermal processing of food are HPP, PEF, ultrasonic waves (U), HIPL, IR, UV, MW, RF, OH and others (Da-Wen, 2005).

9.2 Non-Thermal Technologies for Food Processing

9.2.1 High-Pressure Processing

HPP is a non-thermal processing technology that has shown great potentials in the food industry and is gaining popularity in the food industry because of its ability to inactivate vegetative cells of microorganisms and some enzymes near room temperature, resulting in the almost complete retention of nutritional and sensory characteristics of fresh food without sacrificing shelf life (Heinz & Buckow, 2010). Early work focused on understanding the biological effects of high pressure on food microorganisms and how this could open opportunities for food preservation. Since then, a significant research effort has enabled the development of this tool and its transfer to the industrial scale (Georget *et al.*, 2015). In addition, HPP may cause inactivation or activation of enzymes depending on the applied pressure and the type of enzyme (Terefe *et al.*, 2014) and provides a gentle pasteurisation method in comparison to conventional thermal processing with minimal effects on sensory and nutritional profiles (Sánchez-Moreno *et al.*, 2009; Heinz & Buckow, 2010).

HPP has been used for decades with success in chemical, ceramic, carbon allotropy, steel/alloy, composite materials and plastic industries. HPP of foods uses similar concepts to cold isostatic pressing of metals and ceramics, except that it demands much higher pressure, faster cycling, higher capacity and hygienic design with effective cleaning routines (Mertens & Deplace, 1993; Zimmerman & Bergman, 1993). The process is isostatic, that is, the pressure is transmitted uniformly and instantly, and adiabatic, which means that no matter the food shape or size, there is little variation in temperature with increasing pressure (the temperature increases approximately 3 °C per 100 MPa, depending on the composition of the food) (Otero *et al.*, 2010). This prevents the food from being deformed or heated, which would modify its organoleptic properties.

The first scientific investigation on the application of HPP in food preservation was by Hite in the past decade of the nineteenth century, who reported the application of high pressure for milk preservation (Hite, 1899) and the preservation of fruits and vegetables (Hite *et al.*, 1914). However, it was not until the 1980s that a widespread interest in the technology was renewed leading to several studies on the various aspects of the process (Heinz & Buckow, 2010). The observation that high pressure inactivates microorganisms and reduces the activity of many quality-related food enzymes, while retaining other quality attributes, led to the introduction of a number of high-pressure-processed foods into the market by Japanese and American food companies (Mermelstein, 1997; Hendrickx *et al.*, 1998). The first high-pressure-processed foods introduced into the market were jams, jellies and sauces in Japan in 1990 by the Japanese company Meidi-ya (Thakur & Nelson, 1998). This was followed by fruit preparations, fruit juices, rice cakes and raw squid in Japan; apple and orange juices in France and Portugal; guacamole and oysters in the United States and processed meat products in the United States and Spain (Hugas *et al.*, 2002).

Therefore, among beverages, HPP has been profoundly studied in milk, dairy products and smoothies (Pina-Pérez *et al.*, 2009; Keenan *et al.*, 2012; Devi *et al.*, 2013; Bulut, 2014; Dhakal *et al.*, 2014; Karlović *et al.*, 2014; Martínez-Monteagudo *et al.*, 2014; Nunes de Morais *et al.*, 2014; Sousa *et al.*, 2014; Calzada *et al.*, 2015; Contador *et al.*, 2015; Devi *et al.*, 2015; Li *et al.*, 2015; Rodríguez-Alcalá *et al.*, 2015; Scolari *et al.*, 2015) and milk-based fruit beverages (Sampedro *et al.*, 2009; Cilla *et al.*, 2011; Zulueta *et al.*, 2013; Hernández-Carrión *et al.*, 2014; Rodríguez-Roque *et al.*, 2015).

One of the major advantages of using HPP in fruit and vegetable purees, juices and soups is that the pressure-treated products mimic the sensory properties of fresh produce (Oey *et al.*, 2008; Sánchez-Moreno *et al.*, 2009). It fulfils the consumer demand for 'healthy', nutritious and 'natural' food products besides ensuring their high quality, greater safety and increased shelf life. Consequently, most of the research studies have been done in fruit and vegetable juices and mixed juice beverage (Sánchez-Moreno *et al.*, 2003a, b, 2005a, b; Plaza *et al.*, 2006b; Valdramidis *et al.*, 2009; Cilla *et al.*, 2012; Espina *et al.*, 2013; Abid *et al.*, 2014; Meng-Meng *et al.*, 2014; Yu *et al.*, 2014; Chen *et al.*, 2015; Jayachandran *et al.*, 2015), fruit and vegetable purees (Sánchez-Moreno *et al.*, 2004a, b, 2006b; Oey *et al.*, 2008; Keenan *et al.*, 2011; Fernández-Sestelo *et al.*, 2013; Huang *et al.*, 2014; Palmers *et al.*, 2014; Chakraborty *et al.*, 2015; Marszalek *et al.*, 2015) and nectars (Guerrero-Beltrán *et al.*, 2011; Huang *et al.*, 2013; Wang *et al.*, 2013; Liu *et al.*, 2014).

In addition, the application of HPP for the extraction of bioactive ingredients from plant-derived products is an emergent technique, which has been known as ultra-high-pressure extraction or high hydrostatic pressure extraction (Xi, 2013). However, studies about the effect of HPP in soups are less abundant. Among them, Plaza *et al.* (2006a) studied the carotenoid content and radical scavenging activity in a Mediterranean vegetable soup in Spain known as *gazpacho* treated by high pressure combined with temperature (60 °C) during refrigerated storage. In this study, the *gazpacho* soup showed no significant changes in the total carotenoid content after high-pressure treatments (150 MPa/60 °C/15 min and 350 MPa/60 °C/15 min), although after long-term storage (40 days at 4 °C), a significant decrease of 40% and 46%, respectively, was observed. Regarding the antioxidant capacity, it was not significantly affected after high-pressure treatments. In addition, Muñoz *et al.* (2007) determined the effects of high-pressure treatments and mild temperatures on endogenous microflora and *Escherichia coli* artificially inoculated into the vegetable soup *gazpacho*. The optimum process parameters (pressure and temperature) for a 6-log reduction of *E. coli* were obtained at 269.8 MPa and 59.9 °C.

Nonthermal processing, such as HPP, of fruit and vegetable purees, juices and soups has been revealed as a useful tool to extend their shelf life and quality as well as to preserve their nutritional and functional characteristics. In this sense, the studies by Sánchez-Moreno and coworkers assessing the impact of HPP on the health benefits of fruits, vegetables and derived products are noteworthy (Sánchez-Moreno *et al.*, 2009). There are some specific studies about the effect of consuming *gazpacho* soup, including *gazpacho* treated by high pressure, on vitamin C bioavailability and biomarkers of oxidative stress and inflammation in a healthy human population (Sánchez-Moreno *et al.*, 2004a,b, 2005a, 2006a). *Gazpacho* is a typical Mediterranean dish that can be defined as a ready-to-use vegetable soup containing approximately 80% crude vegetables (50% tomato, 15% cucumber and 10% pepper), 2–10% olive oil and other components (3% onion, 0.8% garlic, 2% wine vinegar and 0.8% sea salt). Daily consumption of 500 ml of *gazpacho* treated by HPP (400 MPa/40 °C/1 min) for 2 weeks was associated with an increased plasma concentration of vitamin C and decreased biomarkers of inflammation in both women and men. In addition, there was an inverse correlation between concentrations of plasma vitamin C and concentrations of 8-*epi*PGF2α, uric acid, PGE_2 and MCP-1, suggesting that vitamin C may play a critical role in reducing the formation of compounds produced by random oxidation of phospholipids by oxygen radicals involved in the development of oxidative processes and inflammation (Sánchez-Moreno *et al.*, 2004a,b).

In line with these studies, Colle *et al.* (2011) prepared a model of tomato soup to evaluate the impact of pilot-scale aseptic processing, including heat treatment and high-pressure homogenisation (HPH), on some selected quality parameters (vitamin C content, lycopene isomer content and lycopene bioaccessibility). A tomato soup without oil as well as a tomato soup containing 5% olive oil was evaluated. With regard to the vitamin C content, thermal processing had a negative effect, especially caused by the breakdown of dehydroascorbic acid. Moreover, HPH caused additional losses of ascorbic acid. Lycopene degradation during thermal processing was limited but was slightly increased by HPH. Furthermore, lycopene isomerisation was restricted, except in the tomato soups containing olive oil. The presence of lipids improved the lycopene bioaccessibility. For all tomato soups, HPH decreased the lycopene bioaccessibility. Increasing the pressure level during homogenisation caused a further decrease in lycopene bioaccessibility. HPH affected the viscosity of the tomato products, whereby increasing the homogenisation pressure resulted in an increased viscosity. Therefore, an inverse relationship between viscosity and lycopene bioaccessibility was observed. Moreover, it is clear that one should endeavour a balance between the viscosity of the tomato product and the lycopene bioaccessibility.

Martínez-Tomás *et al.* (2012) investigated the effect of daily intakes of two differently processed fruit and vegetable soups on β-carotene and lycopene bioavailability, oxidative stress and cardiovascular risk biomarkers. An optimised soup produced using heat treatments and HPH for high nutrient retention and a traditionally produced reference soup were tested. Carrot, tomato and broccoli were used for the optimised and the reference soup (20% of each), but 5% olive oil was added to the optimised soup and 2.5% olive oil to the reference soup. In the reference soup, broccoli, carrots and tomatoes were heated together in one vessel with water, cooked until al dente and then blended with oil. The optimised soup was produced in a 'split-stream process' by separate blanching of tomatoes and broccoli at >85 °C or cooking carrots at 95 °C, followed by pureeing and mixing before HPH at 100 bar. The authors concluded that the daily consumption of both fruit and vegetable soups for 4 weeks enhanced serum β-carotene and lycopene concentrations. The oxidative stress status of the subjects was modulated mainly through a reduction in the activity of antioxidant enzymes. The effect was higher in subjects consuming the optimised soup, suggesting that food processing used may play a large role in enhancing the preservation and bioavailability of bioactive phytochemicals.

In addition to its quality attributes, HPP has the ability to destroy most vegetative microorganisms and inactivate some enzymes. However, in the light of these great attributes and the increasing commercial availability of high-pressure-treated high-acid foods, it has not yet been successfully implemented for processing low-acid foods (i.e. foods with $pH \geq 4.6$). This is mainly due to the limitation of HPP to destroy bacterial spores. In this sense, the overall efficiency of an existing scale-up pressure-assisted thermal sterilisation unit was investigated with regards to the inactivation of *Geobacillus stearothermophilus* spores suspended in pumpkin soup. In this work, the addition of an air line to push the treated liquid food out of the existing pressure-assisted thermal sterilisation unit improved the overall quality of the treated samples, as evidenced by achieving higher log reduction of the spores (Shibeshi & Farid 2010). In addition, the effect of treatment on L-ascorbic acid was similar to that of thermal treatment alone for similar treatment temperatures and time. L-Ascorbic acid analysis indicated that reducing treatment temperature from 121 (thermal) to 115 °C (pressure thermal) could provide significant benefit in retaining nutrients in the pumpkin soup while providing the same degree of sterility (Shibeshi & Farid, 2011).

Bacteria can be injured or inactivated by HPP, depending on the pressure level, holding time and bacterial strains and species. To minimise the outgrowth of HPP-resistant cells, it can be effective to combine HPP with one or several other treatments. Examples can be heat, packaging and antimicrobial agents. In the study by Rode *et al.* (2015), fish soup was treated with high pressure (400 and 600 MPa, 2 min) in combination with different packaging regimes; vacuum, modified atmosphere packaging (MAP), soluble gas stabilisation (SGS) and a combination of the two latter ones (SGS–MAP). The fish soup was, before treatment, inoculated with *Listeria innocua* and stored at 5 °C and further analysed for a period up till 49 days. HPP gave a reduction of *L. innocua* of 3.5 and 7.3 log cfu/g immediately after exposure to 400 and 600 MPa, respectively. Both SGS and MAP showed a significant interaction with pressure. SGS in combination with pressure significantly inhibited the growth of *L. innocua* during storage. The same was observed when packaging in MAP. Bacterial plating on non-selective and selective agar revealed that over 99.9% of the surviving cells after HPP treatment in combination with SGS or SGS–MAP were sublethally injured. In this line, it would be relevant for the food industry to do research also involving HPP of modified atmosphere-packaged products.

In conclusion, it has been scientifically and commercially proven that HPP can produce microbiologically safe and stable products. Research on HPP in liquid foods, apart from milk, has mainly focused on fruits and vegetables juices, mixed juice beverage, purees, nectars and less in soups, with an emphasis on food quality and bioactive components, including strategies for the production of healthy and safe food products and ingredients. In this sense, HPP has shown encouraging potential to manipulate the extractability, allergenicity and functionality of micronutrients and bioactive compounds in foods. Therefore, the growth trend in the commercial application of HPP, including innovations in equipment design, is set to continue in the foreseeable future.

9.2.2 Pulsed Electric Fields

There are several methods of applying electric energy for food pasteurisation as an alternative to thermal pasteurisation. These methods mainly included OH, MW heating and PEFs (Da-Wen, 2005).

PEF is a non-thermal technology for food preservation that uses short pulses of high electric fields with duration of microseconds to milliseconds (1–10 µs) in the order of 20–80 kV/cm. The application of pulses of high voltage to liquid or semi-solid foods placed between two electrodes results in an electric field that produces microbial inactivation and enzymatic activity reduction while better maintaining the original colour, flavour, texture and nutritional value of the unprocessed food. Thus, PEF processing offers to the consumer safe, fresh-like, nutritious products (Martín-Belloso & Elez-Martínez, 2005, 2007a, b; Cortés et al., 2008; Sánchez-Moreno et al., 2009; Morales de la Peña et al., 2011).

The PEF treatment efficacy for microbial and enzyme inactivation is significantly affected by treatment parameters (electric field intensity, form and number of pulses applied, temperature of treatment and distribution of the treatment chamber), the intrinsic properties of bacteria or enzyme and the composition of treatment media, that is, PEF treatment depends on the chemical and physical characteristics of the food (Saldaña et al., 2014).

The results obtained studying the effect of PEF treatments on real foods demonstrated that food components such as fat, protein or carbohydrate content could affect the PEF treatment effectivity. PEF treatments are conducted at ambient, sub-ambient or slightly above ambient temperature for short time (microsecond), so that this technology keeps food below temperature normally used in traditional thermal processing. This would retain the nutritional and functional quality of food including vitamins, minerals, essential flavours and phytochemicals, while consuming less energy losses than traditional thermal processing. Until now, commercial PEF products are restricted to fluid foods that can tolerate high electric fields strength (E), have low electrical conductivity (σ) and do not contain or form bubbles. The particle size of the food is also a limitation. Thus, PEF is applicable to pumpable products with a particle diameter of less than 20 mm.

Microbial inactivation by PEF has been shown to occur due to electrical breakdown of cell membranes that produce the formation of pores caused by the buildup of electrical charges at the cell membrane that ends with membrane disruption and cellular lysis with subsequent leakage out of intracellular compounds which results in cellular death. This is known as electroporation (Mosqueda-Melgar et al., 2008; Martín-Belloso & Soliva-Fortuny, 2011; Morales de la Peña et al., 2011).

In general, PEF treatment has been shown to be an effective means of inactivating bacteria and fungi in field strengths between 20 and 40 kV/cm (energy delivery 100–600 kJ/kg) but has little effect on spores and no effect on virus (Hirneisen *et al.*, 2010). Furthermore, the inactivation of enzymes requires more intense PEF treatments than those required for bacterial inactivation (Elez-Martínez & Martín-Belloso, 2007b).

In recent years, different liquid and semi-liquid foods have been treated with laboratory and pilot plant-scale, continuous-flow PEF system to evaluate the impact of PEF treatments on microbial inactivation, enzyme activity, food physicochemical, nutritional and functional properties and food shelf life (Sánchez-Moreno *et al.*, 2005a, b, 2009; Plaza *et al.*, 2006b; Elez-Martínez & Martín-Belloso, 2007a,b; Mosqueda-Melgar *et al.*, 2008; Cilla *et al.*, 2012; Odriozola-Serrano *et al.*, 2013; Saldaña *et al.*, 2014; Sharma *et al.*, 2014; Rodríguez-Roque *et al.*, 2015).

The most of these PEF treatments assayed in liquid and semi-liquid foods have been done in milk (Sharma *et al.*, 2014), yogurt (Gomes da Cruz *et al.*, 2010), orange juices (Sánchez-Moreno *et al.*, 2005b; Cortés *et al.*, 2008; Plaza *et al.*, 2006a,b, 2011; Buckow *et al.*, 2013), fruit juices (Aguiló-Aguayo *et al.*, 2008, 2010; Odriozola-Serrano *et al.*, 2013), vegetable juices (Odriozola-Serrano *et al.*, 2009, 2013), beverages (Cilla *et al.*, 2012; Rodríguez-Roque *et al.*, 2015), liquid eggs (Espina *et al.*, 2014), beer (Milani *et al.*, 2015) and wine (Puértolas *et al.*, 2010).

However, there are very few published studies on the effect of PEF treatment in soups (Table 9.1). In fact, it has only been found studies about the effect of PEF in pea soup (Vega-Mercado *et al.*, 1996) and *gazpacho*, which is a traditional Mediterranean vegetable soup (Sánchez-Moreno *et al.*, 2005a, 2009).

Pea soup inoculated with *E. coli* and *Bacillus subtilis,* alone or mixed together, was submitted at different PEF treatment conditions. The inactivation of *E. coli* and *B. subtilis* suspended alone in pea soup increased with increases in the intensity of the electric field, number of pulses and pulsing rate. PEF treatment of 30 pulses at 30–33 kV/cm with a pulsed rate of 6.7 Hz reached significantly inactivation levels of 5 $\log_{10}$ (5*D*) and 6.5*D* for *B. subtilis* and *E. coli*, respectively. These results of microbial inactivation using PEF demonstrated the feasibility of the technology for preservation of liquid and semi-liquid foods containing a complex matrix with suspended particles and gelatinised starch (Vega-Mercado *et al.*, 1996).

A Spanish vegetable semi-liquid soup called *gazpacho* and composed of a mixture of tomato, cucumber, green pepper, onion, garlic, salt, virgin olive oil, wine vinegar and sugar was submitted at different PEF-processing parameters such as electric field strength (15–35 kV/cm), treatment time (100–1000 µs), pulse polarity (monopolar vs bipolar), pulse width (2–10 µs) and pulse frequency (50–450 Hz) with the aim to study the effects on vitamin C content and antioxidant activity and compared them to those in heat pasteurisation (90 °C, 1 min) (Elez-Martínez & Martín-Belloso, 2007a).

The more intense PEF treatment to produce microbial inactivation at 35 kV/cm for 1000 µs (bipolar pulse of 4 µs at 200 Hz) reduced the vitamin C content of the vegetable soup gazpacho by 13%, while a decrease of 21% in vitamin C was observed after a pasteurisation treatment (90 °C, 1 min). In general, PEF-treated gazpacho always showed higher vitamin C retention (84–97%) than heat-treated gazpacho. In addition, PEF treatments did not affect the antioxidant activity of freshly prepared soup (Elez-Martínez & Martín-Belloso, 2007a).

PEF treatments of liquid or semi-liquid foods have been revealed as a useful tool to extend their shelf life and quality as well as to preserve their nutritional and functional

Table 9.1 Non-thermal and novel thermal technologies for food processing: soup

Technologies	Product	Source
Non-thermal technologies		
High pressure processing (HPP)	*Gazpacho* (vegetable soup)	Sánchez-Moreno *et al.* (2004a,2004b), Plaza *et al.* (2006a,2006b)
	Pumpkin soup	
	Tomato soup	Shibeshi and Farid (2010, 2011)
	Carrot, tomato and broccoli soup	Colle *et al.* (2011)
	Fish soup	Martínez-Tomás *et al.* (2012)
		Rode *et al.* (2015)
Pulsed electric fields (PEF)	Pea soup	Vega-Mercado *et al.* (1996)
	Gazpacho (vegetable soup)	Sánchez-Moreno *et al.* (2005a, 2009)
Novel thermal technologies		
Ohmic heating (OH)	Multiphase foods	Chen *et al.* (2010)
	Particulate liquid foods	Choi *et al.* (2011)
	Tomato soup	Somavat *et al.* (2012)
	Minestrone soup	Bertolini and Romagnoli (2012)
Microwave heating (MW)	Instant vegetable soup	Wang *et al.* (2009, 2010)
	Vegetable soup enriched with broccoli by-products	Alvarez-Jubete *et al.* (2014)
	Mushroom soup	Li *et al.* (2011)
Radiofrequency heating (RF)	Particulate liquid foods	Zhong *et al.* (2004)

characteristics of these products. In this regard, the research studies carried out on bioavailability and antioxidant properties of PEF-processed vegetable products indicate that the consumption of two glasses (500 ml/day) of PEF-treated gazpacho soup (35 kV/cm for 750 µs in a bipolar mode with pulses of 4 µs at 800 Hz) containing approximately 73 mg of vitamin C caused a better vitamin C bioavailability and antioxidant properties and improve their healthy potential by an increase in plasma vitamin C and a decrease in the oxidative stress and inflammation biomarkers such as the prostaglandin 8-*epi*PGF2α and uric acid in healthy humans (Sánchez-Moreno *et al.*, 2005a, 2009).

Thus, PEF treatment at 35 kV/cm for 750 µs (bipolar pulse of 4 µs at 800 Hz) at room temperature could extend the shelf life of soups such as fresh vegetable *gazpacho* because achieve good microbial and enzymatic inactivation level and their nutritional value and functional properties are better retained than by thermal pasteurisation (Sánchez-Moreno *et al.*, 2005a, 2009).

The commercial success of PEF products is based on selecting appropriate treatment parameters for each food. Actually, the required commercial field strength for microbial inactivation in liquid food as milk and fruit juices is 10–20 kV/cm and the energy delivery is 100–600 kJ/kg. From the point of view of commercialisation, industries could provide to the consumers new safer and cheaper healthy liquid or semi-liquid foods by combining PEF with other physical (low thermal or no-thermal) or chemical (natural antimicrobials) food processing technologies that improve the microbial lethality of PEF technology and reduce the electrical energy costs. Pre-heated soup at 50 °C previous to PEF treatment could enhances PEF process efficiency, reduces treatment time and specific energy requirements to achieve the same microbial inactivation Furthermore, PEF inactivation of microorganism is enhanced by the addition of antimicrobials such as nisin, lysozyme, essential oils (EO) or their active compounds (mandarin EO, lemon EO, citral, limonene, carvacrol, etc.) (Buckow *et al.*, 2013; Espina *et al.*, 2014; Saldaña *et al.*, 2014).

At present, PEF-treated fruit juices and smoothies are on market in Germany, the Netherlands and the United Kingdom using PEF processing equipment with a capacity of 1500–2000 l/h and 5000–8000 l/h. The PEF treatment applied to liquid foods pre-heated at 40 °C required a strength field at 20 kV/cm with an energy input of 120 kJ/kg resulting a final temperature of 60 °C (using intermediate cooling). This treatment is available to achieve 5 $\log_{10}$ (5D) reduction of total bacterial count and also 5D reduction of a pathogen inoculated (*E. coli*) in orange juice according to the US FDA requirements for the implementation of a new non-thermal food processing technology. At this commercial scale, PEF treatments total costs are in a range of 0.02–0.03 US Dollar per litre (Buckow *et al.*, 2013).

9.3 Novel Thermal Technologies for Food Processing

Research in novel thermal technologies for applications such as cooking, pasteurisation/sterilisation, defrosting, thawing and drying have focused on important processing issues such as to achieve high heating rates, significant reduction of processing time, heating uniformity, energy efficiency, microbial inactivation mechanisms and changes in sensorial and nutritional characteristics of foods. These technologies are expected to be less intense than those of conventional thermal processing in terms of nutritional and sensorial changes. Increasing interest has been shown in electroheating in recent years in both scientific literature and commercial heating applications. Electroheating can be subdivided into either direct electroheating where electrical current is applied directly to the food (OH) or indirect electroheating (MW and RF heating) where the electrical energy is firstly converted into electromagnetic radiation which subsequently generates heat within a food. Some of those electroheating processes (OH and RF) are only used industrially but MW can be used on an industrial level as well as domestically (Marra *et al.*, 2009; Chandrasekaran *et al.*, 2013).

9.3.1 Ohmic Heating

OH is one of the thermal processing alternatives that could provide value-added, shelf-stable foods as well as in other applications such as blanching, fermentation, dehydrated, extraction, evaporation sterilisation, pasteurisation and heating food for military and long-duration space missions (Goullieux & Pain, 2005; Knirsch *et al.*, 2010).

OH, also known as Joule heating or electrical resistance heating, is defined as a process where the electric current (alternated) of low frequency (traditionally 50 or 60 Hz) passes through foods and generates thermal energy. Electrical energy dissipated within the food as heat results in remarkable uniform temperature distribution and high-energy conversion efficiency (>90%) (Jun & Sastry, 2005). OH, as well as PEF treatment, is distinguished from other electrical heating methods either by the presence of the electrodes contacting the food (in MW and RF heating electrodes are absent). The critical parameters that influence the rate of OH are the electrical field strengths (E) and electrical conductivities (σ) of the materials which are dependent on temperature, structural and chemical composition of the food components such as proteins and fat (Goullieux & Pain, 2005; Knirsch et al., 2010).

Compared with conventional thermal processing, OH has the advantage that it heats food uniformly, shortening the heating time, saving energy and production cost and improving the quality of the processed product, especially when processing liquid foods containing particulates. In general, a liquid food processed by OH reaches 140 °C in 10 s, whereas a particulate liquid food needs 90 s. In general, the heating rates in OH are significantly higher than in traditional thermal processing (Chen et al., 2010).

OH permits the process of large particulate foods (up to 25 mm) that would be difficult to process using conventional heat exchangers. To achieve a uniform OH process in particulate foods, it is desirable that electrical conductivities of fluid and solid particles are equal. However, significant differences in electrical conductivities between the food components (solid particle and fluid) have been reported. In fact, electric conductivity (σ) is a function of the structure of the material and is often changed by heating. In addition, there are critical σ values, below 0.01 S/m and above 10 S/m, where OH is not applicable.

The main innovation of OH is that it allows the heating of solid and liquid phases at the same rate thus avoiding the over-processing of the liquid phase. OH allows heating the centre of the particles faster than the surrounding liquid by appropriately formulating the ionic contents of the fluid and particulate phase to ensure the appropriate levels of electrical conductivity (σ). In general, non-uniform heating can be reduced by increasing electrolytic contents in food particles by additional pretreatment such as soaking or blanching particles in salt solution. A combination of MW and OH is proposed to improve the uniformity of thermal processing of particulate foods such as soups (Choi et al., 2011).

OH has been employed for microbial and enzyme inactivation of fruit and fruit juices (Evrendilek et al., 2012). In addition, ohmic blanching can be used as an alternative method for vegetable purees (Icier & Ilicali, 2005a). This method results in lower enzyme inactivation times and high retention of colour attributes (Yildiz et al., 2010). Thus, ohmic blanching of pea puree by applying a treatment at 30 V/cm is able to inactivate peroxidase enzyme in less time (201 s) than water blanching (100 °C, 300 s) (Icier et al., 2006). In fact, ohmic blanching at 50 V/cm gives the shortest critical inactivation time of 54 s with the best colour quality. Consequently, the time required to reach 100 °C decreased as the voltage gradient increased (Icier et al., 2006). On the other hand, OH causes browning of preblanched spinach puree more than conventional water heating for the same temperature range (60–90 °C) and the same holding times (Yildiz et al., 2010).

Fruit purees can be used for baby food production and OH could be potentially used to assure their microbiological safety. Therefore, the electrical conductivity changes of fruit purees during OH are the main processing parameters to control. Thus, the electrical conductivity in strawberry-based products decreases when the solids and sugar content are increased (Castro et al., 2004). Similarly, the electrical conductivity values of fruit juice concentrates (apple, sour cherry and orange) showed a decreasing trend when the concentration of soluble solids was increased. However, acidity of fruit products increases the electrical conductivity values (Evrendilek *et al.*, 2012).

However, there are very few published studies on the effect of OH treatment in soups (Table 9.1). Nowadays, OH (60–30 kW) combined with aseptic packaging is an alternative to the traditional thermal treatments for processing viscous and liquid foods containing solid particulates, such as soups, stews, fruit slices in syrups and sauces, fruit purees, jams, fruit juices, baby foods (Icier & Ilicali, 2005a, 2005b; Yildiz *et al.*, 2010; Evrendilek *et al.*, 2012) and milk and milk products (Pereira *et al.*, 2008). This is because it reduces the losses of vitamins and flavours/aromas, while keeping or improving the microbiological safety of the treated products. Moreover, OH is useful for heat-sensitive liquids such as liquid egg that can be ohmically heated in a fraction of a second without coagulation (Knirsch *et al.*, 2010; Chen *et al.*, 2010; Bertolini & Romagnoli, 2012).

The principal microbial inactivation mechanism in OH is the heat as in thermal processing. However, additional non-thermal electroporation mechanism has been reported at low frequency (50–60 Hz), which allows cell walls to build up charges and form pores like in PEF treatments. This does not occur in the MW and RF heating where the electric field is essentially reversed before sufficient charge builds up in the cell walls. The shelf life of OH-processed foods is comparable to that of canned and sterile aseptically processed products but with better nutritional and sensorial characteristics.

The OH process for liquid foods containing large particulates such as soups has been studied and reported by several researchers, covering different aspects from heating behaviour of carrier fluid and particulates (Sarang *et al.*, 2007; Salengke & Sastry, 2007; Choi *et al.*, 2011) to modelling of OH process (Somavat *et al.*, 2012).

Therefore, OH has a great potential as an alternative to thermal processing specially for particulate liquid foods as soups. Thus, tomato soup packaged in a multi-layered pouch has been sterilised by OH showing that this technology useful for long-duration space missions. OH using pulsed alternating current at 80 V with 10 kHz and electrodes inside the package (to test the time required to heat at a constant voltage) has achieved 5-$\log_{10}$ (5D) reduction of inoculated spores of *Geobacillus sterorother-mophilus* in 6 min at 121 °C or 1 min at 130 °C (Somavat *et al.*, 2012).

Minestrone soup, a typical Italian vegetable soup, was used to calculate the production and post-production cost of an OH-treated soup in comparison with a frozen minestrone soup available in the shops. A soup treated by OH combined with aseptic packaging had a production cost of 1.42,306 €/kg, which is 4% higher than that of a frozen soup. However, taking into account the total cost (production and post-production), the cost of both soups, OH and frozen, are similar reaching a value of 2.154,979 €/kg. Therefore, both soups can be sold at the same price. The total cost of both soups is similar due to the fact that frozen soups have greater post-production cost because they need a more expensive cold chain. Furthermore, OH product can be transported and stored at room temperature. Taking into account the superior manageability and nutritional and sensorial quality of OH-treated soups, OH products are competitive compared with traditional thermal technologies (Bertolini & Romagnoli, 2012).

Some major equipment suppliers (APV Baker, Ltd, Crawley, UK; Raztec Corp., Sunnyvale, CA, USA) provide commercial size ohmic heaters for the food industry including Emmepiemme SRL (Italy) and Capenhurst (UK). There are several processing plants currently producing sliced, diced and whole fruit within sauces in various countries, including Italy, Greece, France, Mexico and Japan. In the United States, OH has been used to produce low-acid ready-to-eat meals as well as pasteurised liquid egg (Ramaswamy *et al.*, 2005).

9.3.2 Microwave and Radiofrequency Heating

MW and RF heating refer to the use of electromagnetic waves of certain frequencies to generate heat in a food. MW and RF heating have been widely studied for their applications in the field of food processing such us cooking, pasteurisation/sterilisation, defrosting, thawing and drying (Marra *et al.*, 2009; Chandrasekaran *et al.*, 2013; FDA, 2015).

MWs are electromagnetic waves whose frequency varies between 300 MHz and 300 GHz. MW food processing uses the frequencies of 2450 and 915 MHz. Domestic MW ovens operate generally at a frequency of 2450 MHz, while industrial MW systems operate at frequencies of 915 and 2450 MHz (Table 9.2). Hundreds of 2450 and 915 MHz systems between 10 and 200 kW heating capacities are used in the food industry for precooking bacons, tempering deep frozen meats when making meat patties and precooking many other foods products. Commercial systems performing MW pasteurisation and/or sterilisation of foods are currently available in Europe; however, the use of MWs in the United States to produce shelf-stable, low-acid (pH>4.6) foods requires FDA acceptance (Chandrasekaran *et al.*, 2013).

RF occupies the electromagnetic spectrum between 1 and 300 MHz, and the main frequencies used for industrial heating lie in the range 10–50 MHz. Within the later range, only selected frequencies (13.56 MHz ± 6.68 kHz, 27.12 MHz ± 160 kHz and 40.68 MHz ± 20 kHz) are permitted for industrial, scientific and medical applications (Table 9.2) (Marra *et al.*, 2009).

MW ovens operating at 2450 MHz are common appliances in the households of the major countries around the world. Hundreds of 2450 and 915 MHz systems between 10 and 200 kW heating capacities are used in the food industry for precooking bacons, tempering deep frozen meats when making meat patties and precooking many other foods products. Moreover, MW energy has been widely used for drying and blanching vegetable products without quality degradation as in traditional thermal process (Vadivambal & Jayas, 2007).

Table 9.2 Frequencies assigned for industrial, scientific and medical use (FDA, 2015)

Technology	Frequency
Radio	13.56 MHz ± 6.68 kHz
	27.12 MHz ± 160 kHz
	40.68 MHz ± 20.00 kHz
Microwaves	915 ± 13 MHz
	2450 ± 50 MHz
	5800 ± 75 MHz
	24,125 ± 125 MHz

Commercial systems performing MW pasteurisation and/or sterilisation of foods are currently available in Europe; however, the use of MWs in the United States to produce shelf-stable low-acid (pH>4.6) foods requires FDA acceptance (FDA, 2015). Thus, the FDA acceptance has been recently granted for a sweet potato puree product sterilised using continuous flow MW processing and aseptic packaging. This first industrial implementation of continuous-flow MW sterilisation for low-acid products has been implemented by Yamco in Snow Hill, NC (Coronel *et al.*, 2005; Parrot, 2010).

MW and RF heating of foods mainly occur due to dipolar and ionic mechanisms. Water in the food is often the primary component responsible for dielectric heating. Due to their dipolar nature, water molecules try to follow the electric fields associated with electromagnetic radiations as it oscillates at the very high frequencies that produce heat. The second mechanism of MW and RF heating is through the oscillatory migration of ions in the food that generates heat under the influence of oscillating electric field (Marra *et al.*, 2009; Chandrasekaran *et al.*, 2013; FDA, 2015).

The principal microbial inactivation mechanism in MW and RF is the heat. Time–temperature history of the coldest point determines the microbiological safety of the process as occurred in the traditional thermal processes. MW and RF heating for pasteurisation and sterilisation are preferred to the traditional heating because they are very rapid coming up the desired temperature mainly in solid and semi-solid foods that depend on the slow thermal diffusion process in conventional heating (FDA, 2015). Although it is demonstrated that the heat produced by MW and RF can raise the temperature needed to inactivate foodborne pathogens and enzymes, there are some factors that is necessary to be taking into account, mainly in domestic MW ovens, such as the non-uniform heating distribution and penetration depth (up to 25 mm) (Vadivambal & Jayas, 2010).

The main factors that affect MW and RF heating are the dielectric properties of the food material that represent the ability of food to convert MW or RF energy into heat. The dielectric properties depend on the composition (or formulation), water and salt content, temperature and physical structure of the food material such as the particle size. In general, higher water content and salt improves the dielectric properties of the food but they decrease with fat and oil. In addition, the rate of MW absorption increases when the temperature of the food increases. In the case of MW heating of a multicomponent food like a soup, the temperature distribution could be less uniform than that in a liquid or solid food (Fakhouri & Ramaswamy, 1993).

In recent years, it have been published a great number of studies about the food processing application of MW and RF mainly in the field of food cooking, drying and pasteurisation/sterilisation of food materials. A great number of liquid, semi-liquid and solid foods could be treated by MW and RF. In addition, packaged food products can be treated by MW and RF (Marra *et al.*, 2009; Salazar-González *et al.*, 2012; Chandrasekaran *et al.*, 2013; FDA, 2015).

MW heating for microbial and enzyme inactivation has been largely applied to liquid and semi-liquid foods (apple juice, apple cider, coconut water, grapewater juice, sweet potato puree, kiwi puree, etc.). Furthermore, MW heating has been applied for milk sterilisation (Coronel *et al.*, 2005; Salazar-González *et al.*, 2012; Benlloch-Tinoco *et al.*, 2014).

RF heating of liquid foods such as fruit juices (orange juice, apple juice, apple cider, etc.) have been demonstrated the inactivation of different pathogens and enzymes without vitamin C losses (Marra *et al.*, 2009).

Compared with conventional thermal processing, MW and RF treatments have the advantage that heat food uniformly, shortening the heating time, saving energy and production cost and improving the quality of the processed, especially when processing liquid foods containing particulates. In particulate liquid foods such as soups, MW and RF treatments allow the heating of solid and liquid phases at the same rate avoiding the over-processing of the liquid phase that can also produce the overheating in the outer regions of the solid products. However, there are very few published studies on the effect of MW and RF treatments in soups (Table 9.1).

Some of these studies deal about MW-assisted freeze-dried technology to obtain high-quality instant vegetable soups (Wang *et al.*, 2009, 2010). MW field is applied to supply the heat of sublimation required for freeze drying. Freeze drying combined with MW offers advantages such as reduced processing time and better product quality (Zhang *et al.*, 2006). Thus, a vegetable soup mix contained different ingredients such as cabbage, carrot, tomato, spinach, mushroom, water, salt, sugar and monosodium glutamate in certain proportions. After cooling, the soup was kept in a refrigerator until it reaches a temperature of −30 °C. The optimal drying of vegetable soup with a thickness of 15–20 mm and 450 g can be attained at a temperature of 50–60 °C and at an MW power of 450–675 W. As it was found that sodium chloride and sucrose content had a significant effect on the drying rate, the optimal vegetable soup ingredients required for obtaining better drying characteristics during MW freeze drying was 3.2–5.3 g of sodium chloride per 100 g of water, 2–6.8 g of sucrose per 100 g of water and sodium glutamate of less than 4.5 g per 100 g of water. Under these conditions, the drying time required was less than 6.5 h and the product had a sensory score above 8.0. The combination of food ingredients reduced 40% the time of MW freeze-dried process (Wang *et al.*, 2010).

MW heating of functional soups can be used as a cooking procedure to preserve the bioactive compounds concentration in the final product. The addition of freeze-dried broccoli by-products (florets and stalks) to an instant soup formulation is used to obtain a functional soup rich in bioactive compounds such as isothiocyanates and sulforophane, in particular. The preparation of the soup by adding water and heating using MW energy may enable the conversion of glucosinolates into bioactive isothiocyanates. Thus, MW heating produces the *in situ* activation by the consumer of bioactive compounds in the soup just before consuming, avoiding losses during storage. MW treatments applied in a domestic oven at 400 W for 140 s and 240 W for 210 s produced a temperature increase in the range 63–65 °C that is adequate for consumption and the same time are the optimal conditions for the hydrolysis of glucosinolates by the action of the enzyme myrosinase to produce the bioactive isothiocyanates (Alvarez-Jubete *et al.*, 2014).

MW heating of mushroom soup preserve better the level of free amino acids (61.25 mg/g dw) than traditional heating (56.95 mg/g dw), which might be attributed to the short cooking time of MW cooking. However, mushroom soup treated by mircrowave heating contained less aroma-active compounds and Maillard reaction flavour compounds comparing with traditional heating treatments. This could be explained by short cooking time and different heat penetration of MW heating (Li *et al.*, 2011).

Regarding RF heating, the treatment of carrot and potatoes cubes using a 1% of carboxymethylcellulose solution as carrier in a continuous flow RF unit at 30 kW, 40.68 MHz, has been used to propose RF heating as a potential alternative to convential heating for liquids containing particulates such as soups (Zhong *et al.*, 2004).

In conclusion, OH, MW heating and RF heating are alternatives to the traditional thermal treatments for processing liquid foods containing solid particulates, such as soups. These electroheating technologies have the advantage that heat food uniformly, shortening the heating time, saving energy and production cost and improving sensorial and nutritional quality of the processed product. Meanwhile, more research studies are required to produce electroheating-treated soups at commercial level.

Acknowledgments

This chapter was supported in part by the Project AGL2010-15910 (Spanish Ministry of Science and Innovation) and the Project AGL2013-46326-R (Spanish Ministry of Economy and Competitiveness).

References

Abid, M., Jabbar, S., Hu, B., Hashim, M.M., Wu, T., Wu, Z., Khan, M.A. & Zeng, X. (2014) Synergistic impact of sonication and high hydrostatic pressure on microbial and enzymatic inactivation of apple juice. *LWT – Food Science and Technology* **59**, 70–76.

Aguiló-Aguayo, I., Sobrino-López, A., Soliva-Fortuny, R. & Martín-Belloso, O. (2008) Influence of high-intensity pulsed electric field processing on lipoxygenase and b-glucosidase activities in strawberry juice. *Innovative Food Science and Emerging Technologies* **9**, 455–462.

Aguiló-Aguayo, I., Soliva-Fortuny, R. & Martín-Belloso, O. (2010) High-intensity pulsed electric fields processing parameters affecting polyphenoloxidase activity of strawberry juice. *Journal of Food Science* **79**, C641–C646.

Alvarez-Jubete, L., Valverde, J., Kehoe, K., Reilly, K., Rai, D.K. & Barry-Ryan, C. (2014) Development of a novel functional soup rich in bioactive sulforaphane using broccoli (*Brassica oleracea* L. ssp. *italica*) florets and byproducts. *Food and Bioprocess Technology* **7**, 1310–1321.

Baxter Food Group (2015) Available: http://www.just-food.com/news/baxters-launches-functional-soups_id116168.aspx [16 Mar 2015].

Benlloch-Tinoco, M., Igual, M., Salvador, A., Rodrigo, D. & Martínez-Navarrete, N. (2014) Quality and acceptability of microwave and conventionally pasteurised kiwifruit puree. *Food and Bioprocess Technology* **7**, 3282–3292.

Bertolini, M. & Romagnoli, G. (2012) An Italian case study for the process-target-cost evaluation of the ohmic treatment and aseptic packaging of a vegetable soup (minestrone). *Journal of Food Engineering* **110**, 214–219.

Buckow, R., Ng, S. & Toepfl, S. (2013) Pulsed electric field processing of orange juice: a review on microbial, enzymatic, nutritional, and sensory quality and stability. *Comprehensive Reviews in Food Science and Food Safety* **12**, 455–467.

Bulut, S. (2014) Inactivation of *Escherichia coli* in milk by high pressure processing at low and subzero temperatures. *High Pressure Research* **34**, 439–446.

Calzada, J., del Olmo, A., Picon, A., Gaya, P. & Nuñez, M. (2015) Effect of high-pressure processing on the microbiology, proteolysis, biogenic amines and flavour of cheese made from unpasteurized milk. *Food and Bioprocess Technology* **8**, 319–332.

Castro, I, Teixeira, J.A., Salengke, S., Sastry, S.K. & Vicente, A.A. (2004) Ohmic heating of strawberry products: electrical conductivity measurements and ascorbic acid degradation kinetics. *Innovative Food Science and Emerging Technologies* **5**, 27–36.

Chakraborty, S., Rao, P.S. & Mishra, H.N. (2015) Effect of combined high pressure–temperature treatments on color and nutritional quality attributes of pineapple (*Ananas comosus* L.) puree. *Innovative Food Science and Emerging Technologies* **28**, 10–21.

Chandrasekaran, S., Ramanthan, S. & Basak, T. (2013) Microwave food processing: a review. *Food Research International* **52**, 243–261.

Chen, C., Abdelrahim, K. & Beckerich, I. (2010) Sensitivity analysis of continuous ohmic heating process for multiphase foods. *Journal of Food Engineering* **98**, 257–265.

Chen, J., Zheng, X., Dong, J., Chen, Y. & Tian, J. (2015) Optimization of effective high hydrostatic pressure treatment of *Bacillus subtilis* in Hami melon juice. *LWT – Food Science and Technology* **60**, 1168–1173.

Choi, W., Nguyen, L.T., Lee, S.H. & Jun S. (2011) A microwave and ohmic heating combination heater for uniform heating of liquid-particle food mixtures. *Journal of Food Science* **76**, E576-E584.

Cilla, A., Lagarda, M.J., Alegría, A., de Ancos, B., Cano, M.P., Sánchez-Moreno, C., Plaza, L. & Barberá, R. (2011) Effect of processing and food matrix on calcium and phosphorus bioavailability from milk-based fruit beverages in Caco-2 cells. *Food Research International* **44**, 3030–3038.

Cilla, A., Alegría, A., De Ancos, B., Sánchez-Moreno, C., Cano, M.P., Plaza, L., Clemente, G., Lagarda, M.J. & Barberá, R. (2012) Bioaccessibility of tocopherols, carotenoids and ascorbic acid from milk and soya-based fruit beverages: influence of food matrix and processing. *Journal of Agricultural and Food Chemistry* **60**, 7282–7290.

Colle, I.J.P., Andrys, A., Grundelius, A., Lemmens, L., Löfgren, A., Van Buggenhout, S., Van Loey, A. & Hendrickx, M. (2011) Effect of pilot-scale aseptic processing on tomato soup quality parameters. *Journal of Food Science* **76**, C714-C723.

Contador, R., Delgado, F.J., García-Parra, J., Garrido, M. & Ramírez, R. (2015) Volatile profile of breast milk subjected to high-pressure processing or thermal treatment. *Food Chemistry* **180**, 17–24.

Coronel, P., Truong, V.D., Sumunovic, J., Sandeep, K.P. & Cartwright, G.D. (2005). Aseptic processing of sweet potato purees using a continuous flow microwave system. *Journal of Food Science* **70**, 531–536.

Cortés, C. Esteve, M.J. & Frígola, A. (2008) Color of orange juice treated by high intensity pulsed electric fields during refrigerated storage and comparison with pasteurized juice. *Food Control* **19**, 151–158.

Da-Wen, S. (ed) (2005) *Emerging Technologies for Food Processing*. Food Science and Technology International Series. Elsevier Academic Press, London, UK.

Devi, A.F., Buckow, R., Hemar, Y. & Kasapis, S. (2013) Structuring dairy systems through high pressure processing. *Journal of Food Engineering* **114**, 106–122.

Devi, A.F., Buckow, R., Singh, T., Hemar, Y. & Kasapis, S. (2015) Colour change and proteolysis of skim milk during high pressure thermal–processing. *Journal of Food Engineering* **147**, 102–110.

Dhakal, S., Liu, C., Zhang, Y., Roux, K.H., Sathe, S.K., Balasubramaniam, V.M. (2014) Effect of high pressure processing on the immunoreactivity of almond milk. *Food Research International*, **62**, 215–222.

Elez-Martínez, P. & Martín-Belloso, O. (2007a) Effects of high intensity pulsed electric field processing conditions on vitamin C and antioxidant capacity of orange juice and "gazpacho", a cold vegetable soup. *Food Chemistry* **102**, 201–209.

Elez-Martínez, P. & Martín-Belloso, O. (2007b) Impact of pulsed electric fields on food enzymes and shelf-life. In *Food Preservation by Pulsed Electric Fields* (eds H.L.M. Lelieveld, S. Notermans & S.W.H. de Haan), CRC-Press, New York, USA.

Espina, L., García-Gonzalo, D., Laglaoui, A., Mackey, B.M. & Pagán, R. (2013) Synergistic combinations of high hydrostatic pressure and essential oils or their constituents and their use in preservation of fruit juices. *International Journal of Food Microbiology* **161**, 23–30.

Espina, L., Monfort, S., Álvarez, I, García-Gonzalo, D. & Pagán, R. (2014) Combination of pulsed electric fields, mild heat and essential oils as an alternative to the ultrapasteurization of liquid whole egg. *International Journal of Food Microbiology* **189**, 119–125.

Evrendilek, G.A., Baysal, T., Icier, F., Yildiz, H., Demirdoven, A. & Bozkurt, H. (2012) Processing of fruits and fruit juices by novel electrotechnologies. *Food Engineering Reviews* **4**, 68–87.

Fakhouri, M.O. & Ramaswamy, H.S. (1993). Temperature uniformity of microwave heated foods as influence by product type and composition. *Food Research International* **26**, 89–95.

FDA (2015) *Kinetics of Microbial Inactivation for Alternative Food Processing Technologies: Microwave and Radio Frequency Processing*, [Online], Available: http://www.fda .gov/Food/FoodScienceResearch/SafePracticesforFoodProcesses/ucm100250.htm [16 Mar 2015].

Fernández-Sestelo, A., Sendra de Saá, R., Pérez-Lamela, C., Torrado-Agrasar, A., Rúa, M.L. & Pastrana-Castro, L. (2013) Overall quality properties in pressurized kiwi purée: Microbial, physicochemical, nutritive and sensory tests during refrigerated storage. *Innovative Food Science and Emerging Technologies* **20**, 64–72.

Floros J.D., Newsome R., Fisher, W., Barbosa-Cánovas, G.V., Chen, H., Dunne, C.P., German, J.B., Richard, L.H., Heldman, D.R., Karwe, M.V., Knabel, S.J., Labuza, T.P., Lund, D.B., Newell-McGloughlin, M., Robinson, J.L., Sebranek, J.G., Shewfelt, R.L., Tracy, W.F., Weaver, C.A. & Ziegler, G.R. (2010) Feeding the world today and tomorrow: the importance of food science and technology. *Comprehensive Reviews in Food Science and Food Safety* **9**, 572–598.

Gelski, J. (2014) New technologies for heating up for soup industry. *Food Business News*, [Online]. Available: http://www.foodbusinessnews.net/articles/news_home/ Research [16 Mar 2015].

Georget, E., Sevenich, R., Reineke, K., Mathys, A., Heinz, V., Callanan, M., Rauh, C. & Knorr, D. (2015) Inactivation of microorganisms by high isostatic pressure processing in complex matrices: A review. *Innovative Food Science and Emerging Technologies* **27**, 1–14.

Gomes da Cruz, A., de Assis Fonseca Fariaa, J.A., Isay Saadb, S.M., André Bolinia, H.M., Souza Santa'Ana, A. & Cristianinia, M. (2010) High pressure processing and pulsed electric fields: potential use in probiotic dairy foods processing. *Trends in Food Science and Technology* **21**, 483–493.

Goullieux, A. & Pain, J.P. (2005) Omhic heating. In *Emerging Technologies for Food Processing* (ed S. Da-Wen), pp. 469–505. Elsevier Academic Press, London, UK.

Guerrero-Beltrán, J.A., Barbosa-Cánovas, G.V. & Welti-Chanes J. (2011) High hydrostatic pressure effect on *Saccharomyces cerevisiae*, *Escherichia coli* and *Listeria innocua* in pear nectar. *Journal of Food Quality* **34**, 371–378.

Hain Pure Foods (2015) Available: http://www.hainpurefoods.com [16 Mar 2015].

Heinz, V. & Buckow, R. (2010) Food preservation by high pressure. *Journal für Verbraucherschutz und Lebensmittelsicherheit* **5**, 73–81.

Hendrickx, M., Ludikhuyze, L., Vanden Broeck, I. & Weemaes, C. (1998) Effect of high pressure on enzymes related to food quality. *Trends in Food Science and Technology* **9**, 197–203.

Hernández-Carrión, M., Tárrega, A., Hernando, I., Fiszman, S.M. & Quiles, A. (2014) High hydrostatic pressure treatment provides persimmon good characteristics to formulate milk-based beverages with enhanced functionality. *Food and Function* **5**, 1250–1260.

Hirneisen, K.A., Black, E.P., Cascarino, J.L., Fino, V.R., Hoorver, D.G. & Kniel, K.E. (2010) Viral inactivation in foods: a review of traditional and novel food-processing technologies. *Comprehensive Reviews in Food Science and Food Safety* **9**, 3–20.

Hite, B.H. (1899) The effect of pressure on the preservation of milk. *West Virginia University. Agricultural Experiment Station Bulletin* **58**, 15–35.

Hite, B.H., Giddings, N.J. & Weakly, C.E. (1914) The effects of pressure on certain microorganisms encountered in the preservation of fruits and vegetables. *West Virginia University. Agricultural Experiment Station Bulletin* **146**, 1–67.

Huang, W., Bi, X., Zhang, X., Liao, X., Hu, X. & Wu, J. (2013) Comparative study of enzymes, phenolics, carotenoids and color of apricot nectars treated by high hydrostatic pressure and high temperature short time. *Innovative Food Science and Emerging Technologies* **18**, 74–82.

Huang, R., Li, X., Huang, Y. & Chen, H. (2014) Strategies to enhance high pressure inactivation of murine norovirus in strawberry puree and on strawberries. *International Journal of Food Microbiology* **185**, 1–6.

Hugas, M., Garcia, M. & Monfort, J.M. (2002) New mild technologies in meat processing: high pressure treatment on the sarcoplasmic reticulum of red and white muscles. *Meat Science* **62**, 359–371.

Icier, F. & Ilicali, C. (2005a) Tempeature dependent electrical conductivities of fruit purees during ohmic heating. *Food Research International* **38**, 1135–1142.

Icier, F. & Ilicali, C. (2005b) The effects of concentration on electrical conductivity of orange juice concentrates during ohmic heating. *European Food Research and Technology* **220**, 406–414.

Icier, F., Yildiz, H. & Baysal, T. (2006) Peroxidase inactivation and colour changes during ohmic blanching of pea puree. *Journal of Food Engineering* **74**, 424–429.

Jayachandran, L.E., Chakraborty, S. & Rao, P.S. (2015) Effect of high pressure processing on physicochemical properties and bioactive compounds in litchi based mixed fruit beverage. *Innovative Food Science and Emerging Technologies* **28**, 1–9.

Jun, S. & Sastry, S.K. (2005) Modeling and optimizing of pulsed ohmic heating of foods inside the flexible package. *Journal of Food Process Engineering* **28**, 417–436.

Karlović, S., Bosiljkov, T., Brnčić, M., Semenski, D., Dujmić, F., Tripalo, B. & Ježek, D. (2014) Reducing fat globules particle-size in goat ilk: Ultrasound and high hydrostatic pressures approach. *Chemical and Biochemical Engineering Quarterly* **28**, 499–507.

Keenan, D.F., Brunton, N., Butler, F., Wouters, R. & Gormley, R. (2011) Evaluation of thermal and high hydrostatic pressure processed apple purees enriched with prebiotic inclusions. *Innovative Food Science and Emerging Technologies* **12**, 261–268.

Keenan, D.F., Rößle, C., Gormley, R., Butler, F. & Brunton, N.P. (2012) Effect of high hydrostatic pressure and thermal processing on the nutritional quality and enzyme activity of fruit smoothies. *LWT – Food Science and Technology* **45**, 50–57

Knirsch, M.C., Santos, C., Vicente, A.A., Penna, T.C. (2010) Ohmic heating: a review. *Trends in Food Science & Technology* **21**, 436–441.

Li, Q., Zhang, H.H., Claver, I.P., Zhu, K.X., Peng, W. & Zhou, H.M. (2011) Effect of different cooking methods on the flavour constituents of mushroom (*Agaricus bisporus* (Lange) Sing) soup. *International Journal of Food Science and Technology* **46**, 1100–1108.

Li, R., Wang, Y., Wang, S. & Liao, X. (2015) Comparative study of changes in microbiological quality and physicochemical properties of N_2-infused and N_2-degassed banana smoothies after high pressure processing. *Food and Bioprocess Technology* **8**, 333–342.

Liu, F., Grauwet, T., Kebede, B.T., Van Loey, A., Liao, X. & Hendrickx, M. (2014) Comparing the effects of high hydrostatic pressure and thermal processing on blanched and unblanched mango (*Mangifera indica* L.) nectar: Using headspace fingerprinting as an untargeted approach. *Food and Bioprocess Technology* **7**, 3000–3011.

Marra, F., Zhang, L. & Lyng, J. (2009) Radio frequency treatment of foods: Review of recent advances. *Journal of Food Engineering* **91**, 497–508.

Marszalek, K., Mitek, M., & Skapska, S. (2015) The effect of thermal pasteurization and high pressure processing at cold and mild temperatures on the chemical composition, microbial and enzyme activity in strawberry purée. *Innovative Food Science and Emerging Technologies* **27**, 48–56.

Martín-Belloso, O. & Elez-Martínez, P. (2005) Food safety aspects of pulsed electric fields. In *Emerging Technologies for Food Processing* (ed S. Da-Wen), pp. 183–217. Elsevier Academic Press, London, UK.

Martín-Belloso, O. & Soliva-Fortuny, R. (2011) Pulsed electric field processing basics. In *Nonthermal Processing Technologies for Food* (eds H. Zhang, V.M. Barbosa-Cánovas, V.M. Balasubramaniam, C.P. Dunne, D.F. Farkas & J.T.C. Yuan) Chapter 11, pp. 157–175. Wiley-Blackwell and IFT-Press, Ames, Iowa, USA.

Martínez-Monteagudo, S.I., Gänzle, M.G. & Saldaña, M.D.A. (2014) High-pressure and temperature effects on the inactivation of *Bacillus amyloliquefaciens*, alkaline phosphatase and storage stability of conjugated linoleic acid in milk. *Innovative Food Science and Emerging Technologies* **26**, 59–66.

Martínez-Tomás, R., Pérez-Llamas, F., Sánchez-Campillo, M., González-Silvera, D., Cascales, A.I., García-Fernández, M., López-Jiménez, J.A., Zamora Navarro, S., Burgos, M.I., López-Azorín, F., Wellner, A., Avilés Plaza, F., Bialek, L., Alminger, M. & Larqué, E. (2012) Daily intake of fruit and vegetable soups processed in different ways increases human serum β-carotene and lycopene concentrations and reduces levels of several oxidative stress markers in healthy subjects. *Food Chemistry* **134**, 127–133.

Meng-Meng, L., Chang-Xin, L. & Xu-Qiao, F. (2014) Status quo and trend of fruit and vegetable juice processing by application of ultra high pressure. *Journal of Food Safety and Quality* **5**, 567–576.

Mermelstein, N.H. (1997) High pressure processing reaches the US market. *Food Technology* **51**, 95–96.

Mertens, B. & Deplace, G. (1993) Engineering aspect of high pressure processing in food industry. *Food Technology* **47**, 164–167.

Milani E.A., Alkhafaji, S. & Silva F.V.M. (2015) Pulsed electric field continuous pasteurization of different types of beers. *Food Control* **50**, 223–229.

Morales de la Peña, M., Elez-Martínez, P. & Martín-Belloso, O. (2011) Food preservation by pulsed electric fields: an engineering perspective. *Food Engineering Reviews* **3**, 94–107.

Mosqueda-Melgar, J., Elez-Martínez, P., Raybaudi-Massilia, R.M. & Martín-Belloso, O. (2008) Effects of pulsed electric fields on pathogenic microorganisms of major concern in fluid foods: a review. *Critical Reviews in Food Science and Nutrition* **48**, 747–759.

Muñoz, M., de Ancos, B., Sánchez-Moreno, C. & Cano, M.P. (2007) Effects of high pressure and mild heat on endogenous microflora and on the inactivation and sublethal injury of *Escherichia coli* inoculated into fruit juices and vegetable soup. *Journal of Food Protection* **70**, 1587–1593.

Nunes de Morais, A.C., da Rocha Ferreira, E.H. & Rosenthal, A. (2014) High isostatic pressure application in dairy products: a review. *Revista do Instituto de Laticínios Cândido Tostes* **69**, 357–374.

NutraHelix Biotech Pvt. Ltd. (2012) Available: http://www.nutrahelix.com/pdf/Functional%20Soups.pdf [16 Mar 2015].

Odriozola-Serrano, I., Soliva-Fortuny, R., Hernández-Jover, T. & Martín-Belloso, O. (2009) Carotenoid and phenolic profile of tomato juices processed by high intensity pulsed electric fields compared with conventional thermal treatments. *Food Chemistry* **112**, 258–266.

Odriozola-Serrano, I., Aguiló-Aguayo, I., Soliva-Fortuny, R. & Martín-Belloso, O. (2013) Pulsed electric fields processing effects on quality and health-related constituents for plant-based foods. *Trends in Food Science and Technology* **29**, 98–107.

Oey I., Van der Plancken I., Van Loey A., & Hendrickx, M.E. (2008) Does high-pressure processing influence nutritional aspects of plant based food systems? *Trends in Food Science and Technology* **19**, 300–308.

Oms-Oliu, G., Rojas-Graü, M.G., Alandes González, L., Varela, P., Soliva-Fortuny, R., Hernando Hernando, M.I., Pérez Munuera, I., Fisman, S. & Martín-Belloso, O. (2010) Recent approaches using chemical treatments to preserve quality of fresh-cut fruit: a review. *Postharvest Biology and Technology* **57**,139-148.

Otero, L., Guignon, B., Aparicio, C. & Sanz, P.D. (2010) Modeling thermophysical properties of food under high pressure. *Critical Reviews in Food Science and Nutrition* **50**, 344–368.

Palmers, S., Grauwet, T., Kebede, B.T., Hendrickx, M.E. & Van Loey, A. (2014). Reduction of furan formation by high-pressure high-temperature treatment of individual vegetable purées. *Food and Bioprocess Technology* **7**, 2679–2693.

Parrot, D. (2010) Microwave technology sterilizes sweet potato puree. *Food Technology* **64**, 66.

Pereira, R., Martins, R. & Vicente, A. (2008) Goat milk free fatty acids characterization during conventional and ohmic heating pasteurization. *Journal of Dairy Science* **91**, 2925–2937.

Pina-Pérez, M.C., Silva-Angulo, A.B., Muguerza-Marquínez, B., Aliaga, D.R. & López, A.M. (2009) Synergistic effect of high hydrostatic pressure and natural antimicrobials on inactivation kinetics of *Bacillus cereus* in a liquid whole egg and skim milk mixed beverage. *Foodborne Pathogens and Disease* **6**, 649–656.

Plaza, L., Sánchez-Moreno, C., de Ancos, B. & Cano, M.P. (2006a) Carotenoid content and antioxidant capacity of Mediterranean vegetable soup (gazpacho) treated by high-pressure/temperature during refrigerated storage. *European Food Research and Technology* **223**, 210–215.

Plaza, L., Sánchez-Moreno, C., Elez-Martínez, P., De Ancos, B., Martín-Belloso, O. & Cano, M.P. (2006b) Effect of refrigerated storage on vitamin C and antioxidant activity of orange juice processed by high-pressure or pulsed electric fields with regard to low pasteurization. *European Food Research and Technology* **223**, 487–493.

Plaza, L., Sánchez-Moreno, C., de Ancos, B, Elez-Martínez, P., Martín-Belloso, O. & Cano, M.P. (2011) Carotenoid and flavanone content during refrigerated storage of orange juice processed by high-pressure, pulsed electric fields and low pasteurization. *LWT – Food Science and Technology* **44**, 834–839.

Podravka, D.D. (2015) Available: http://www.podravka.com/media/news/fini-mini-vital-new-design-for-functional-soups [16 Mar 2015].

Puértolas, E., López, N., Condón, S., Álvarez, I. & Raso J (2010) Potential applications of PEF to improve red wine quality. *Trends in Food Science and Technology* **21**, 247–255.

Ramaswamy, R., Balasubramaniam, V.M., & Sastry, S.K. (2005) Ohmic heating of foods. Fact sheet for food processors. FSE 3–05. Ohio State University Extension Fact Sheet, Columbus, OH.

Rode, T.M., Hovda, M.B. & Rotabakk, B.T. (2015) Favourable effects of soluble gas stabilisation and modified atmosphere for suppressing regrowth of high pressure treated *Listeria innocua*. *Food Control* **51**, 108–113.

Rodríguez-Alcalá, L.M., Castro-Gómez, P., Felipe, X., Noriega, L. & Fontecha, J. (2015) Effect of processing of cow milk by high pressures under conditions up to 900 MPa on the composition of neutral, polar lipids and fatty acids. *LWT – Food Science and Technology* **62**, 265–270.

Rodríguez-Roque, M.J., De Ancos, B., Sánchez-Moreno, C., Cano, M.P., Elez-Martínez, P. & Martín-Belloso, O. (2015) Impact of food matrix and processing on the in vitro bioaccessibility of vitamin C, phenolic compounds, and hydrophilic antioxidant activity from fruit juice-based beverages. *Journal of Functional Foods* **14**, 33–43.

Salazar-González, C., San Martín-González, M.F., López-Malo, A., & Sosa-Morales, M.E. (2012) Recent studies related to microwave processing of fluid foods. *Food and Bioprocess Technology* **5**, 31–46.

Saldaña, G., Álvarez, I., Condón, S. & Raso, J. (2014) Microbial aspects related to the feasibility of PEF technology for food pasteurization. *Critical Reviews in Food Science and Nutrition* **54**, 1415–1426.

Salengke, S., & Sastry, S.K. (2007) Experimental investigation of Ohmic heating of solid–liquid mixtures under worst–case heating scenarios. *Journal of Food Engineering* **83**, 324–336.

Sampedro, F., Geveke, D.J., Fan, X. & Zhang, H.Q. (2009) Effect of PEF, HHP and thermal treatment on PME inactivation and volatile compounds concentration of an orange juice–milk based beverage. *Innovative Food Science and Emerging Technologies* **10**, 463–469.

Sánchez-Moreno, C., Plaza, L., de Ancos, B. & Cano, M.P. (2003a) Vitamin C, provitamin A arotenoids, and other arotenoids in high-pressurized orange juice during refrigerated storage. *Journal of Agricultural and Food Chemistry*, **51**, 647–653.

Sánchez-Moreno, C., Plaza, L., de Ancos, B. & Cano, M.P. (2003b) Effect of high-pressure processing on health-promoting attributes of freshly squeezed orange juice (*Citrus sinensis* L.) during chilled storage. *European Food Research and Technology* **216**, 18–22.

Sánchez-Moreno, C., Cano, M.P., de Ancos, B., Plaza, L., Olmedilla, B., Granado, F. & Martín, A. (2004a) Consumption of high-pressurized vegetable soup increases plasma vitamin C and decreases oxidative stress and inflammatory biomarkers in healthy humans. *Journal of Nutrition* **134**, 3021–3025.

Sánchez-Moreno, C., Plaza, L., de Ancos, B. & Cano, M.P. (2004b) Effect of combined treatments of high-pressure and natural additives on carotenoid extractability and antioxidant activity of tomato puree (*Lycopersicum esculentum* Mill.). *European Food Research and Technology* **219**, 151–160.

Sánchez-Moreno, C., Cano, M.P., de Ancos, B., Plaza, L., Olmedilla, B. Granado, F., Elez-Martínez, P., Martín-Belloso, O. & Martín, A. (2005a) Intake of Mediterranean vegetable soup treated by pulsed electric fields affects plasma vitamin C and antioxidant biomarkers in humans. *International Journal of Food Sciences and Nutrition* **56**, 115–124.

Sánchez-Moreno, C., Plaza, L., Elez-Martínez, P., de Ancos, B., Martín-Belloso, O. & Cano, M.P. (2005b) Impact of high pressure and pulsed electric fields on bioactive compounds and antioxidant activity of orange juice in comparison with traditional thermal processing. *Journal of Agricultural and Food Chemistry* **53**, 4403–4409.

Sánchez-Moreno, C., Cano, M.P., de Ancos, B., Plaza, L., Olmedilla, B., Granado, F. & Martín, A. (2006a) Mediterranean vegetable soup consumption increases plasma vitamin C and decreases F_2-isoprostanes, prostaglandin E_2 and monocyte chemotactic protein-1 in healthy humans. *Journal of Nutritional Biochemistry* **17**, 183–189.

Sánchez-Moreno, C., Plaza, L., de Ancos, B. & Cano, M.P. (2006b) Impact of high-pressure and traditional thermal processing of tomato puree on carotenoids, vitamin C and antioxidant activity. *Journal of the Science of Food and Agriculture* **86**, 171–179.

Sánchez-Moreno, C., de Ancos, B., Plaza, L., Elez-Martínez, P. & Cano, M.P. (2009) Nutritional approaches and health-related properties of plant foods processed by high pressure and pulsed electric fields. *Critical Reviews in Food Science and Nutrition* **49**, 552–576.

Sarang, S., Sastry, S.K., Gaines, J., Yang, T.C.S. & Dunne, P. (2007) Product formulation for Ohmic heating: blanching as a pretreatment method to improve uniformity in heating of solid–liquid food mixtures. *Journal of Food Engineering* **87**, 227–234.

Scolari, G., Zacconi, C., Busconi, M. & Lambri, M. (2015) Effect of the combined treatments of high hydrostatic pressure and temperature on *Zygosaccharomyces bailii* and *Listeria monocytogenes* in smoothies. *Food Control* **47**, 166–174.

Sharma, P., Oey, I., Bremer, P. & Everett, D.W. (2014) Reduction of bacterial counts and inactivation of enzymes in bovine whole milk using pulsed electric fields. *International Dairy Journal* **39**, 146–156.

Shibeshi, K. & Farid, M.M. (2010) Pressure-assisted thermal sterilization of soup. *High Pressure Research* **30**, 530–537.

Shibeshi, K. & Farid, M.M. (2011) Scale-up unit of a unique moderately high pressure unit to enhance microbial inactivation. *Journal of Food Engineering* **105**, 522–529.

Siró, I., Kápolna, E., Kápolna, B. & Lugasi, A. (2008). Functional food. Product development, marketing and consumer acceptance. A review. *Appetite* **51**, 456–467.

Somavat, R., Kamonpatan, P., Mohamed, H.M.H., Sastry, S.K. (2012) Ohmic sterilization inside a multi-layered laminate pouch for long-duration space missions. *Journal of Food Engineering* **112**, 134–143.

Sousa, S.G., Delgadillo, I. & Saraiva, J.A. (2014) Effect of thermal pasteurisation and high-pressure processing on immunoglobulin content and lysozyme and lactoperoxidase activity in human colostrum. *Food Chemistry* **151**, 79–85.

Terefe, N.S., Buckow, R. & Versteeg, C. (2014) Quality-related enzymes in fruit and vegetable products: effects of novel food processing technologies, Part 1: High-pressure processing. *Critical Reviews in Food Science and Nutrition* **54**, 24–63.

Thakur, B.R. & Nelson, P.E. (1998). High pressure processing and preservation of foods. *Food Reviews International* **14**, 427–447.

Vadivambal, R. & Jayas, D.S. (2007) Changes in quality of microwave-treated agricultural products: a review. *Biosystems Engineering* **98**, 1–16

Vadivambal, R. & Jayas, D.S. (2010) Non-uniform temperature distribution during microwave heating of food materials: A review. *Food and Bioprocess Technology* **3**, 161–171.

Valdramidis, V.P., Graham, W.D., Beattie, A., Linton, M., McKay, A., Fearon, A.M. & Patterson, M.F. (2009) Defining the stability interfaces of apple juice: implications on the optimisation and design of high hydrostatic pressure treatment. *Innovative Food Science and Emerging Technologies* **10**, 396–404.

Vega-Mercado, H., Martín-Belloso, O., Chang, F.J. Barbosa-Cánovas, G.V. & Swanson, B.G. (1996) Inactivation of *E.coli* and *B. subtilis* suspended in pea soup using pulsed electric fields. *Journal of Food Processing and Preservation* **20**, 501–510.

Wang, R., Zhang, M., Mujumdar, A.S. & Sun, J.C. (2009) Microwave freeze-drying characteristics and sensory quality of instant vegetable soup. *Drying Technology* **27**, 962–968.

Wang, R., Zhang, M. & Mujumdar, A.S. (2010) Effect of food ingredient on microwave freeze-drying of instant vegetable soup. *LWT – Food Science and Technology* **43**, 1144–1150.

Wang, Y., Yi, J., Yi, J., Dong, P., Hu, X. & Liao, X. (2013) Influence of pressurization rate and mode on inactivation of natural microorganisms in purple sweet potato nectar by high hydrostatic pressure. *Food and Bioprocess Technology* **6**, 1570–1579.

Weaver, C.A., Dwyer, J., Fulgoni III,, V.L., King, J.C., Leville, G.A., MacDonald, R.S., Ordovás, J. & Schakenberg, D. (2014) Processed Foods: contributions to nutrition. *American Journal of Clinical Nutrition* **99**, 1525–1542.

Xi, J. (2013) High-pressure processing as emergent technology for the extraction of bioactive ingredients from plant materials. *Critical Reviews in Food Science and Nutrition* **53**, 837–852.

Yildiz, H., Icier, F. & Baysal, T. (2010) Changes in beta-carotene, chlorophyll and color of spinach puree during ohmic heating. *Journal of Food Process Engineering* **33**, 763–779.

Yu, Y., Xu, Y., Wu, J., Xiao, G., Fu, M. & Zhang, Y. (2014) Effect of ultra-high pressure homogenisation processing on phenolic compounds, antioxidant capacity and anti-glucosidase of mulberry juice. *Food Chemistry* **153**, 114–120.

Zhang, M., Tang, J., Mujumdar, A.S. & Wang, S. (2006) Trends in microwave-related drying of fruits and vegetables. *Trends in Food Science and Technology* **17**, 524–534.

Zhong, Q., Sandeep, K.P. & Swartzel, K.R. (2004) Continuous flow ratio frequency heating of particulate foods. *Innovative Food Science and Emerging Technologies* **5**, 474–483.

Zimmerman, F. & Bergman, C. (1993) Isostatic high-pressure equipment for food preservation. *Food Technology* **47**, 162–163.

Zulueta, A., Barba, F.J., Esteve, M.J. & Frígola, A. (2013) Changes in quality and nutritional parameters during refrigerated storage of an orange juice milk beverage treated by equivalent thermal and non-thermal processes for mild pasteurization. *Food and Bioprocess Technology* **6**, 2018–2030.

10
Functional Beverages

Francesc Puiggròs[1], Begoña Muguerza[2], Anna Arola-Arnal[2], Gerard Aragonès[2], Susana Suárez-Garcia[2], Cinta Bladé[2], Lluís Arola[1,2], and Manuel Suárez[2]*

[1]*Technological Unit of Nutrition and Health, EURECAT-Technological Center of Catalonia, Reus, Spain*
[2]*Department of Biochemistry and Biotechnology, Nutrigenomics Research Group, Universitat Rovira i Virgili, Tarragona, Spain*

10.1   Introduction

The origin of functional foods is Japan, in the mid-1930s when, after intense research, Dr Shirota identified some bacteria able to improve consumer's health and included them in the formulation of a new product called Yakult, a based-fermented milk drink that claimed to keep intestines healthy (Heasman & Mellentin, 2001). Some decades after that, Japan experienced a new interest in this field and launched to the market several products identified as functional foods. Basically, foods fortified with bioactive compounds that had beneficial physiological effects in the organism. The ultimate objective of these foods was to protect the consumer's health and therefore reduce the health care cost, which was continuously increasing (Hardy, 2002). These types of products expanded to other countries although their introduction and acceptance in the markets differed among them. While in the United States of America and Canada their development and acceptance was very rapid, in Europe the interest for functional foods did not start up to the latter half of the 1990s (Menrad, 2003).

Nowadays, the concept of functional foods is generally recognized by consumers, who identify them as healthier and more beneficial products. In addition, consumers have become more aware of the relationship between diet and disease prevention, thus increasing the demand of these products. Consequently, the food industry is continuously innovating and developing new items within this field, aiming to catch their attention (Mollet & Rowland, 2002). However, although these products have gained great importance in the world, the definition of functional food presents some ambiguities and differs among countries, still lacking a global consensus (Lau *et al.*, 2012). For example, the FUFOSE, an European Commission concerted action held between

*Corresponding author: Manuel Suárez, manuel.suarez@urv.cat

Innovative Technologies in Beverage Processing, First Edition.
Edited by Ingrid Aguiló-Aguayo and Lucía Plaza.
© 2017 John Wiley & Sons Ltd. Published 2017 by John Wiley & Sons Ltd.

1995 and 1997 and coordinated by the International Life Sciences Institute (ILSI), established the functional foods concept as "those that is satisfactorily shown that beneficially affects one or more body functions, improving the health and welfare and/or reducing the risk of disease." It is said that functional foods must be in the form of food and its effectiveness must be obtained by consumption of normal amounts within the framework of a balanced diet (Diplock *et al.*, 1999). On the other hand, Health Canada defines functional food as "food that is similar in appearance to, or may be, a conventional food that is consumed as part of the usual diet, with demonstrated physiological benefits and/or that reduces the risk of chronic disease beyond basic nutritional functions." Under this broad definition, several products are commercialized as functional foods. Thus, natural products (without modification), fortified foods (with increased amount of a component that is naturally present in the product), enriched foods (in which a component that is not naturally found has been added), and altered foods (in which a component has been replaced by another) are all sold as functional foods (Siró *et al.*, 2008; Lau *et al.*, 2012).

From an economic perspective, the importance of functional foods is continuously increasing in the entire world. By countries, Japan and the United States have the biggest market of functional foods. On the other hand, European countries are steadily increasing their consumption but still are far behind them (Ozen *et al.*, 2012). According to a research carried out by Global Industry Analysts, it was estimated that the functional food market would reach 100,000 million of Euros by 2015 and from these 40% belong to the US market, while a 25% belong to Europe (Murcia, 2013). These data show the actual and future importance of this market within the food industry, being one of the sectors that are growing more in a relatively short period of time.

Within this framework, beverages are one of the functional food matrices that have contributed more to the development of new functional foods. Functional beverages are ready-to-consume, thus becoming a simple and easy way to ingest healthy compounds by the consumers who are continuously looking for products that could rapidly meet their needs in a manner compatible with their increasingly busy life. In the functional beverage formulation, fruits and vegetables have been considered the ideal candidates for bioactive compounds. On the one hand, fruits are described as health-promoting products. A general overview of their composition reveals a broad range of healthy compounds such as vitamins, minerals, fibers, antioxidants, and phenolic compounds, sufficient to confirm their bioactivity (Sun-Waterhouse, 2011). This fact agrees with the epidemiological data and recommendations made by experts about the need to enhance the daily consumption of some bioactives. As a response, the food industry has launched a broad range of new functional beverages (Joshipura *et al.*, 2001), and several products such as fortified milk products or juices, either combined with milk or alone, aromatized water, and energy drinks, among others, have emerged in the market (Sun-Waterhouse, 2011). In this chapter, a complete revision of the state of the art of the functional beverages in all its versions and the latest trends in their formulation is presented.

10.2   Functional Food Regulation

As it was stated earlier, there is still not an international consensus definition for the functional food concept. In fact, there are several regulations and laws that lay down their own rules for the production, commercialization, and about the health claims that those foods can bear. In addition, this fact is entailed with inherent problems such as inter-Frontiers product's distribution that increase the need for a unified definition and control.

The first country to legislate functional foods was Japan. In 1952, a regulation about the improvement of nutrition throughout food (248/1952) was proclaimed. Later on, the terms FOSHU, "Foods for Specified Health Uses" (legislation 41/1991) and FHC, "Foods with Health Claims" (2001) were developed. Nowadays, a revised form of the FOSHU and FHC system published in 2005 is under use (Ohama *et al.*, 2014). Regarding the United States, the organization responsible for the control of food safety issues, including those regarding health claims, is the Food and Drug Administration (FDA). Since 1993, it is allowed to include food messages claiming health-promoting properties, but industrial operators are requested to deliver the FDA scientific data about the health claim. Finally, in the European Union, a step forward in the control and regulation of food and feed safety issues was done in 2002 with the creation of the European Food Safety Authority (EFSA). This organization evaluates all the aspects related with food in Europe and provides scientific support in the legislation process. The EFSA assessment process involves a set of scientific guidances based on some FUFOSE results, obtained in the PASSCLAIM (Process for the Scientific Support for Claims on Foods). These guidances provide a set of consensus criteria to be taken into account when assessing validity of health claims made on foods. However, very few dossiers submitted to the EFSA have reached this goal, mainly due to the high scientific standards of the process of health-claim assessment (Bech-Larsen & Scholderer, 2007; Juarez-Iglesias, 2010).

Besides these, other countries follow their own regulations to control the functional foods market within their frontiers (Jew *et al.*, 2008), thereby increasing all the problematic issues related with such a broad number of differences. To overcome this situation, efforts have been made during the past decade to define international guidelines in the process of acceptance of health claims within the context of the Codex Alimentarius (Kwak & Jukes, 2001).

10.3 Natural Ingredients in the Formulation of Functional Beverages

The awareness of consumers about the beneficial effects of bioactive compounds has had a direct influence on the success of functional foods. To satisfy consumer's requirements, the food industry continuously innovates and develops new functional products, using novel sources of bioactive compounds. Nowadays, R&D departments from food industries and recognized research groups are working together to identify and evaluate bioactive compounds that may exert physiological activities in the organism. The complete process includes characterization of food matrices, extraction of its bioactive compounds, *in vitro* and *in vivo* evaluation of their bioactivity using cell lines and animal models, and, finally, the validation of the results in intervention studies in humans. Only when all these aspects are overcome, bioactive compounds can be used in the formulation of functional foods (Hasler, 1998).

In the development of functional beverages, fruits and vegetables are especially good sources of bioactive compounds due to the diverse range of molecules that can be found as part of its composition. Indeed, epidemiological studies have suggested associations between the consumption of polyphenolic-rich foods or beverages and the prevention of diseases related to oxidative stress, such as cancer or neurodegenerative disease (Scalbert & Williamson, 2000; Rimm, 2002). Many studies suggest that some of these beneficial effects can be due to the presence of flavonoids (Hertog et al., 1994; Kris-Etherton & Keen, 2002), which are the most abundant group of polyphenols in the human diet (Scalbert & Williamson, 2000). However, it is important to take into account that lots of these compounds remain in the by-products and wastes generated in the industrial processing of these products, which not only are discarded without use but in turn represent a problem for the food industries. Therefore, the use of these by-products as source of bioactive compounds allows to solve a problem while an economical benefit is obtained (Fernández-Ginés et al., 2008).

In addition, it has been demonstrated that although bioactive compounds exert beneficial activities by themselves, its combination provides enhanced results. Consequently, it can be postulated that supplementation with a combination of some of them may be more beneficial than ingestion of only one. This fact was observed in the study carried out by Huang et al. (2000) in which they compared supplementation with vitamin C, vitamin E, their combination, or three servings of fruits and vegetables. They observed that the most beneficial effects were obtained with fruits and vegetables rather than the pure bioactive compounds pointing to these synergistic effects.

Some fruits have revealed the ability to act against infections caused by microorganisms. For example, it is well known that cranberries can help to overcome urinary tract infections. This fruit is usually sold as juice, either natural or enriched with other vitamins or bioactive compounds that could complement its activity. Researchers have demonstrated that its functionality can be attributed to its phenolic content, specifically proanthocyanidins (also known as condensed tannins). It has been observed that these compounds prevent the adhesion of *Escherichia coli* to the epithelial cells of the urinary tract (Howell et al., 1998). Further studies supported these results, confirming the potential use of cranberry for antimicrobial purpose (Leahy et al., 2001). Other fruits with antimicrobial activity are grapes. Several studies have demonstrated that extracts from different parts of grapes such as seeds, skins, and juice successfully inhibit the growth of both Gram-positive and Gram-negative bacteria, although with different intensities (Jayaprakasha et al., 2003). Among the Gram-positive bacteria, the inhibition of *Listeria monocytogenes* can be highlighted. This can be attributed to its high content of phenolic compounds (Georgiev et al., 2014). In addition, it has been observed that the inhibition of *L. monocytogenes* follows different patterns of action depending on the bioactive compound (Rhodes et al., 2006). In grape juice and skin extracts, both rich in anthocyanins, the anti-inhibitory activity was dependent on the pH. On the contrary, the activity of grape seeds extracts, rich in catechin, epicatechine, and epicatechine galate, was pH independent. One of the most important consequences from this antibacterial activity is that flavonoids from grape extracts can modulate the composition of human gut microflora. In an *in vitro* experiment, it was observed that red wine extract was able to reduce the ratio of firmicutes/bacteroidetes (Kemperman et al., 2013). This modification was related with a reduction of body weight pointing out that these compounds are excellent candidates to be used in the formulation of functional beverages addressed to control the mass body weight. It is important to note that, in addition to the antibacterial activity, grape-derived products have also shown antiviral (Bekhit et al., 2011) and antifungal activities (*Candida albicans*) highlighting the potential of these products (Papadopoulou et al., 2005).

Other products rich in phenolic compounds different from grapes have also shown antimicrobial activity. For example, Ferrer *et al.* (2009) demonstrated that an olive powder was able to inhibit *Bacillus cereus* growth. This is especially suitable for beverage industry because it can be used to control cold chain break in vegetables juices and purees. Yerba Mate, another product rich in polyphenols, had been used as an antimicrobial food preservative and, while the need for more research exists, evidence seems to demonstrate that it is a plant with a variety of compounds that can be applied for use in human health (Burris *et al.*, 2012).

10.4   New Trends in the Formulation of Functional Beverages

As consumers continue to demand healthy alternatives, beverage manufacturers have developed new line extensions or products that target not only the overall wellness but also specific health conditions. In the following sections, novel trends in the formulation of functional beverages are described.

10.4.1   Tendencies in Fruit Ingredients

Apart from the traditional fruits (such as berries, grapes, apples, and oranges) that are used in the formulation of functional beverages, new fruit sources are currently under study trying to search for new compounds and activities. For example, in the review of Gruenwald (2009), a list of African fruits that can be potentially used in these products can be found. This list includes the fruit of the baobab, which has antioxidant, anti-inflammatory, and prebiotic effects. Baobab (*Adansonia digitata*), used in traditional medicine by Egyptians, can be used in the formulation of juices, soft drinks, and smoothies (Osman, 2004). Other African fruits and herbs traditionally used in medicine, which are being introduced in beverage formulation, are Devil's claw (*Harpagophytum procumbens*) and the sausage tree fruit (*Kigelia pinnata*).

Following the search of new compounds, other interesting sources are the fruits from the Asian countries, for example, mangosteen (*Garcinia mangostana*), noni (*Morinda citrifolia*), durian (*Durio zibethinus*), and goji (*Lycium barbarum*), which exert a broad range of activities such as antioxidant, anticancer, and anti-inflammatory (Amagase & Nance, 2008; Akihisa *et al.*, 2007; Haruenkit *et al.*, 2007). From these fruits, Durian represents an especially complicated case due to its disgusting odor and flavour, which prevents its use in the beverage industry without a mask.

10.4.2   Green Botanicals in Beverages

Recent trends in the formulation of functional beverages include the use of green botanicals, which are a rich source of bioactive compounds such as vitamins, minerals, and antioxidants. This group of vegetables is still unexploited in the functional food sector and represents a promising alternative in the search and development of new products. The main challenge in the use of these sources is to mask the taste associated with them in order to ease their acceptance by consumers. To do so, new technologies of extraction, for example, using low temperatures, are under study (Fitzgerald *et al.*, 2011).

Among green botanicals, seaweeds have a long tradition in Asian countries and have been recently introduced in the Western countries due to its image as natural and healthy source of compounds. Seaweeds are rich in minerals, vitamins, peptides, lecithin, and prebiotics. In addition, microalgae have polyunsaturated fatty acids (PUFAs), carotenoids, and polyphenols (Dhargalkar & Verlecar, 2009). Different studies have demonstrated their beneficial effects as hypotensive, antioxidant, and preventive of diabetes, cancer, and cardiovascular disease (CVD) (Burtin, 2003; Yang et al., 2010; Lee et al., 2010).

Cereal grasses are other products that are currently investigated in order to determine their potential application in beverages. These include alfalfa, wheat, barley, oat, and rye and are rich in amino acids, vitamins, and phytonutrients (Gruenwald, 2009). The consumption of juices from these products seems to act against ulcerative colitis and anemia (Ben-Arye et al., 2002; Marawaha et al., 2004).

Another source of bioactive compounds that is currently experiencing a high interest is *Aloe vera*. This plant has been used in cosmetics long ago but recently has started its inclusion in beverages. This is due to its broad range of bioactive compounds (more than 200) that can exert health-promoting activities such as digestive, antidiabetic, anticholesterolemic, antioxidant, and prevention of kidney stones, among others (Bolkent et al., 2004; Tanaka et al., 2006; Hu et al., 2003).

On the other hand, the recent trends in the formulation of functional beverages include the use of medicinal herbs, for example, chamomile, feverfew, meadowsweet, and willow (Harbourne et al., 2013). One more time, their bioactivity, basically anti-inflammatory (Blumenthal et al., 2000; Bone & Mills, 2000), can be attributed to its richness in phenolic compounds such as flavonoids and phenolic acids (Mulinacci et al., 2000; Cai et al., 2004). In addition, their essential oils, rich in terpenes such as sesquiterpenes, have also been described as health promoting (Medić-Šarić et al., 1997).

Finally, one new recruit in the field of functional food development are bamboo shoots. Again, these products come from the traditional medicine of Asian countries. Due to the high activities described from the consumption of bamboo shoots, such as digestive, hypotensive, and preventive of CVD and cancer, its introduction in the formulation of beverages has been evaluated and is currently used in some functional foods (Nirmala et al., 2014; Lobovikov, 2003; Chongtham et al., 2011).

10.4.3   By-Products in Beverage Formulation

Once the great potential of bioactive compounds as health-promoting molecules was demonstrated, food industries tried to find different sources in order to use them in the formulation of functional beverages. Their purposes were to obtain rich sources of these compounds at the lowest cost that is possible and at the same time to discover new compounds with unknown activities. In this search, the by-products generated during the industrial processing of fruits and vegetables turned out to be one of the most suitable sources (Mirabella et al., 2014). These by-products, which are obtained in high quantities, retain high amounts of bioactive compounds such as phenolic compounds, vitamins, and minerals. The main by-products generated in the food industry are the pomace, peels, seeds, oils, leaves, and stones from fruits such as oranges, apples, grapes, peaches, apricots, lemons, and pineapple, among others (Sun-Waterhouse, 2011).

The high content of bioactive compounds of these by-products can be explained by several factors. For example, in the case of fruits that are used to produce juices, these compounds are retained within the fiber, the main component of the fruit pomace. Bioactive compounds interact with fiber and establish strong bonds that cannot be cleaved during the industrial processing, and therefore fiber ends up acting as scavenger. In addition, it has been demonstrated that this fiber has also health-promoting activity, increasing the importance of pomace as one of the most promising sources of bioactive compounds (Nawirska & Kwaśniewska, 2005; Larrauri, 1999).

Apart from fruits, new trends include the search of bioactive compounds from vegetables by-products. For example, broccoli by-products have shown to be rich in flavonoids, phenolic acids, glucosinolates, minerals, and vitamin C (Domínguez-Perles et al., 2010). In their manuscript, Domínguez-Perles et al. postulated that these compounds can be used in the design of novel functional beverages, for example, using green tea as matrix. Thus, a product in which a combination of complementary bioactives from both green tea and broccoli is obtained. Another study evaluated the use of carrot pomace as source of bioactives. After hydrolysis, they obtained an extract that could be used as an additive that not only enhanced the organoleptic properties of the juices but also significantly increased its phenolic content and its antioxidant capacity (Stoll et al., 2003).

The interest in by-products can be seen by the high number of research manuscripts published during the past decade. Even official organizations from the countries aware of the repercussion that these by-products have within the industry, not only as source of bioactive compounds but also due to their environmental impact, are promoting their study by means of grant programs such as the Horizon 2020 calls.

10.5   Novel Infusions (Tea and Tea Alternatives)

In the beverage sector, tea is, after water, the most widely consumed nonalcoholic beverage in the world well ahead of coffee, carbonated soft drinks, and other popular beverages (Ashurst, 2004). Native to China, tea infusions, produced from fermented (black tea) and wild leaves (green tea) of *Camellia sinensis*, are well accepted worldwide due to its taste, aroma, and bioactive compounds (Khan *et al.*, 2007). This contributes to reach the 20 billion of cups consumed every day worldwide (Marcos *et al.*, 1998). In addition, there are lots of infusions made of leaves or flowers that were known in Europe long before tea was introduced such as *Lamiaceae* (*Mentha*, *Thymus*), *Asteraceae* (*Matricaria*), *Rosaceae* (*Rubus*), and *Malvaceae* (*Tilia*) (Sõukand *et al.*, 2013) that are still consumed. Nowadays, teas and similar infusions are receiving great attention within the health and wellness food market. Thus, innovations in ready-to-drink teas and more elaborate packaging are being designed to comply with brewing requirements of premium leaf while responding as a convenient vehicle for fortification that can add even more health benefits to this drink. Hence, herbal extracts can be incorporated into soft drinks, mineral water-based drinks, and energy drinks (Ansari & Kumar, 2012).

Regarding beneficial effects, tea is associated with reduction of serum cholesterol, prevention of low-density lipoprotein oxidation, decreased risk of CVD, and inhibitory activity against tumorigenesis (Chung *et al.*, 2003). These effects are mainly attributed to its flavonoid content. In fact, black tea is the major single contributor to monomeric flavonoids both in the Europe and US diets (Song & Chun,

2008; Vogiatzoglou *et al.*, 2014). The main green tea polyphenols are epicatechin, epicatechin-3-gallate (ECG), epigallocatechin (EGC), and epigallocatechin-3-gallate (EGCG) (Yang *et al.*, 2002). In addition, other tea compounds that have been reported to be beneficial to human health are fluoride, caffeine, and essential minerals (Cabrera *et al.*, 2003). Epidemiological data support a link between black tea consumption and reduced risk of CVD possibly through a blood-pressure (BP) lowering effect (Hodgson & Croft, 2010). However, the results from intervention studies are inconsistent and either found no significant effect (Hooper *et al.*, 2008; Taubert *et al.*, 2007) or included low number of trials (Hartley *et al.*, 2013). Even so, meta-analysis performed through 11 high-quality randomized controlled trials found a beneficial effect of regular tea ingestion for both systolic and diastolic BP, and this was more pronounced in those with a higher baseline BP (Greyling *et al.*, 2014) and with low doses and long interventions (Peng *et al.*, 2014). From a mechanistic point of view, this effect has been attributed to the influence of flavonoids on endothelial function, which is considered as an early marker for BP changes (Greyling *et al.*, 2014). Human data indicate that black tea consumption improves endothelial function (Ras *et al.*, 2011). Furthermore, other reports underlie the beneficial effect of green tea on BP through the maintenance of the vascular tone by balancing vasoconstricting and vasodilating substances (Bhardwaj & Khanna, 2013) through an improvement of ventricular function that exerts beneficial effects via increasing nitric oxide (NO) production from endothelium in PI3-kinase-dependent pathways (Grassi *et al.*, 2008) and improving the regulation of eNOS activation and ROS production, thereby increasing the production of NO (Babu & Liu, 2008). Green tea is able to reduce oxidative stress and manage the generation of ROS by inhibiting pro-oxidant enzymes and inducing antioxidant (Cabrera *et al.*, 2006). Moreover, green tea catechins can induce an anti-inflammatory effect by suppression of several inflammatory factors, such as cytokines, adhesion molecules, and NF-κB (Lin & Lin, 1997).

Other CVD risk factors are reported to be positively affected by tea infusions. Green tea consumption might significantly reduce plasma low-density-lipoprotein cholesterol and total cholesterol concentrations (Zheng *et al.*, 2011) and may have a favorable effect on glucose control in adults (Liu *et al.*, 2013). However, these outcomes have been inconsistent, probably due to uncontrolled confounding factors caused by the lifestyle habits. Furthermore, regular black tea consumption improves body weight and body fat distribution, waist circumference, and waist-to-hip ratio compared to a caffeine-matched control beverage, after 3 months of regular consumption (Bøhn *et al.*, 2014).

The positive effects of tea in cancer have been partially attributed to the antioxidative property of tea polyphenols (Wiseman *et al.*, 1997). In this sense, tea consumption (1–6 cups/day) consistently leads to a significant increase in the antioxidant capacity of the blood to reduce the damage of important biomolecules, such as DNAs and lipids (Hakim *et al.*, 2004). There are some reports that show inhibitory effect of tea on *in vivo* models of lung cancer formation induced by cigarette smoking (Xu *et al.*, 1992; Hakim *et al.*, 2004). However, the quality of the studies is insufficient to draw strong conclusions. For example, due to the lack of intervention adherence, it may be difficult to estimate real tea dosage, which is especially prone to bias and imprecision (Rietveld & Wiseman, 2003).

Regarding the bioactivity of other infusions, it has been suggested that yerba mate exerts some benefits such as antioxidant, anti-inflammatory, antiobesity, and anticancer (Burris *et al.*, 2012). In fact, a 60-day intervention pilot study about the effects of mate consumption promoted a significant increase in GSH concentration and a decrease in LOOH levels in diabetic and prediabetic subjects, attenuating oxidative stress and preventing diabetes complications (Boaventura et al., 2013). Mate infusions have polyphenolic content comparable to tea and orange juice, and their antioxidant activity is slightly higher than wines, orange juice, and black tea but lower than green tea. Recently, it has been reported that the consumption of mate infusions would significantly contribute to the overall antioxidant intake, providing high amounts of caffeoylquinic acid derivatives, with biological effects potentially beneficial for human health (Bravo et al., 2007).

10.6 Fortified Beverages

There is no global consensus about the products included within the fortified beverages group. Some authors consider that fortified beverages include four subsegments, namely sports drinks, enriched beverages, nutraceutical, and energy drinks (Ashurst, 2004), while others consider that fortified beverages are achieved by just adding a given nutrient to a beverage, which serves as a vehicle of this nutrient. Recent definitions consider fortified beverages as those that had been added as beneficial compounds, including phytochemicals derived from natural products (Ansari & Kumar, 2012).

Milk and, more recently, fruit juices have revealed appropriate food matrices to develop fortified beverages. As dairy products supply high amounts of the daily requirement of minerals (calcium, magnesium, iron, and potassium) and vitamins (A, D, B2, and B12), it is used as the basic building block of many products to deliver other nutritional products that bear health claims. Whey proteins, hydrolyzed proteins, and low glycemic carbohydrates are also used as important matrices for new products addressed to different health condition (Özer & Kirmaci, 2010). Recently, fermented processes had become recognized as a relatively easy and natural technology to raise vitamin levels in beverages (Hugenholtz, 2013). Numerous examples, especially involving dairy applications, have been natural enrichment with riboflavin (Burgess et al., 2004), folate (Sybesma et al., 2003), vitamin B12 (Taranto et al., 2003), vitamin K2 (Tani, 1992), and, sometimes, several of these vitamins, simultaneously (Sybesma et al., 2004). In addition, successful applications in dairy products, such as folate enrichment by using Lactobacillus plantarum (Hugenschmidt et al., 2011), are examples of other fermented food products (Table 10.1).

Fortification of products is a simple method to overcome nutritional deficiencies and related pathologies. For example, it is well known that age-related bone loss and osteoporosis are linked to low dietary calcium uptake and inadequate vitamin D storage in the body. Thus, fortification of semiskimmed or nonfat milks with vitamin D is required. The consumption of dairy enriched with calcium and vitamin D-3 is a simple strategy to reduce age-related bone loss in the elderly (Daly et al., 2006) and to reduce the risk of pre-eclampsia and maternal mortality (Hofmeyr et al., 2010). Furthermore, it is reported that the bioavailability of calcium added as gluconate to buffalo milk at a level of 50 mg/100 ml is higher than with other calcium salts (Ranjan et al., 2005). Another case of interest is iron because it is one of the most critical nutritional deficiencies worldwide, especially in underdeveloped and developing countries. Among the strategies to prevent and treat anemia, food fortification is a low-cost alternative, easily accepted into the food habits of the population (Yang & Huffman, 2011). For

Table 10.1 Matrix and bioactive compounds used in the development of fortified beverages

Matrix	Bioactive compound	References
Fermented dairy products	Riboflavin	Burgess *et al.* (2004)
	Folate	Sybesma *et al.* (2003); Hugenschmidt *et al.* (2011)
	Vitamin B12	Taranto *et al.* (2003)
	Vitamin K2	Tani (1992)
	Iron	Silva *et al.* (2008)
Milk	Vitamin D	Daly *et al.* (2006)
	Calcium gluconate	Ranjan *et al.* (2005)
	Essential fatty acids	Yang and Huffman (2011)
Thickened beverage	Inulin	Dahl *et al.* (2005)
Orange-flavored beverage	ß-Glucan	Temelli *et al.* (2004)

example, it seems that iron absorption is enhanced by fermented milks due to lactic acid and other organic acids (Silva *et al.*, 2008). However, although there are loads of papers about the health-promoting effect of iron-fortified milks, scientific studies of the technological strategies in milk and dairy beverages are limited (Özer & Kirmaci, 2010). The main technological problems of the iron fortification of dairy beverage are the development of undesirable color and its poor solubility, as well as the stimulation of protein instability.

Regarding the beneficial effects of functional foods, a positive correlation of fortified food supplementation during pregnancy with the offspring's learning capability throughout childhood has been seen. A few studies have also shown that fortified foods have impacts on increasing birth length and reducing preterm delivery. Thus, fortified beverages and supplements containing milk and essential fatty acids offer benefits, improving maternal status and pregnancy outcome (Yang & Huffman, 2011). Zinc, copper, iodine, selenium, and vitamins E, C, and B group are other important nutrients used in fortified beverages. However, due to their negative impact in the organoleptic properties, many tested products have not included these nutrients or have done it in a limited way. In fact, although minerals such as magnesium and selenium are potentially useful for functional dairy beverages, to the best of our knowledge, no commercial dairy beverage enriched with selenium is currently on the market (Foster *et al.*, 1998). Recent efforts to improve the selenium level in cow's and sheep's milk through natural measures instead of adding selenite to milk have been successful.

Finally, dietary fiber enrichment of food beverages with inulin, β-glucan, or pectins is a very promising strategy to effectively exert positive effects on cardiovascular and digestive metabolisms. Studies conducted in a double-blind, three-week study, testing an inulin-fortified beverage against a starch-thickened one, showed that the test beverage was well accepted by the subjects (Dahl *et al.*, 2005). In another study, an orange-flavored beverage containing β-glucan extracted from barley was developed, thus showing the interest of the industry in the development of these products (Temelli *et al.*, 2004).

10.7 Cocoa-Based Beverages

Cocoa is very rich in flavanols (Lee *et al.*, 2003; Manach *et al.*, 2004), particularly catechin, epicatechin, and procyanidin (Wollgast & Anklam, 2000). Although other sources of these compounds such as wine and tea exist, cocoa is believed to have the highest content of flavanols (Arts *et al.*, 1999; Lee *et al.*, 2003). In fact, it is due to this composition that cocoa and cocoa derivatives are considered good candidates for functional foods. Early coca research was related with the role of these compounds in the development of flavor and aroma in chocolate, but posteriorly work has changed its focus on their potential health benefits. There is a growing body of evidence that cocoa may have beneficial effects on health, particularly in relation to the CVD. Indeed, the cardiovascular benefits of cocoa and cocoa derivatives have been extensively studied (Ding *et al.*, 2006; Cooper *et al.*, 2008; Galleano *et al.*, 2009). It was reported that cocoa consumption reduced cardiovascular mortality in the indigenous populations of Kuna islands (Hollenberg *et al.*, 1997), and more recently, it has also been associated with lower cardiovascular mortality (Buijsse, 2006; Mink *et al.*, 2007; Fraga *et al.*, 2011). Actually, an epidemiological long-term study has reported a lowering effect of cocoa intake on cardiovascular mortality in elderly men (Buijsse, 2006). In addition, a prospective study in postmenopausal women demonstrated a borderline inverse correlation of chocolate intake and CVD (Mink *et al.*, 2007). The antioxidant action of cocoa polyphenols can reduce the susceptibility of blood cholesterol to oxidation and protect the blood vessel walls against oxidative damage (Pearson *et al.*, 2001; Osakabe *et al.*, 2004), which are important aspects in the development of atherosclerosis. Interestingly, the protective effect of epicatechin and procyanidins on LDL-oxidation can occur at physiologically relevant concentrations, ranging from 0.1 to 0.5 μM (Steinberg *et al.*, 2002). Inflammation is recognized as another independent mechanism in the pathogenesis of atherosclerosis. Some data from human studies suggest that cocoa polyphenols may affect the inflammatory response via eicosanoid modulation (Schramm *et al.*, 2001; Holt *et al.*, 2002; Sies *et al.*, 2005), increasing the levels of anti-inflammatory prostacyclines and decreasing the levels of the proinflammatory leukotrienes (Rein *et al.*, 2000; Schramm *et al.*, 2001; Holt *et al.*, 2002; Pearson *et al.*, 2002). In addition, cocoa polyphenols also appear to exert beneficial effects on blood platelets (Pearson *et al.*, 2005), having an aspirin-like effect that may reduce thrombosis risk (Pearson *et al.*, 2002). They may also induce endothelium-dependent vessel relaxation improving the endothelial function (Heiss *et al.*, 2003, 2005; Engler *et al.*, 2004; Vlachopoulos *et al.*, 2005; Schroeter *et al.*, 2006). Indeed, different studies have shown significant decrements of BP in elderly, normotensive, and hypertensive subjects (Taubert *et al.*, 2003; Grassi *et al.*, 2005a,b). With regard to the potential mechanisms involved, numerous studies indicate that cocoa polyphenols might exert nitric oxide (NO)-mediated cardioprotective effects (Keen *et al.*, 2005; Cooper *et al.*, 2008). In fact, results from our group have demonstrated that the NG-nitro-L-arginine methyl ester (L-NAME), a nonselective NO synthesis inhibitor, completely inhibited the antihypertensive effects of cocoa polyphenols (Quiñones *et al.*, 2011). It is important to point out that beneficial effects of NO modulation include regulation of blood pressure and also lowering NO affected hypercholesterolemia, inhibition of platelet aggregation, and monocyte adhesion, all of which are involved in the progression of atherosclerosis. Hence, events related to reduce cardiovascular risk with NO levels as a target include lipoprotein oxidation, inflammation, platelet aggregation, and endothelial dysfunction (Cooper *et al.*, 2008).

Therefore, the large concentration of beneficial effects attributed to cocoa polyphenols makes it an ideal ingredient for functional beverages. However, an important point to consider in the use of cocoa in functional food is the polyphenols bioavailability from food matrix. In concrete, the milk has been considered a controversial matrix. Some authors reported that milk had a negative impact on the bioavailability of cocoa, and also tea, polyphenols as milk proteins can cause binding of these molecules, the milk–polyphenol complexes could resist gastric breakdown, and therefore, the polyphenols would not be available for absorption (Leenen *et al.*, 2000). Other possible explanation is that milk might hinder polyphenol absorption by increasing gastric pH (Serafini *et al.*, 1996). In fact, a crossover study concluded that the interaction between milk proteins and cocoa polyphenols inhibited the absorption of epicatechin (Serafini & Crozier, 2003). However, other authors demonstrated that the epicatechin concentration in plasma was the same after the consumption of milk- or water-based cocoa beverages under isocaloric and isolipidemic conditions (Schroeter *et al.*, 2003; Roura *et al.*, 2007). In addition, another study concluded that the consumption of flavanol-rich cocoa with milk did not present effects on catechin and epicatechin absorption (Schramm *et al.*, 2003). Moreover, the use of high polyphenols cocoa powder dissolved in milk leads to an increase in the cocoa metabolites measured in plasma and urine showing a higher bioavailability (Tomas-Barberán *et al.*, 2007).

10.8   Energy Drinks

Within the functional beverages framework, a new field emerged some years ago with the onset of energy drinks. These drinks are defined as nonalcoholic beverages that are commercialized highlighting their energizing effects due to its composition rich in caffeine, taurine, and vitamins, among other ingredients (Schaffer *et al.*, 2014). As can be stated from an external scientific report commissioned by the EFSA (Zucconi *et al.*, 2013), these products have steadily gained consumers, especially among adolescents, who are attracted by their stimulating effects. It is important to state that energy drinks should not be confused with isotonic beverages that provide minerals, carbohydrates, and water lost during intense sports activities.

Industrial stakeholders agree to consider these beverages as members of the family of functional foodstuffs. In fact, their health claims are under the scope of the European Nutrition and Health Claims Regulation (NHCR) by the EC Regulation 1924/2006. However, most energy drinks contain high doses of caffeine (ranging from 70 to 400 mg/l), compound which content in food products is highly regulated in some countries. Consequently, the energy drinks must accomplish supplementary regulations depending on the country of production and commercialization (Zucconi *et al.*, 2013).

Regarding their composition, caffeine is the most abundant ingredient. Caffeine is a xanthine alkaloid, widely distributed in the plant kingdom, present in many products of the diet such as coffee, cocoa, and tea. Caffeine is an adenosine receptor antagonist and its main effect is to act as central nervous system stimulant, reducing the sleep sensation and boosting the overall activity of the organism (Jones, 2008). Its known drawbacks are related to sleep disorders, anxiety, cardiac arrest, and a general overstimulation of the central nervous system, which can have serious consequences after chronic intake of high doses (Higgins *et al.*, 2010). In addition, it is also known that its effect increases the basal energy expenditure that can be related with the weight lost due to a greater amount of expended calories (Klepacki, 2010; Dulloo *et al.*, 1989). Although the European Union has not approved any caffeine health claims, the EFSA has delivered three positive opinions linking the consumption of 3–4 g of caffeine 1 h before performing exercise with the reduction of the fatigue perception and the increase in both endurance capacity and performance. The last trends in the energy drinks include the use of caffeine from natural sources such as guarana. Guarana is a plant from the Amazon basin that is known to be one of the richest sources of caffeine in the plant kingdom. Its consumption is widely spread in Brazil and its introduction in the formulation of the energy drinks allowed to commercialize these products as "more naturals." In addition, guarana is also rich in a diverse range of vitamins such as vitamins A, E, and B-families, minerals, and phenolic compounds, which are health-promoting molecules that complement and increase the functional effect of these beverages.

Apart from caffeine, taurine and D-glucuronolactone are the other most common compounds included in energy drinks. Taurine is a nonproteinic amino acid naturally produced by human body and also a component of the diet. It is essential for the cardiovascular function and the development and function of the skeletal muscle, the central nervous system, and the retina. For its part, D-glucurono-γ-lactone is a structural component of connective tissues. Regarding these two ingredients, a scientific opinion by the EFSA in 2009 concluded that the exposure to taurine and D-glucurono-γ-lactone at the doses currently used in the energy drinks are not sufficient to be considered as a safety concern, thus contradicting complaints from some consumer organizations. Finally, other bioactive compounds included in the formulation of energy drinks are sugars (basically maltodextrins, which are good source of energy rapidly used after ingestion), antioxidants (which protect the organism against oxidative stress caused from free radicals), B-vitamins (coenzymes needed to convert the sugars included within the energy drinks into energy), ginseng (a Chinese medicinal herb that it is said to increase memory and relieve stress, anxiety, and fatigue), L-carnitine (an amino acid that stimulates lipid metabolism), and gingko biloba extract (a traditional antioxidant from Chinese medicine) (Higgins *et al.*, 2010). The broad range of compounds that nowadays are included in the energy drink reflects that it is a market still in expansion.

10.9 Beverage Emulsions

Regarding innovation in beverage emulsions, following up the known as "healthy-style pattern," last trends include the replacement of synthetic surfactants by others with a natural origin. Among these, the use of saponins has emerged as a very interesting alternative. Saponins are a heterogenic group of glycosidic secondary metabolites widely distributed in plants that exert a broad range of pharmacological activities including anti-inflammatory, hypocholesterolemic, vasoprotective, immunomodulatory, hypoglycemic, and antifungal, among others (Podolak *et al.*, 2010). As an example, saponins from the bark of *Quillaja saponaria* have been used as surfactant molecule in beverage emulsions instead of Tweens (Piorkowski & McClements, 2013). In order to evaluate their suitability as additives in beverages, some authors have determined its most important parameters. For example, Yang *et al.* (2013) evaluated their stability and effectiveness in the formation of edible delivery systems, therefore considering it suitable for encapsulation in beverages.

10.10   Conclusions and Future Trends

Although it has suffered an intense development in the past years, functional food science is still in early stages. Future development includes discovery of new phytonutrients and their activities as well as technological improvement in order to introduce them in an effective way. Everything points out that optimized nutrition will be the major challenge of nutrition in the twenty first century and functional foods have their own role to play in it. These food compounds or food with health-enhancing properties have provided immense commercial opportunities to food and drink industry, which has spent millions of euros to deliver new foodstuffs bearing health claims to the market. In fact, functional food and nutraceutical market in the European Union have grown from about $1.8 billion out of a $5.7 billion global market in 1999 to $8 billion out of a global market of $167 billion in 2010. This expansion has led to the development of auxiliary services based on R&D. As an example, over 140 European companies from more than 20 countries participated in the European FF Network. Due to the economic importance of this area (current global turnover estimated to be $200 billion USD) and the potential to exponentially increase markets, there is a strong interest from both industry and governments to capitalize on these opportunities. Future trends also include the inclusion of foodomics on functional foods. Thus, genomics, proteomics, and metabolomics improvement and development will play a crucial role in this field allowing developing the concept of personalized nutrition in a more proper way and approaching this concept in the food industry.

References

Akihisa, T. *et al.* (2007) Anti-inflammatory and potential cancer chemopreventive constituents of the fruits of Morinda citrifolia (Noni). *Journal of Natural Products* **70**, 754–757.

Amagase, H. & Nance, D.M. (2008) A randomized, double-blind, placebo-controlled, clinical study of the general effects of a standardized *Lycium barbarum* (Goji) juice, GoChi. *Journal of Alternative and Complementary Medicine* **14**, 403–412.

Ansari, M.M. & Kumar, D.S. (2012) Fortification of food and beverages with phytonutrients. *Food and Public Health* **2**, 241–253.

Arts, I.C., Hollman, P.C. & Kromhout, D. (1999) Chocolate as a source of tea flavonoids. *Lancet* **354**, 488.

Ashurst, P.R. (2004) *Chemistry and Technology of Soft Drinks and Fruit Juices* (ed. P.R. Ashurst). Wiley-Blackwell, ISBN: 978-1-4443-3381-7.

Babu, P.V.A. & Liu, D. (2008) Green tea catechins and cardiovascular health: an update. *Current Medicinal Chemistry* **15**, 1840–1850.

Bech-Larsen, T. & Scholderer, J. (2007) Functional foods in Europe: consumer research, market experiences and regulatory aspects. *Trends in Food Science and Technology* **18**, 231–234.

Bekhit, A.E.D. *et al.* (2011) Antioxidant activities, sensory and anti-influenza activity of grape skin tea infusion. *Food Chemistry* **129**, 837–845.

Ben-Arye, E. *et al.* (2002) Wheat grass juice in the treatment of active distal ulcerative colitis: a randomized double-blind placebo-controlled trial. *Scandinavian Journal of Gastroenterology* **37**, 444–449.

Bhardwaj, P. & Khanna, D. (2013) Green tea catechins: defensive role in cardiovascular disorders. *Chinese Journal of Natural Medicines* **11**, 345–353.

Blumenthal, M., Goldberg, A. & Brinckmann, J. (2000) *Herbal Medicine: Expanded Commission E Monographs*. Integrative Medicine Communications, American Botanical Council, Newton, MA.

Boaventura, B.C.B. *et al.* (2013) Antioxidant potential of mate tea (Ilex paraguariensis) in type 2 diabetic mellitus and pre-diabetic individuals. *Journal of Functional Foods* **5**, 1057–1064.

Bøhn, S.K. *et al.* (2014) Effects of black tea on body composition and metabolic outcomes related to cardiovascular disease risk: a randomized controlled trial. *Food & Function* **5**, 1613–1620.

Bolkent, S. *et al.* (2004) Effect of *Aloe vera* (L.) Burm. fil. leaf gel and pulp extracts on kidney in type-II diabetic rat models. *Indian Journal of Experimental Biology* **42**, 48–52.

Bone, K. & Mills, S. (2000) *Principles and Practice of Phytotherapy: Modern Herbal Medicine*, Elsevier Ltd.

Bravo, L., Goya, L. & Lecumberri, E. (2007) LC/MS characterization of phenolic constituents of mate (Ilex paraguariensis, St. Hil.) and its antioxidant activity compared to commonly consumed beverages. *Food Research International* **40**, 393–405.

Buijsse, B. (2006) Cocoa intake, blood pressure, and cardiovascular mortality: the zutphen elderly study. *Archives of Internal Medicine* **166**, 411–417.

Burgess, C. *et al.* (2004) Riboflavin production in *Lactococcus lactis*: Potential for *in situ* production of vitamin-enriched foods. *Applied and Environmental Microbiology* **70**, 5769–5777.

Burris, K.P. *et al.* (2012) Composition and bioactive properties of yerba mate (llex paraguariensis A. St.-Hil.): a review. *Chilean Journal of Agricultural Research* **72**, 268–275.

Burtin, P. (2003) Nutritional value of seaweeds. *Electronic Journal of Environmental Agricultural and Food Chemistry (EJEAFChe)* **2**, 498–503.

Cabrera, C., Artacho, R. & Giménez, R. (2006) Beneficial effects of green tea – a review. *Journal of the American College of Nutrition* **25**, 79–99.

Cabrera, C., Giménez, R. & López, M.C. (2003) Determination of tea components with antioxidant activity. *Journal of Agricultural and Food Chemistry* **51**, 4427–4435.

Cai, Y. *et al.* (2004) Antioxidant activity and phenolic compounds of 112 traditional Chinese medicinal plants associated with anticancer. *Life Sciences* **74**, 2157–2184.

Chongtham, N., Bisht, M.S. & Haorongbam, S. (2011) Nutritional properties of bamboo shoots: potential and prospects for utilization as a health food. *Comprehensive Reviews in Food Science and Food Safety* **10**, 153–168.

Chung, F.-L. *et al.* (2003) Tea and cancer prevention: studies in animals and humans. *Journal of Nutrition* **133**, 3268S–3274S.

Cooper, K. *et al.* (2008) Cocoa and health: a decade of research. *British Journal of Nutrition* **99**, 1–11.

Dahl, W.J. *et al.* (2005) Effects of thickened beverages fortified with inulin on beverage acceptance, gastrointestinal function, and bone resorption in institutionalized adults. *Nutrition* **21**, 308–311.

Daly, R.M. *et al.* (2006) Calcium- and vitamin D3-fortified milk reduces bone loss at clinically relevant skeletal sites in older men: a 2-year randomized controlled trial. *Journal of Bone and Mineral Research: The Official Journal of the American Society for Bone and Mineral Research* **21**, 397–405.

Dhargalkar, V.K. & Verlecar, X.N. (2009) Southern ocean seaweeds: a resource for exploration in food and drugs. *Aquaculture* **287**, 229–242.

Ding, E.L. *et al.* (2006) Chocolate and prevention of cardiovascular disease: a systematic review. *Nutrition & Metabolism* **3**, 2.

Diplock, A.T. *et al.* (1999) Scientific concepts of functional foods in Europe: consensus document. *British Journal of Nutrition* **81**(Suppl. 1), S1–S27.

Domínguez-Perles, R. *et al.* (2010) Broccoli-derived by-products – a promising source of bioactive ingredients. *Journal of Food Science* **75**, 383–392.

Dulloo, A.G. *et al.* (1989) Normal caffeine consumption: influence on thermogenesis and daily energy expenditure in lean and postobese human volunteers. *American Journal of Clinical Nutrition* **49**, 44–50.

Engler, M.B. *et al.* (2004) Flavonoid-rich dark chocolate improves endothelial function and increases plasma epicatechin concentrations in healthy adults. *Journal of the American College Nutrition* **23**, 197–204.

Fernández-Ginés, J.M. *et al.* (2008) Production of functional ingredients from by-products. *Alimentaria: Investigación, Tecnología y Seguridad* **398**, 76–79.

Ferrer, C. *et al.* (2009) Effect of olive powder on the growth and inhibition of *Bacillus cereus*. *Foodborne Pathogens and Disease* **6**, 33–37.

Fitzgerald, C. *et al.* (2011) Heart health peptides from macroalgae and their potential use in functional foods. *Journal of Agricultural and Food Chemistry* **59**, 6829–6836.

Foster, L.H., Chaplin, M.F. & Sumar, S. (1998) The effect of heat treatment on intrinsic and fortified selenium levels in cow's milk. *Food Chemistry* **62**, 21–25.

Fraga, C.G. *et al.* (2011) Cocoa flavanols: effects on vascular nitric oxide and blood pressure. *Journal of Clinical Biochemistry and Nutrition* **48**, 63–67.

Galleano, M., Oteiza, P.I. & Fraga, C.G. (2009) Cocoa, chocolate, and cardiovascular disease. *Journal of Cardiovascular Pharmacology* **54**, 483–490.

Georgiev, V., Ananga, A. & Tsolova, V. (2014) Recent advances and uses of grape flavonoids as nutraceuticals. *Nutrients* **6**, 391–415.

Grassi, D., Lippi, C. *et al.* (2005a) Short-term administration of dark chocolate is followed by a significant increase in insulin sensitivity and a decrease in blood pressure in healthy persons. *American Journal of Clinical Nutrition* **81**, 611–614.

Grassi, D., Necozione, S. *et al.* (2005b) Cocoa reduces blood pressure and insulin resistance and improves endothelium-dependent vasodilation in hypertensives. *Hypertension* **46**, 398–405.

Grassi, D. *et al.* (2008) Tea, flavonoids, and nitric oxide-mediated vascular reactivity. *Journal of Nutrition* **138**, 1554S–1560S.

Greyling, A. *et al.* (2014) The effect of black tea on blood pressure: a systematic review with meta-analysis of randomized controlled trials. *PLoS ONE*, **9**(7), e103247. doi: 10.1371/journal.pone.0103247. eCollection 2014.

Gruenwald, J., 2009. Novel botanical ingredients for beverages. *Clinics in Dermatology* **27**, 210–216.

Hakim, I.A. *et al.* (2004) Effect of a 4-month tea intervention on oxidative DNA damage among heavy smokers: role of glutathione S-transferase genotypes. *Cancer Epidemiology Biomarkers and Prevention* **13**, 242–249.

Harbourne, N. *et al.* (2013) Stability of phytochemicals as sources of anti-inflammatory nutraceuticals in beverages – a review. *Food Research International* **50**, 480–486.

Hardy, G. (2002) Nutraceuticals and functional foods: introduction and meaning a european consensus of scientific concepts of functional foods. *Nutrition* **16**, 688–697.

Hartley, L. *et al.* (2013) Green and black tea for the primary prevention of cardiovascular disease. *Cochrane Database of Systematic Review* **6**, CD009934.

Haruenkit, R. *et al.* (2007) Comparative study of health properties and nutritional value of durian, mangosteen, and snake fruit: experiments *in vitro* and *in vivo*. *Journal of Agricultural and Food Chemistry* **55**, 5842–5849.

Hasler, C.M. (1998) Functional foods: their role in disease prevention and health promotion. *Food Technology* **52**, 63–70.

Heasman, M. & Mellentin, J. (2001) *The Functional Foods Revolution: Healthy People, Healthy Profits?*, Paperback.

Heiss, C. *et al.* (2003) Vascular effects of cocoa rich in flavan-3-ols. *JAMA* **290**, 1030–1031.

Heiss, C. *et al.* (2005) Acute consumption of flavanol-rich cocoa and the reversal of endothelial dysfunction in smokers. *Journal of the American College of Cardiology* **46**, 1276–1283.

Hertog, M.G. *et al.* (1994) Dietary flavonoids and cancer risk in the Zutphen elderly study. *Nutrition and Cancer* **22**, 175–184.

Higgins, J.P., Tuttle, T.D. & Higgins, C.L. (2010) Energy beverages: content and safety. *Mayo Clinic Proceedings* **85**, 1033–1041.

Hodgson, J.M. & Croft, K.D. (2010) Tea flavonoids and cardiovascular health. *Molecular Aspects of Medicine* **31**, 495–502.

Hofmeyr, G.J. *et al.* (2010) Calcium supplementation during pregnancy for preventing hypertensive disorders and related problems. *Cochrane Database of Systematic Reviews (Online)* **8**, CD001059.

Hollenberg, N.K. *et al.* (1997) Aging, acculturation, salt intake, and hypertension in the Kuna of Panama. *Hypertension* **29**, 171–176.

Holt, R.R. *et al.* (2002) Chocolate consumption and platelet function. *Journal of the American Medical Association* **287**, 2212–2213.

Hooper, L. *et al.* (2008) Flavonoids, flavonoid-rich foods, and cardiovascular risk: a meta-analysis of randomized controlled trials. *American Journal of Clinical Nutrition* **88**, 38–50.

Howell, A.B. *et al.* (1998) Inhibition of the adherence of P-fimbriated *Escherichia coli* to uroepithelial-cell surfaces by proanthocyanidin extracts from cranberries. *New England Journal of Medicine* **339**, 1085–1086.

Hu, Y., Xu, J. & Hu, Q. (2003) Evaluation of antioxidant potential of *Aloe vera* (*Aloe barbadensis* Miller). *Journal of Agricultural Food Chemistry* **51**,7788–7791.

Huang, H.Y., Helzlsouer, K.J. & Appel, L.J. (2000) The effects of vitamin C and vitamin E on oxidative DNA damage: results from a randomized controlled trial. *Cancer Epidemiology, Biomarkers & Prevention: A Publication of the American Association for Cancer Research, Cosponsored by the American Society of Preventive Oncology* **9**, 647–652.

Hugenholtz, J. (2013) Traditional biotechnology for new foods and beverages. *Current Opinion in Biotechnology* **24**, 155–159.

Hugenschmidt, S., Schwenninger, S.M. & Lacroix, C. (2011) Concurrent high production of natural folate and vitamin B12 using a co-culture process with *Lactobacillus plantarum* SM39 and *Propionibacterium freudenreichii* DF13. *Process Biochemistry* **46**, 1063–1070.

Jayaprakasha, G.K., Selvi, T. & Sakariah, K.K. (2003) Antibacterial and antioxidant activities of grape (*Vitis vinifera*) seed extracts. *Food Research International* **36**, 117–122.

Jew, S. *et al.* (2008) Generic and product-specific health claim processes for functional foods across global jurisdictions. *Journal of Nutrition* **138**, 1228S–1236S.

Jones, G. (2008) Caffeine and other sympathomimetic stimulants; modes of action and effects on sports performance. *Essays in Biochemistry* **44**, 109–123.

Joshipura, K.J. *et al.* (2001) The effect of fruit and vegetable intake on risk for coronary heart disease. *Annals of Internal Medicine* **134**, 1106–1114.

Juarez-Iglesias, M. (2010) La normativa europea para la evaluación de las declaraciones nutricionales y propiedades saludables de los alimentos. In *Alimentos Saludables y de Diseño Específico. Alimentos Funcionales*, pp. 151–159. Instituto Tomás Pascual.

Keen, C.L. *et al.* (2005) Cocoa antioxidants and cardiovascular health. *American Journal of Clinical Nutrition* **81**(1 Suppl).

Kemperman, R. *et al.* (2013) Impact of polyphenols from black tea and red wine/grape juice on a gut model microbiome. *Food Research International* **53**, 659–669.

Khan, S.A. *et al.* (2007) Influence of green tea on enzymes of carbohydrate metabolism, antioxidant defense, and plasma membrane in rat tissues. *Nutrition* **23**, 687–695.

Klepacki, B. (2010) Energy drinks: a review article. *Strength and Conditioning Journal* **32**, 37–41.

Kris-Etherton, P.M. & Keen, C.L. (2002) Evidence that the antioxidant flavonoids in tea and cocoa are beneficial for cardiovascular health. *Current Opinion in Lipidology* **13**, 41–49.

Kwak, N.S. & Jukes, D. (2001) Issues in the substantiation process of health claims. *Critical Reviews in Food Science and Nutrition* **41**, 465–479.

Larrauri, J.A. (1999) New approaches in the preparation of high dietary fibre powders from fruit by-products. *Trends in Food Science and Technology* **10**, 3–8.

Lau, T.C. *et al.* (2012) Functional food: a growing trend among the health conscious. *Asian Social Science* **9**, 198–208.

Leahy, M., Roderick, R. & Brilliant, K. (2001) The cranberry-promising health benefits, old and new. *Nutrition Today* **36**, 254–265.

Lee, K.W. *et al.* (2003) Cocoa has more phenolic phytochemicals and a higher antioxidant capacity than teas and red wine. *Journal of Agricultural and Food Chemistry* **51**, 7292–7295.

Lee, H.J. *et al.* (2010) Algae consumption and risk of type 2 diabetes: Korean national health and nutrition examination survey in 2005. *Journal of Nutritional Science and Vitaminology* **56**, 13–18.

Leenen, R. *et al.* (2000) A single dose of tea with or without milk increases plasma antioxidant activity in humans. *European Journal of Clinical Nutrition* **54**, 87–92.

Lin, Y.L. & Lin, J.K. (1997) (−)-Epigallocatechin-3-gallate blocks the induction of nitric oxide synthase by down-regulating lipopolysaccharide-induced activity of transcription factor nuclear factor-kappaB. *Molecular Pharmacology* **52**, 465–472.

Liu, K. *et al.* (2013) Effect of green tea on glucose control and insulin sensitivity: a meta-analysis of 17 randomized controlled trials. *American Journal of Clinical Nutrition* **98**, 340–348.

Lobovikov, M. (2003) Bamboo and rattan products and trade. *Journal of Bamboo and Rattan* **2**, 397–406.

Manach, C. *et al.* (2004) Polyphenols: food sources and bioavailability. *American Journal of Clinical Nutrition* **79**, 727–747.

Marawaha, R.K. *et al.* (2004) Wheat grass juice reduces transfusion requirement in patients with thalassemia major: a pilot study. *Indian Pediatrics* **41**, 716–720.

Marcos, A. *et al.* (1998) Preliminary study using trace element concentrations and a chemometrics approach to determine the geographical origin of tea. *Journal of Analytical Atomic Spectrometry* **13**, 521–525.

Medić-Šarić, M. *et al.* (1997) Application of numerical methods to thin-layer chromatographic investigation of the main components of chamomile (*Chamomilla recutita (L.) Rauschert*) essential oil, *Journal of Chromatography A* **776**, 355–360.

Menrad, K. (2003) Market and marketing of functional food in Europe. *Journal of Food Engineering* **56**, 181–188.

Mink, P.J. *et al.* (2007) Flavonoid intake and cardiovascular disease mortality: a prospective study in postmenopausal women. *American Journal of Clinical Nutrition* **85**, 895–909.

Mirabella, N., Castellani, V. & Sala, S. (2014) Current options for the valorization of food manufacturing waste: a review. *Journal of Cleaner Production* **65**, 28–41.

Mollet, B. & Rowland, I. (2002) Functional foods: at the frontier between food and pharma Editorial overview Beat Mollet and Ian Rowland. *Current Opinion in Biotechnology* **13**, 483–485.

Mulinacci, N. *et al.* (2000) Characterization of Matricaria recutita L. flower extracts by HPLC-MS and HPLC-DAD analysis. *Chromatographia* **51**, 301–302.

Murcia, J.L. (2013) Alimentos funcionales. *Distribución y Consumo*.

Nawirska, A. & Kwaśniewska, M. (2005) Dietary fibre fractions from fruit and vegetable processing waste. *Food Chemistry* **91**, 221–225.

Nirmala, C., Bisht, M.S. & Laishram, M. (2014) Review bioactive compounds in bamboo shoots: health benefits and prospects for developing functional foods. *International Journal of Food Science and Technology* **49**, 1425–1431.

Ohama, H., Ikeda, H. & Moriyama, H. (2014) Health foods and foods with health claims in Japan. *Toxicology* **221**, 95–111

Osakabe, N. *et al.* (2004) Dose–response study of daily cocoa intake on the oxidative susceptibility of low-density lipoprotein in healthy human volunteers. *Journal of Health Science* **50**, 679–684.

Osman, M. (2004) Chemical and nutrient analysis of baobab (*Adansonia digitata*) fruit and seed protein solubility. *Plant Foods for Human Nutrition* **59**, 29–33.

Ozen, A.E., Pons, A. & Tur, J. (2012) Worldwide consumption of functional foods: a systematic review. *Nutrition Reviews* **70**, 472–481.

Özer, B.H. & Kirmaci, H.A. (2010) Functional milks and dairy beverages. *International Journal of Dairy Technology* **63**, 1–15.

Papadopoulou, C., Soulti, K. & Roussis, I.G. (2005) Potential antimicrobial activity of red and white wine phenolic extracts against strains of *Staphylococcus aureus*, *Escherichia coli* and *Candida albicans*. *Food Technology and Biotechnology* **43**, 41–46.

Pearson, D.A. *et al.* (2001) *Flavonoids and Other Polyphenols*, Elsevier.

Pearson, D. *et al.* (2002) The effects of flavanol-rich cocoa and aspirin on *ex vivo* platelet function. *Thrombosis Research* **106**, 191–197.

Pearson, D. *et al.* (2005) Flavanols and platelet reactivity. *Clinical & Developmental Immunology* **12**, 1–9.

Peng, X. *et al.* (2014) Effect of green tea consumption on blood pressure: a meta-analysis of 13 randomized controlled trials. *Scientific Reports* **6251**,1–7.

Piorkowski, D.T. & McClements, D.J. (2013) Beverage emulsions: recent developments in formulation, production, and applications. *Food Hydrocolloids*, 1–37

Podolak, I., Galanty, A. & Sobolewska, D. (2010) Saponins as cytotoxic agents: a review. *Phytochemistry Reviews* **9**, 425–474.

Quiñones, M. *et al.* (2011) Evidence that nitric oxide mediates the blood pressure lowering effect of a polyphenol-rich cocoa powder in spontaneously hypertensive rats. *Pharmacological Research* **64**, 478–481.

Ranjan, P. *et al.* (2005) Bioavailability of calcium and physicochemical properties of calcium-fortified buffalo milk. *International Journal of Dairy Technology* **58**, 185–189.

Ras, R.T., Zock, P.L. & Draijer, R. (2011) Tea consumption enhances endothelial-dependent vasodilation; a meta-analysis. *PLoS ONE*, **6**(3), e16974. doi: 10.1371/journal.pone.0016974.

Rein, D. *et al.* (2000) Cocoa and wine polyphenols modulate platelet activation and function. *Journal of Nutrition* **130**, 2120S–2126S.

Rhodes, P.L. *et al.* (2006) Antilisterial activity of grape juice and grape extracts derived from *Vitis vinifera* variety Ribier. *International Journal of Food Microbiology* **107**, 281–286.

Rietveld, A. & Wiseman, S. (2003) Antioxidant effects of tea: evidence from human clinical trials. *Journal of Nutrition* **133**, 3285S–3292S.

Rimm, E.B. (2002) Fruit and vegetables – building a solid foundation. *American Journal of Clinical Nutrition* **76**(1), 1–2.

Roura, E. *et al.* (2007) Human urine: epicatechin metabolites and antioxidant activity after cocoa beverage intake. *Free Radical Research* **41**, 943–949.

Scalbert, A. & Williamson, G. (2000) Chocolate: modern science investigates an ancient medicine. *Journal of Medicinal Food* **3**, 121–125.

Schaffer, S.W. *et al.* (2014) Effect of taurine and potential interactions with caffeine on cardiovascular function. *Amino Acids* **46**, 1147–1157.

Schramm, D.D. *et al.* (2001) Chocolate procyanidins decrease the leukotriene prostacyclin ratio in humans and human aortic endothelial cells. *American Journal of Clinical Nutrition* **71**, 36–40.

Schramm, D.D. *et al.* (2003) Food effects on the absorption and pharmacokinetics of cocoa flavanols. *Life Sciences* **73**, 857–869.

Schroeter, H. *et al.* (2003) Milk and absorption of dietary flavanols. *Nature* **426**, 787–788.

Schroeter, H. *et al.* (2006) (−)-Epicatechin mediates beneficial effects of flavanol-rich cocoa on vascular function in humans. *Proceedings of the National Academy of Sciences of the United States of America* **103**, 1024–1029.

Serafini, M. & Crozier, A. (2003) Nutrition: milk and absorption of dietary flavanols. *Nature* **426**, 788.

Serafini, M., Ghiselli, A. & Ferro-Luzzi, A. (1996) *In vivo* antioxidant effect of green and black tea in man. *European Journal of Clinical Nutrition* **50**, 28–32.

Sies, H. *et al.* (2005) Cocoa polyphenols and inflammatory mediators. *American Journal of Clinical Nutrition* **81**(1 Suppl), 304S–312S.

Silva, M.R. *et al.* (2008) Growth of preschool children was improved when fed an iron-fortified fermented milk beverage supplemented with *Lactobacillus acidophilus*. *Nutrition Research* **28**, 226–232.

Siró, I. *et al.* (2008) Functional food. Product development, marketing and consumer acceptance – a review. *Appetite* **51**, 456–467.

Song, W.O. & Chun, O.K. (2008) Tea is the major source of flavan-3-ol and flavonol in the U.S. diet. *Journal of Nutrition* **138**, 1543S–1547S.

Sõukand, R. *et al.* (2013) Plants used for making recreational tea in Europe: a review based on specific research sites. *Journal of Ethnobiology and Ethnomedicine* **9**, 58.

Steinberg, F.M. *et al.* (2002) Cocoa procyanidin chain length does not determine ability to protect LDL from oxidation when monomer units are controlled. *Journal of Nutritional Biochemistry* **13**, 645–652.

Stoll, T. *et al.* (2003) Application of hydrolyzed carrot pomace as a functional food ingredient to beverages. *Food, Agriculture and Environment* **1**, 88–92.

Sun-Waterhouse, D. (2011) The development of fruit-based functional foods targeting the health and wellness market: a review. *International Journal of Food Science and Technology* **46**, 899–920.

Sybesma, W. *et al.* (2003) Effects of cultivation conditions on folate production by lactic acid bacteria. *Metabolic Engineering* **69**, 4542–4548.

Sybesma, W. *et al.* (2004) Multivitamin production in *Lactococcus lactis* using metabolic engineering. *Metabolic Engineering* **70**, 109–115.

Tanaka, M. *et al.* (2006) Identification of five phytosterols from *Aloe vera* gel as anti-diabetic compounds. *Biological & Pharmaceutical Bulletin* **29**, 1418–1422.

Tani, Y. (1992) Microbial process of menaquinone production. *Journal of Nutritional Science and Vitaminology* **38**, 251–254.

Taranto, M.P. *et al.* (2003) *Lactobacillus reuteri* CRL1098 produces cobalamin. *Journal of Bacteriology* **185**, 5643–5647.

Taubert, D., Roesen, R. & Schömig, E. (2007) Effect of cocoa and tea intake on blood pressure: a meta-analysis. *Archives of Internal Medicine* **167**, 626–634.

Taubert, D. *et al.* (2003) Chocolate and blood pressure in elderly individuals with isolated systolic hypertension. *JAMA* **290**, 1029–1030.

Temelli, F., Bansema, C. & Stobbe, K. (2004) Development of an orange-flavored barley β-glucan beverage. *Cereal Chemistry* **81**, 499–503.

Tomas-Barberán, F.A. *et al.* (2007) A new process to develop a cocoa powder with higher flavonoid monomer content and enhanced bioavailability in healthy humans. *Journal of Agricultural and Food Chemistry* **55**, 3926–3935.

Vlachopoulos, C. *et al.* (2005) Effect of dark chocolate on arterial function in healthy individuals. *American Journal of Hypertension* **18**, 785–791.

Vogiatzoglou, A. *et al.* (2014) Assessment of the dietary intake of total flavan-3-ols, monomeric flavan-3-ols, proanthocyanidins and theaflavins in the European Union. *British Journal of Nutrition* **111**, 1463–1473.

Wiseman, S.A., Balentine, D.A. & Frei, B. (1997) Antioxidants in tea. *Critical Reviews in Food Science and Nutrition* **37**, 705–718.

Wollgast, J. & Anklam, E. (2000) Review on polyphenols in *Theobroma cacao*: changes in composition during the manufacture of chocolate and methodology for identification and quantification. *Food Research International* **33**, 423–447.

Xu, Y. *et al.* (1992) Inhibition of tobacco-specific nitrosamine-induced lung tumorigenesis in A/J mice by green tea and its major polyphenol as antioxidants. *Cancer Research* **52**, 3875–3879.

Yang, Z. & Huffman, S.L. (2011) Review of fortified food and beverage products for pregnant and lactating women and their impact on nutritional status. *Maternal and Child Nutrition* **7**, 19–43.

Yang, C.S., Maliakal, P. & Meng, X. (2002) Inhibition of carcinogenesis by tea. *Annual Review of Pharmacology and Toxicology* **42**, 25–54.

Yang, Y.J. *et al.* (2010) A case–control study on seaweed consumption and the risk of breast cancer. *British Journal of Nutrition* **103**, 1345–1353.

Yang, Y. *et al.* (2013) Formation and stability of emulsions using a natural small molecule surfactant: Quillaja saponin (Q-Naturale). *Food Hydrocolloids* **30**, 589–596.

Zheng, X.X. *et al.* (2011) Green tea intake lowers fasting serum total and LDL cholesterol in adults: a meta-analysis of 14 randomized controlled trials. *American Journal of Clinical Nutrition* **94**, 601–610.

Zucconi, S. *et al.* (2013) *Gathering consumption data on specific consumer groups of energy drinks*, Available at: www.efsa.europa.eu/publications.

Part III

Waste in the Juice and Non-Alcoholic Beverage Sector

11

Waste/By-Product Utilisations

Ciaran Fitzgerald, Mohammad Hossain, and Dilip K. Rai*

Department of Food BioSciences, Teagasc Food Research Centre Ashtown, Dublin, Ireland

11.1 Introduction

The non-alcoholic juice beverage industry is a massive sector of the global food market estimated to be worth $92 billion dollars by 2015 (Clark, 2009). One of the major challenges of this industry is the disposal of large amounts of by-products in an environmentally safe and economically viable manner. While the by-products are traditionally used as a low-cost supplement to livestock feed or fertilisers, many by-products have further biotechnological uses such as generating biofuel or are used as biosorbents to clean up surface waters polluted with heavy metals. With the functional food market now estimated to be worth €167 billion worldwide, natural sources of bioactive ingredients are become increasingly important (Granato *et al.*, 2010). Waste streams of the fruit juice industry are a rich source of essential oils, phenolics and other bioactive compounds that have great value to the functional food industry. A major advantage to using bioactive compounds from natural sources is that consumers react positively to them over synthetically manufactured ingredients (Dickson-Spillmann *et al.*, 2011). This chapter aims to outline the many ways in which the by-products of the fruit beverage industry can be recycled and revalued. It also attempts to highlight some novel technologies that may be used to increase the value of these by-products (Fig. 11.1).

11.2 Major Waste and By-Products Generated from the Juice and Non-Alcoholic Beverage Sector

Fruit juice production involves the mechanical separation of the natural liquid from the solid matrix. Therefore, in essence, the production of juice generates high amounts

*Corresponding author: Dilip K. Rai, Dilip.Rai@teagasc.ie

Innovative Technologies in Beverage Processing, First Edition.
Edited by Ingrid Aguiló-Aguayo and Lucía Plaza.
© 2017 John Wiley & Sons Ltd. Published 2017 by John Wiley & Sons Ltd.

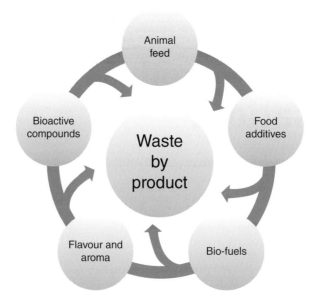

Figure 11.1   Waste by-product utilisation.

of by-products in the form of skin, pomace, wastewater and other materials resulting from the extraction of the juice. The amount of by-products produced varies depending on the source, for example, the yield of juice from oranges is around 50% (Clark, 2009), apple juice generates 25% of its original weight in pomace and tomatoes yield 30% secondary product during juice production. As is the case of any industry where large amounts of secondary by-product are generated, utilisation of these by-products becomes critical to the economics of the industry itself (Clark, 2009). Figure 11.2 displays the process of orange juice production and sources of by-products.

As can be seen from Table 11.1, millions of tons of by-products are produced annually from the juice and non-alcoholic beverage sector. Often, fruit is processed fresh in the same country as they are grown, juices and other products may then be concentrated which reduces storage and transportation costs. Orange juice as one of the most popular fruit beverages generates the largest amount of by-products per annum generating 32 million tons of pomace, skin and pulp globally per year (Martín *et al.*, 2010). The largest producers are Brazil, which produces around 58% of the worlds' orange juice, followed by the United States (23%), while the European Union (EU) produces 6.1% annually (USFDA, 2014). The apple juice industry also produces a significant amount of secondary products every year. China is the largest producer of apples, growing half the world's apple harvest annually. Consequently, it is also the leading producer of apple juice and apple juice concentrate, mainly utilising apples that are too small or misshapen to be used for the fresh apple market (USDA, 2010). Argentina is the biggest producer and processor of lemons and limes processing up to 1.15 million tons of lemons generating over half a million tons of by-products yearly (FAO, 2012).

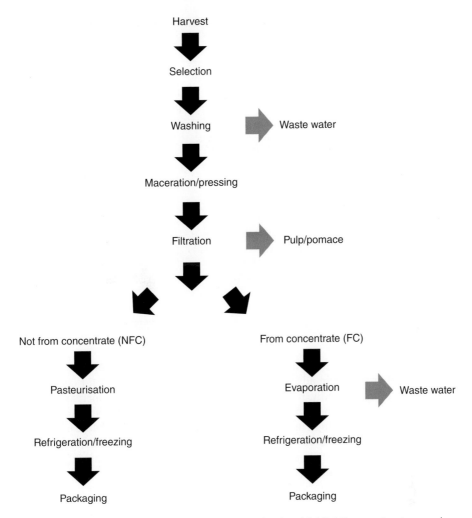

Figure 11.2 General process of orange juice production highlighting waste streams (grey arrows).

11.3 Utilisation of By-Products from the Non-Alcoholic Beverage Sector

11.3.1 Animal Feed

The feeding of agricultural by-product to livestock dates back hundreds of years to the beginning of farming itself. It is now widely practised globally in practically all areas of the agri-food sector. In fact, in the United States alone, there have been 355 different

Table 11.1   Commonly juiced fruit and vegetables with their approximate amount of by-products produced per annum

Source	Product	By-product	Percentage of whole	Global production (million tons per annum)	References
Fruits					
Orange	Orange juice	Peel, membrane and pulp	50	32	Clark (2009) and Martín *et al.* (2010)
Apple	Juice, cider	Pomace	25	17.5	Rha *et al.* (2011)
Lemon and lime	Juice, essential oil	Peel, membrane and pulp	50	1.3	Lario *et al.* (2004) and FAO (2012)
Grape	Juice	Skin, pomace	20	10	Kammerer *et al.* (2004)
Grapefruit	Juice	Peel, membrane and pulp	50	0.6	Brown (2011)
Vegetables					
Tomato	Juice, paste	Peel, seeds and pomace	5	1.52	Del Valle *et al.* (2006)
Carrot	Juice	Pomace	40	–	Baljeet *et al.* (2014)

by-product materials documented as sources of livestock feed (DePeters *et al.*, 1997). This practice has two main advantages: firstly, it eliminates the need for costly waste disposal strategies; secondly, it also reduces dependency on grains increasing cereal availability for human consumption. Due to the high plant cell wall content of all fruit pulps, they are naturally high in dietary fibre (mainly pectin and cellulose), which are seen as a good source of feed for ruminants. Ruminant digestion systems are inherently capable of fermenting high-fibre material like that created by the non-alcoholic beverage sector (Bampidis & Robinson, 2006).

Apart from the high dietary fibre content, fruit beverage by-products are nutritionally beneficial to livestock in numerous other ways. Dried citrus pulp is a common by-product fed to livestock and is seen as a high-energy inexpensive supplement to regular grain rations (Grasser *et al.*, 1995). It has a comparable energy content to that of fresh grass, hay and silage (Tamminga, 1996). It can contain protein levels as high as 12.87% as well as being low in fat and rich in flavonoids (Marín *et al.*, 2007). Citrus pulp feed is also known to be rich in calcium (Arosemena *et al.*, 1995). As well as having nutritional advantages, citrus by-products have been shown to have little or no effect on milk yield and composition in ruminants such as sheep (Fegeros *et al.*, 1995). Other studies have shown when dairy cows' feed is substituted with citrus pulp that milk fat content and yield was significantly increased (Belabasakis & Tsirgogianni, 1996). Similarly, fermented apple pomace was used to supplement the dietary in-take of protein in pigs by 36% (Ajila *et al.*, 2015). The fermented apple pomace also serve as a source of minerals (i.e. zinc, copper, magnesium and iron) and vitamins in the animal feed. An increase in mean body weight of broiler chickens from 270 to 537 g was observed when apple pomace was mixed in equal amounts to standard poultry feed (Joshi *et al.*, 2000).

11.3.2 Pectin

Pectin is amply available from fruit beverage waste streams in particular as by-products of apple and orange juice production. For example, apple pomace can contain as high as 9.73% pectin and is easily and inexpensively extracted using citric acid (Canteri-Schemin *et al.*, 2005). Pectin extracted from apple pomace and citrus peels has huge application as a natural food ingredient owing to its various technofunctional properties. Pectins are used in food products as gelling agents, thickeners, emulsifiers, stabilisers, and texturisers. Another potential usage of pectins is to separate heavy metals from water sources. Heavy metals, as such, if untreated in the environment, persistently recirculate until they end up in the food chain. When consumed, heavy metals such as mercury, cadmium, lead and arsenic can cause disruption to the nervous, cardiovascular, respiratory, gastrointestinal, hepatic, renal, hematopoietic, immunological and dermatologic systems. Furthermore, due to their bioaccumalatory nature post exposure, low levels of these molecules can develop into a variety of diseases over a prolonged period (Inoue, 2013). The process termed 'biosorption' exploits the ability of biological material to bind contaminants such as heavy metals via physicochemical processes (Volesky & Holan, 1995). This is seen as a much more cost-effective method compared to synthetic sorbent materials and has proven to be quite effective at absorbing lead, copper, zinc and cadmium from polluted waters (Schiewer & Patil, 2008). Over the past few decades, many applications have been developed to utilise waste streams of the non-alcoholic beverage industries for biotechnological purposes. One such example is the potential usage of pectin extracted from waste streams such as apple pomace and citrus peels to extract heavy metals from surface water due to activities such as mining and metal processing.

11.3.3 Biofuel

As the cost of fossil fuels rises and oil fields begin to dry up, the area of biofuels will become increasingly important to meet global fuel supplies. Through a process of anaerobic digestion, the peel of citrus fruit may be turned into biogas consisting of methane and carbon dioxide. The anti-microbial effect of most citrus peels' essential oils necessitate a pretreatment process to ensure efficient digestion (essential oils, another important source of valorisation from fruit beverages, will be discussed further in the chapter) (Martín *et al.*, 2010). The most common source of bio-methane is animal manure and sewage sludge produced through liquid anaerobic digestion. However, lignocellulosic solid waste such as fruit peel and pressings can be processed in to bio-methane through a process called solid-state anaerobic digestion. This process is seen as a greener technique as it requires less energy due to less heat being needed for heating smaller reactors, and also there is no need for constant agitation (Li *et al.*, 2011). Orange peel can produce as much as 455 ml of methane per gram of solid material used (Raposo *et al.*, 2012).

Bioethanol is the most common renewable fuel produced nowadays and is primarily produced using the starch from corn grain or the sucrose from sugar cane as a energy sources for the anaerobic fermentation of ethanol (Gray *et al.*, 2006). However, as the supply of these substrates decreases, other sources are needed to meet the demand for the fuel. The peel of citrus fruit is rich in fermentable sugars such as glucose, fructose and sucrose as well as polysaccharides such as cellulose and pectin. These sugars are ideal for fermentation by microorganisms such as *Saccharomyces cerevisiae* to produce ethanol (Oberoi *et al.*, 2010). Recent efforts by car manufacturers have seen a growth in the production of vehicles that run on E85, which is a blend of ethanol (85%) and petroleum (15%). This fuel not only massively reduces the demand for fossil fuels but also reduces the amount of greenhouse gases emitted to the atmosphere dramatically (Sun & Cheng, 2002). Furthermore, the resulting digestate of these fermentations may be used as a high-quality fertiliser making the process even more profitable and environmentally friendly (Tsagarakis & Georgantzs, 2003).

11.3.4 Flavour and Aroma

Citrus peel is rich in a terpene compound known as limonene, which is responsible for the orange aroma found in the fruits' skin. Limonene acts as an anti-microbial agent, and therefore, citrus wastes need to be pretreated prior to the process of creating biofuels to maximise yield (Pourbafrani *et al.*, 2010). Limonene may be isolated during this pretreatment process and its distinctive orange flavour and aroma is used in the cosmeceutical and food industries and also as an environmental friendly solvent (Lohrasbi *et al.*, 2010).

Essential oils are also common by-products of the juice industry. During the removal of water to produce fruit juice concentrates essential oils are extracted along with the aqueous portion and can be easily removed through centrifugation (Högnadóttir & Rouseff, 2003). Essential oils can also be easily extracted from the peels of citrus fruit waste and have numerous health-promoting applications. Apart from being used in the cosmeceutical and food industries for their distinct aromas and flavours, essential oils have shown to have potent anti-microbial, anti-oxidant and anti-inflammatory properties (Yang *et al.*, 2009). Because essential oils consist of a wide spectrum of compounds, bacterial resistance is rare, which makes them ideal candidates for anti-microbial agents and as anti-spoilage ingredients for the food industry (Maggi *et al.*, 2009).

11.3.5 Food Additives

The by-products of the fruit juice industry also have many applications as food additives. During the industrial production of fruit juices, tartaric acid is partially removed to stabilise the product using ion exchange towers. After this process, there are huge amounts of tartaric acid washed out and lost in waste waters. Through electrodialysis, evaporation and reverse osmosis, this waste stream may be revalued and the tartaric acid recycled as an acidulant compound for soft drinks or as an additive in medicines and cosmetics, as an anti-oxidant or emulsifying agent (Andrés *et al.*, 1997).

Colourants are the most used type of additives in the food industry. As consumers become more discerning when it comes to the use of synthetic compounds as colourants natural sources of colour compounds are becoming increasingly prevalent (Ayala-Zavala *et al.*, 2011). Food colourants sourced from fruit beverage waste are not only cheaply sourced but are also seen as possessing high purity and colour stability, an essential attribute for products with long shelf lives (Pszczola, 1998). Pigments such as delphinidin-type anthocyanins can be easily extracted from the waste of grape juice production for inclusion as colourants for foods that require blue–violet to shades (Stintzing & Carle, 2004). Another widely used food colourant pigment, β-carotene can be extracted from orange peel and not only is it a safe food colourant but also associated with anti-cancer properties (Ghazi, 1999).

11.4   Potential Sources of Bioactive Compounds

11.4.1   Phenolic Compounds

Phenolic compounds are one of the most abundant phytochemicals found in plants. Due to their anti-oxidant activity, consumption of these compounds has been associated with health benefits when eaten regularly. Beverage processing activities produce huge amounts of waste rich in phenolic compounds (Balasundram *et al.*, 2006). In fact, the peel of citrus fruits such as lemons, oranges and grapefruit has proven to contain 15% more phenolic compounds than the edible portion itself (Gorinstein *et al.*, 2001). The main polyphenol is hesperidin which accounts for 50% of the phenolic compounds found in citrus peel (Fernández-López *et al.*, 2009). Similarly, the peels of apples, peaches and pears are known to contain twice as much phenolic compounds as the edible portion of the fruits (Gorinstein *et al.*, 2002). Apple pomace has been proposed as a potential source of phloridzin (Lavelli & Corti, 2011), a polyphenol considered as anti-diabetic compound (Ehrenkranz *et al.*, 2005). Grape seeds and skins as by-products of the grape juice industry contain high amounts of the phenolic compounds known as anthocyanins (Shrikhande, 2000). The main challenge is to safely remove and concentrate these compounds from their source material, so that they may be re-incorporated into food stuffs which may incur a positive health effect on the consumer in the form of functional food products.

11.4.2   Bioactive Peptides

The area of mining fruit processing waste for bioactive peptides has barely been touched upon. This is surprising as with protein levels as high as 12.87%, waste from the fruit beverage industry may be seen as a potentially rich source of bioactive proteins and peptides (Marín *et al.*, 2007). The seeds of fruits have a particularly high protein content, which could well be a source for sustainable nutritional protein as an alternative to meat proteins. Anti-fungal proteins have been found in the seeds of plums (Lam & Ng, 2009) and kiwi fruits (Wang & Ng, 2002).

11.5   Novel Technologies Involved in the Processing of Fruit Beverage Waste

11.5.1   Pulsed Electric Field

Pulsed electric field (PEF) is an exciting new technology when considering the processing of waste from the fruit beverage industry. It is an environmentally friendly and energy-efficient process whereby short bursts of electricity are transferred in to a medium. By using PEF, electroporation of plant cell membranes occurs rapidly, improving tissue softness and influencing textural properties (Toepfl *et al.*, 2006).

When used during the juicing process, it enhances juice extraction and, in some cases, can replace the use of enzymes for maceration purposes (Eshtiaghi & Knorr, 2000). PEF is also used on the waste of the beverage industry where it enhanced the extraction yield of bioactive compounds. Corrales *et al.* (2008) showed that by using PEF, the extraction of anti-oxidant anthocyanins from grape by-products can not only be enhanced but can also be used to selectively extract compounds of interest (Corrales *et al.*, 2008).

11.5.2   Ultrasonication

Another physical technology being introduced to food processing is ultrasonication. Ultrasonication has for some time been used in laboratories to assist in dissolving chemicals and degassing of solvents prior to chromatography experiments. However, it is also useful for enhancing extraction of compounds from plant cell material including fruit beverage by-products. Ultrasonication induces cavitation by forming collapsing bubbles in the solvent generating localised pressure which causes plant material to rupture and release intracellular contents in to the solvent. This technology also has the advantage of being environmentally friendly and energy efficient (Knorr, 1994). Ghafoor *et al.* (2009) utilised this technology to enhance the extraction of anti-oxidant compounds from grape seeds, showing that using ultrasonication the extract had significantly increased total phenolic contents, anti-oxidant activities and anthocyanin contents (Ghafoor *et al.*, 2009).

References

Ajila, C.M., Sarma, S.J., Brar, S.K., Godbout, S., Cote, M., Guay, F., Verma, M. & Valéro J.R. (2015) Fermented apple pomace as a feed additive to enhance growth performance of growing pigs and its effects on emissions. *Agriculture* **5**(2), 313–329.

Andrés, L.J., Riera, F.A. & Alvarez, R. (1997) Recovery and concentration by electro-dialysis of tartaric acid from fruit juice industries waste waters. *Journal of Chemical Technology and Biotechnology* **70**, 247–252.

Arosemena, A., Depeters, E.J. & Fadel, J.G. (1995) Extent of variability in nutrient composition within selected by-product feedstuffs. *Animal Feed Science and Technology* **54**, 103–120.

Ayala-Zavala, J., Vega-Vega, V., Rosas-Domínguez, C., Palafox-Carlos, H., Villa-Rodrígyez, J., Siddiqui, M., Dávila-Aviña, J. & González-Aguilar, G. (2011) Agro-industrial potential of exotic fruit byproducts as a source of food additives. *Food Research International* **44**, 1866–1874.

Balasundram, N., Sundram, K. & Samman, S. (2006) Phenolic compounds in plants and agri-industrial by-products: antioxidant activity, occurrence, and potential uses. *Food Chemistry* **99**, 191–203.

Baljeet, S., Ritika, B. & Reena, K. (2014) Effect of incorporation of carrot pomace powder and germinated chickpea flour on the quality characteristics of biscuits. *International Food Research Journal* **21**, 217–222.

Bampidis, V. & Robinson, P. (2006) Citrus by-products as ruminant feeds: a review. *Animal Feed Science and Technology* **128**, 175–217.

Belabasakis, N.G. & Tsirgogianni, D. (1996) Effects of dried citrus pulp on milk yield, milk composition and blood components of dairy cows. *Animal Feed Science and Technology* **60**, 87–92.

Brown, M.G. (2011) *Impacts of the European Union Tariff On the Florida Price for Grapefruit Juice*. Florida Department of Citrus.

Canteri-Schemin, M.H., Fertonani, H.C.R., Waszczynkyj, N. & Wosiacki, G. (2005) Extraction of pectin from apple pomace. *Brazilian Archives of Biology and Technology* **48**, 259–266.

Clark, J.P. (2009) Fruit and vegetable juice processing. In *Case Studies in Food Engineering*. Springer.

Corrales, M., Toepfl, S., Butz, P., Knorr, D. & Tauscher, B. (2008) Extraction of anthocyanins from grape by-products assisted by ultrasonics, high hydrostatic pressure or pulsed electric fields: a comparison. *Innovative Food Science & Emerging Technologies* **9**, 85–91.

Del Valle, M., Cámara, M. & Torija, M.E. (2006) The nutritional and functional potential of tomato by-products. In *X International Symposium on the Processing Tomato* 758, 165–172.

Depeters, E., Fadel, J. & Arosemena, A. (1997) Digestion kinetics of neutral detergent fiber and chemical composition within some selected by-product feedstuffs. *Animal Feed Science and Technology* **67**, 127–140.

Dickson-Spillmann, M., Siegrist, M. & Keller, C. (2011) Attitudes toward chemicals are associated with preference for natural food. *Food Quality and Preference* **22**, 149–156.

Ehrenkranz, J.R.L., Lewis, N.G., Ronald Kahn, C., & Roth, J. (2005) Phlorizin: a review. *Diabetes/Metabolism Research and Reviews* **21**(1), 31–38.

Eshtiaghi, M. & Knorr, D. (2000) Anwendung elektrischer Hochspannungsimpulse zum Zellaufschluss bei der Saftgewinnung am Beispiel von Weintrauben. *LVT* **45**, 23–27.

FAO (2012) Citrus fruit fresh and processed annual statistics 2012. Food and Agriculture Organization of the United Nations.

Fegeros, K., Zervas, G., Sstamouli, S. & Apostolaki, E. (1995) Nutritive value of dried citrus pulp and its effect on milk yield and milk composition of lactating ewes. *Journal of Dairy Science* **78**, 1116–1121.

Fernández-López, J., Ssendra-Nadal, E., Navarro, C., Sayas, E., Viuda-Martos, M. & Alvarez, J.A.P. (2009) Storage stability of a high dietary fibre powder from orange by-products. *International Journal of Food Science & Technology* **44**, 748–756.

Ghafoor, K., Choi, Y.H., Jeon, J.Y. & Jo, I.H. (2009) Optimization of ultrasound-assisted extraction of phenolic compounds, antioxidants, and anthocyanins from grape (vitis vinifera) seeds. *Journal of Agricultural and Food Chemistry* **57**, 4988–4994.

Ghazi, A. (1999) Extraction of beta-carotene from orange peels. *Nahrung* **43**, 274–277.

Gorinstein, S., Martín-Belloso, O., Park, Y.-S., Haruenkit R., Lojek, A., Ĉ́ Ž, M., Caspi, A., Libman, I. & Trakhtenberg, S. (2001) Comparison of some biochemical characteristics of different citrus fruits. *Food Chemistry* **74**, 309–315.

Gorinstein, S., Martín-Belloso, O., Lojek, A., ČÍŽ, M., Soliva-Fortuny, R., Park, Y.S., Caspi, A., Libman, I. & Trakhtenberg, S. (2002) Comparative content of some phytochemicals in Spanish apples, peaches and pears. *Journal of the Science of Food and Agriculture* **82**, 1166–1170.

Granato, D., Branco, G.F., Nazzaro, F., Cruz, A.G. & Faria, J.A. (2010) Functional foods and nondairy probiotic food development: trends, concepts, and products. *Comprehensive Reviews in Food Science and Food Safety* **9**, 292–302.

Grasser, L., Fadel, J., Garnett, I. & Ddepeters, E. (1995) Quantity and economic importance of nine selected by-products used in California dairy rations. *Journal of Dairy Science* **78**, 962–971.

Gray, K.A., Zhao, L. & Emptage, M. (2006) Bioethanol. *Current Opinion in Chemical Biology* **10**, 141–146.

Högnadóttir, Á. & Rousseff, R.L. (2003) Identification of aroma active compounds in orange essence oil using gas chromatography–olfactometry and gas chromatography–mass spectrometry. *Journal of Chromatography. A* **998**, 201–211.

Inoue, K. (2013) Heavy metal toxicity. *Journal of Clinical Toxicology* **3**, doi: 10.4172/2161-0495.S3-007.

Joshi, V.K., Gupta, K., Devrajan, A., Lal, B.B., & Arya, S.P. (2000) Production and evaluation of fermented apple pomace in the feed of broilers. *Journal of Food Science and Technology* **37**(6), 609–612.

Kammerer, D., Claus, A., Carle, R. & Schieber, A. (2004) Polyphenol screening of pomace from red and white grape varieties (*Vitis vinifera* L.) by HPLC-DAD-MS/MS. *Journal of Agricultural and Food Chemistry* **52**, 4360–4367.

Knorr, D. (1994) Plant cell and tissue cultures as model systems for monitoring the impact of unit operations on plant foods. *Trends in Food Science & Technology* **5**, 328–331.

Lam, S. K. & Ng, T.B. (2009) Passiflin, a novel dimeric antifungal protein from seeds of the passion fruit. *Phytomedicine* **16**, 172–180.

Lario, Y., Sendra, E., García-Pérez, J., Fuentes, C., Sayas-Barberá, E., Fernández-López, J. & Pérez-Álvarez, J. (2004) Preparation of high dietary fiber powder from lemon juice by-products. *Innovative Food Science & Emerging Technologies* **5**, 113–117.

Lavelli, V. & Corti, S. (2011) Phloridzin and other phytochemicals in apple pomace: stability evaluation upon dehydration and storage of dried product. *Food Chemistry* **129**(4), 1578–1583.

Li, Y., Park, S.Y. & Zhu, J. (2011) Solid-state anaerobic digestion for methane production from organic waste. *Renewable and Sustainable Energy Reviews* **15**, 821–826.

Lohrasbi, M., Pourbafrani, M., Niklasson, C. & Taherzadeh, M.J. (2010) Process design and economic analysis of a citrus waste biorefinery with biofuels and limonene as products. *Bioresource Technology* **101**, 7382–7388.

Maggi, F., Bramucci, M., Cecchini, C., Coman, M.M., Cresci, A., Cristalli, G., Lupidi, G., Papa, F., Quassinti, L. & Sagratini, G. (2009) Composition and biological activity of essential oil of *Achillea ligustica* (Asteraceae) naturalized in central Italy: Ideal candidate for anti-cariogenic formulations. *Fitoterapia* **80**, 313–319.

Marín, F.R., Soler-Rivas, C., Benavente-García, O., Castillo, J. & Pérez-Álvarez, J.A. (2007) By-products from different citrus processes as a source of customized functional fibres. *Food Chemistry* **100**, 736–741.

Martín, M., Siles, J., Chica, A. & Martín, A. (2010) Biomethanization of orange peel waste. *Bioresource Technology* **101**, 8993–8999.

Oberoi, H.S., Vadlani, P.V., Madl, R.L., Saida, L. & Abeykoon, J.P. (2010) Ethanol production from orange peels: two-stage hydrolysis and fermentation studies using optimized parameters through experimental design. *Journal of Agricultural and Food Chemistry* **58**, 3422–3429.

Pourbafrani, M., Forgás, G., Horváth, I.S., Niklasson, C. & Taherzadeh, M.J. (2010) Production of biofuels, limonene and pectin from citrus wastes. *Bioresource Technology* **101**, 4246–4250.

Pszczola, D. (1998) Natural colors: pigments of imagination. *Food Technology (USA)* **52**.

Raposo, F., De La Rubia, M., Fernández-Cegrí, V. & Borja, R. (2012) Anaerobic digestion of solid organic substrates in batch mode: an overview relating to methane yields and experimental procedures. *Renewable and Sustainable Energy Reviews* **16**, 861–877.

Rha, H.J., Bae, I.Y., Lee, S., Yoo, S.H., Chang, P.S. & Lee, H.G. (2011) Enhancement of anti-radical activity of pectin from apple pomace by hydroxamation. *Food Hydrocolloids* **25**, 545–548.

Schiewer, S. & Patil, S.B. (2008) Pectin-rich fruit wastes as biosorbents for heavy metal removal: equilibrium and kinetics. *Bioresource Technology* **99**, 1896–1903.

Shrikhande, A.J. (2000) Wine by-products with health benefits. *Food Research International* **33**, 469–474.

Stintzing, F.C. & Carle, R. (2004) Functional properties of anthocyanins and betalains in plants, food, and in human nutrition. *Trends in Food Science & Technology* **15**, 19–38.

Sun, Y. & Cheng, J. (2002). Hydrolysis of lignocellulosic materials for ethanol production: a review. *Bioresource Technology* **83**, 1–11.

Tamminga, S. (1996) A review on environmental impacts of nutritional strategies in ruminants. *Journal of Animal Science* **74**, 3112–3124.

Toepfl, S., Mathys, A., Heinz, V. & Knorr, D. (2006) Review: potential of high hydrostatic pressure and pulsed electric fields for energy efficient and environmentally friendly food processing. *Food Reviews International* **22**, 405–423.

Tsagarakis, K. & Georgantzs, N. (2003) The role of information on farmers' willingness to use recycled water for irrigation. *Water Supply* **3**, 105–113.

USDA (2010) Investment in Processing Industry Turns Chinese Apples Into Juice Exports. *United States Department of Agriculture.*

USFDA (2014) Citrus: World Markets and Trade. *United States Department of Agriculture.*

Volesky, B. & Holan, Z. (1995) Biosorption of heavy metals. *Biotechnology Progress* **11**, 235–250.

Wang, H. & Ng, T.B. (2002) Isolation of an antifungal thaumatin-like protein from kiwi fruits. *Phytochemistry* **61**, 1–6.

Yang, E.J., Kim, S.S., Oh, T.H., Baik, J.S., Lee, N.H. & Hyun, C.G. (2009) Essential oil of citrus fruit waste attenuates LPS-induced nitric oxide production and inhibits the growth of skin pathogens. *International Journal of Agriculture and Biology* **11**, 791–794.

Index

Innovative Technologies in Beverage Processing, First Edition.
Edited by Ingrid Aguiló-Aguayo and Lucía Plaza.
© 2017 John Wiley & Sons Ltd. Published 2017 by John Wiley & Sons Ltd.

Food Science and Technology Books

WILEY Blackwell

GENERAL FOOD SCIENCE & TECHNOLOGY, ENGINEERING AND PROCESSING

Food Texture Design and Optimization	Dar	9780470672426
Nano- and Microencapsulation for Foods	Kwak	9781118292334
Extrusion Processing Technology: Food and Non-Food Biomaterials	Bouvier	9781444338119
Food Processing: Principles and Applications, 2nd Edition	Clark	9780470671146
The Extra-Virgin Olive Oil Handbook	Peri	9781118460450
Mathematical and Statistical Methods in Food Science and Technology	Granato	9781118433683
The Chemistry of Food	Velisek	9781118383841
Dates: Postharvest Science, Processing Technology and Health Benefits	Siddiq	9781118292372
Resistant Starch: Sources, Applications and Health Benefits	Shi	9780813809519
Statistical Methods for Food Science: Introductory 2nd Edition	Bower	9781118541647
Formulation Engineering of Foods	Norton	9780470672907
Practical Ethics for Food Professionals: Research, Education and the Workplace	Clark	9780470673430
Edible Oil Processing, 2nd Edition	Hamm	9781444336849
Bio-Nanotechnology: A Revolution in Food, Biomedical and Health Sciences	Bagchi	9780470670378
Dry Beans and Pulses : Production, Processing and Nutrition	Siddiq	9780813823874
Genetically Modified and non-Genetically Modified Food Supply Chains: Co-Existence and Traceability	Bertheau	9781444337785
Food Materials Science and Engineering	Bhandari	9781405199223
Handbook of Fruits and Fruit Processing, second edition	Sinha	9780813808949
Tropical and Subtropical Fruits: Postharvest Physiology, Processing and Packaging	Siddiq	9780813811420
Food Biochemistry and Food Processing, 2nd Edition	Simpson	9780813808741
Dense Phase Carbon Dioxide	Balaban	9780813806495
Nanotechnology Research Methods for Food and Bioproducts	Padua	9780813817316
Handbook of Food Process Design, 2 Volume Set	Ahmed	9781444330113
Ozone in Food Processing	O'Donnell	9781444334425
Food Oral Processing	Chen	9781444330120
Food Carbohydrate Chemistry	Wrolstad	9780813826653
Organic Production & Food Quality	Blair	9780813812175
Handbook of Vegetables and Vegetable Processing	Sinha	9780813815411

FUNCTIONAL FOODS, NUTRACEUTICALS & HEALTH

Antioxidants and Functional Components in Aquatic Foods	Kristinsson	9780813813677
Food Oligosaccharides: Production, Analysis and Bioactivity	Moreno-Fuentes	9781118426494
Novel Plant Bioresources: Applications in Food, Medicine and Cosmetics	Gurib-Fakim	9781118460610
Functional Foods and Dietary Supplements: Processing Effects and Health Benefits	Noomhorm	9781118227879
Food Allergen Testing: Molecular, Immunochemical and Chromatographic Techniques	Siragakis	9781118519202
Bioactive Compounds from Marine Foods: Plant and Animal Sources	Hernández-Ledesma	9781118412848
Bioactives in Fruit: Health Benefits and Functional Foods	Skinner	9780470674970
Marine Proteins and Peptides: Biological Activities and Applications	Kim	9781118375068
Dried Fruits: Phytochemicals and Health Effects	Alasalvar	9780813811734
Handbook of Plant Food Phytochemicals	Tiwari	9781444338102
Analysis of Antioxidant-Rich Phytochemicals	Xu	9780813823911
Phytonutrients	Salter	9781405131513
Coffee: Emerging Health Effects and Disease Prevention	Chu	9780470958780
Functional Foods, Nutraceuticals & Disease Prevention	Paliyath	9780813824536
Nondigestible Carbohydrates and Digestive Health	Paeschke	9780813817620
Bioactive Proteins and Peptides as Functional Foods and Nutraceuticals	Mine	9780813813110
Probiotics and Health Claims	Kneifel	9781405194914
Functional Food Product Development	Smith	9781405178761

INGREDIENTS

Fats in Food Technology, 2nd Edition	Rajah	9781405195423
Processing and Nutrition of Fats and Oils	Hernandez	9780813827674
Stevioside: Technology, Applications and Health	De	9781118350669
The Chemistry of Food Additives and Preservatives	Msagati	9781118274149
Sweeteners and Sugar Alternatives in Food Technology, 2nd Edition	O'Donnell	9780470659687
Hydrocolloids in Food Processing	Laaman	9780813820767
Natural Food Flavors and Colorants	Attokaran	9780813821108
Handbook of Vanilla Science and Technology	Havkin-Frenkel	9781405193252
Enzymes in Food Technology, 2nd edition	Whitehurst	9781405183666
Food Stabilisers, Thickeners and Gelling Agents	Imeson	9781405132671
Glucose Syrups - Technology and Applications	Hull	9781405175562
Dictionary of Flavors, 2nd edition	DeRovira	9780813821351

FOOD SAFETY, QUALITY AND MICROBIOLOGY

Practical Food Safety	Bhat	9781118474600
Food Chemical Hazard Detection	Wang	9781118488591
Food Safety for the 21st Century	Wallace	9781118897980
Guide to Foodborne Pathogens, 2nd Edition	Labbe	9780470671429
Improving Import Food Safety	Ellefson	9780813808772
Food Irradiation Research and Technology, 2nd Edition	Fan	9780813802091
Food Safety: The Science of Keeping Food Safe	Shaw	9781444337228

For further details and ordering information, please visit www.wiley.com/go/food

Food Science and Technology from Wiley Blackwell

Decontamination of Fresh and Minimally Processed Produce	Gomez-Lopez	9780813823843
Progress in Food Preservation	Bhat	9780470655856
Food Safety for the 21st Century: Managing HACCP and Food Safety throughout the Global Supply Chain	Wallace	9781405189118
The Microbiology of Safe Food, 2nd edition	Forsythe	9781405140058

SENSORY SCIENCE, CONSUMER RESEARCH & NEW PRODUCT DEVELOPMENT

Olive Oil Sensory Science	Monteleone	9781118332528
Quantitative Sensory Analysis: Psychophysics, Models and Intelligent Design	Lawless	9780470673461
Product Innovation Toolbox: A Field Guide to Consumer Understanding and Research	Beckley	9780813823973
Sensory and Consumer Research in Food Product Design and Dev, 2nd Ed	Moskowitz	9780813813660
Sensory Evaluation: A Practical Handbook	Kemp	9781405162104
Statistical Methods for Food Science	Bower	9781405167642
Concept Research in Food Product Design and Development	Moskowitz	9780813824246
Sensory and Consumer Research in Food Product Design and Development	Moskowitz	9780813816326

FOOD INDUSTRY SUSTAINABILITY & WASTE MANAGEMENT

Food and Agricultural Wastewater Utilization and Treatment, 2nd Edition	Liu	9781118353974
Sustainable Food Processing	Tiwari	9780470672235
Food and Industrial Bioproducts and Bioprocessing	Dunford	9780813821054
Handbook of Sustainability for the Food Sciences	Morawicki	9780813817354
Sustainability in the Food Industry	Baldwin	9780813808468
Lean Manufacturing in the Food Industry	Dudbridge	9780813810072

FOOD LAWS & REGULATIONS

Guide to US Food Laws and Regulations, 2nd Edition	Curtis	9781118227787
Food and Drink - Good Manufacturing Practice: A Guide to its Responsible Management (GMP6), 6th Edition	Manning	9781118318201
The BRC Global Standard for Food Safety: A Guide to a Successful Audit, 2nd Edition	Kill	9780470670651
Food Labeling Compliance Review, 4th edition	Summers	9780813821818

DAIRY FOODS

Lactic Acid Bacteria: Biodiversity and Taxonomy	Holzapfel	9781444333831
From Milk By-Products to Milk Ingredients: Upgrading the Cycle	de Boer	9780470672228
Milk and Dairy Products as Functional Foods	Kanekanian	9781444336832
Milk and Dairy Products in Human Nutrition: Production, Composition and Health	Park	9780470674185
Manufacturing Yogurt and Fermented Milks, 2nd Edition	Chandan	9781119967088
Sustainable Dairy Production	de Jong	9780470655849
Advances in Dairy Ingredients	Smithers	9780813823959
Membrane Processing: Dairy and Beverage Applications	Tamime	9781444333374
Analytical Methods for Food and Dairy Powders	Schuck	9780470655986
Dairy Ingredients for Food Processing	Chandan	9780813817460
Processed Cheeses and Analogues	Tamime	9781405186421
Technology of Cheesemaking, 2nd edition	Law	9781405182980

SEAFOOD, MEAT AND POULTRY

Seafood Processing: Technology, Quality and Safety	Boziaris	9781118346211
Should We Eat Meat? Evolution and Consequences of Modern Carnivory	Smil	9781118278727
Handbook of Meat, Poultry and Seafood Quality, second edition	Nollet	9780470958322
The Seafood Industry: Species, Products, Processing, and Safety , 2nd Edition	Granata	9780813802589
Organic Meat Production and Processing	Ricke	9780813821269
Handbook of Seafood Quality, Safety and Health Effects	Alasalvar	9781405180702

BAKERY & CEREALS

Oats Nutrition and Technology	Chu	9781118354117
Cereals and Pulses: Nutraceutical Properties and Health Benefits	Yu	9780813818399
Whole Grains and Health	Marquart	9780813807775
Gluten-Free Food Science and Technology	Gallagher	9781405159159
Baked Products - Science,Technology and Practice	Cauvain	9781405127028

BEVERAGES & FERMENTED FOODS/BEVERAGES

Encyclopedia of Brewing	Boulton	9781405167444
Sweet, Reinforced and Fortified Wines: Grape Biochemistry, Technology and Vinification	Mencarelli	9780470672242
Technology of Bottled Water, 3rd edition	Dege	9781405199322
Wine Flavour Chemistry, 2nd edition	Bakker	9781444330427
Wine Quality: Tasting and Selection	Grainger	9781405113663

PACKAGING

Handbook of Paper and Paperboard Packaging Technology, 2nd Edition	Kirwan	9780470670668
Food and Beverage Packaging Technology, 2nd edition	Coles	9780813812748
Food and Package Engineering	Morris	9780813814797
Modified Atmosphere Packaging for Fresh-Cut Fruits and Vegetables	Brody	9780813812748

For further details and ordering information, please visit www.wiley.com/go/foo